SECOND EDITION Communication Networks
A First Course

Jean Walrand
University of California at Berkeley

WCB
McGraw-Hill

Boston   Burr Ridge, IL   Dubuque, IA   Madison, WI   New York   San Francisco   St. Louis
Bangkok   Bogotá   Caracas   Lisbon   London   Madrid
Mexico City   Milan   New Delhi   Seoul   Singapore   Sydney   Taipei   Toronto

This book is for my wife Annie who made it possible and my children Isabelle and Julie who made it necessary.

WCB/McGraw-Hill

*A Division of The **McGraw·Hill** Companies*

COMMUNICATION NETWORKS

Copyright © 1998 by The McGraw-Hill Companies, Inc. All rights reserved. Previous edition © 1991 by Richard D. Irwin, a Times Mirror Higher Education Group, Inc. company. Printed in the United States of America. Except as permitted under the United States Copyright Act of 1976, no part of this publication may be reproduced or distributed in any form or by any means, or stored in a data base or retrieval system, without the prior written permission of the publisher.

This book is printed on acid-free paper.

1 2 3 4 5 6 7 8 9 0 DOC/DOC 1 0 9 8

ISBN 0-256-17404-0

Vice president and editorial director:	*Kevin T. Kane*
Publisher:	*Tom Casson*
Executive editor:	*Elizabeth A. Jones*
Senior developmental editor:	*Kelley Butcher*
Marketing manager:	*John T. Wannemacher*
Project manager:	*Pat Frederickson*
Production supervisor:	*Heather D. Burbridge*
Designer:	*Michael Warrell*
Compositor:	*Techsetters, Inc.*
Typeface:	*10/12 Times Roman*
Printer:	*R. R. Donnelley & Sons Company*

Library of Congress Cataloging-in-Publication Data

Walrand, Jean.
    Communication networks: a first course / Jean Walrand. — 2nd ed.
        p.   cm.
    Includes bibliographical references and index.
    ISBN 0-256-17404-0
    1. Computer networks.   I. Title
TK5105.5.W35   1998
004.6—dc21   97-45604

Printed in the United States of America

http://www.mhhe.com

Contents

Chapter 3   Internet   47

Preface

A universal revolution is underway in telecommunications. The changes taking place are having a dramatic impact on how individuals and institutions communicate and do business with one another. Government at all levels, retailing and finance, health care, education, and entertainment are among the areas of human activity profoundly affected by the technological advances now underway.

In order to adapt and contribute effectively to these changes engineers and computer scientists need to acquire a solid foundation and understanding of communication networks. The purpose of this text is to help both students and practicing professionals master the background knowledge necessary to participate in and influence ongoing developments in telecommunications.

This book has been written and designed to be accessible to junior-level engineering and computer science students as well as to advanced undergraduate and graduate students with no prior knowledge of communication theory or computer networks. Several of the chapters make use of elementary probability concepts and calculations, which are explained in an appendix. The last chapter of the book uses results from queueing theory, derivations of which are included in another appendix.

Throughout the text, to the extent possible, descriptions of communication networks have been separated from their mathematical analysis to accommodate differing levels of reader comprehension and preparedness.

Many of the implementation details of current networks are transitory; only a few basic principles are likely to survive the current changes in telecommunication technology. This text has been organized and written with a focus on principles—such as the management of complexity, standardized connectivity, and resource sharing—which will continue to guide the development of networks. These principles motivate the layering of software, standardization, hierarchical addressing and routing, and the multiplexing and switching methods implemented in networks. Although many specific networks are discussed in the book, a continuous effort has been made along the way to emphasize the underlying structure of the field.

Since the first edition was published in 1991, networks have experienced an explosive growth fueled largely by the World Wide Web. Although the mechanisms of networks and their design principles have not substantially changed, the level of interest in the Internet and networking possibilities has increased substantially.

Yet another significant development is the progress made in the deployment of Asynchronous Transfer Mode (ATM) networks. ATM is used for local area network (LAN) backbones, inside LAN switches and Internet protocol (IP) routers, and as an Internet technology.

Other networking technologies are also being pushed forward: fast access links on cable television networks and on telephone lines, wireless access with cellular data networks, better compression for multimedia, and improved security.

This second edition has been rewritten to integrate these developments. The book is now organized by network technology: Internet, LANs, and ATM are the main themes. After explaining how these networks work, we explore more specialized topics. Thus, the book is organized "big picture first" then more specific details.

Organization of the Book

In a nutshell, the book explains the Internet, Ethernet, and a few other local area networks, Asynchronous Transfer Mode, some concepts of digital communication links, and elements of security and compression. You learn how networks work and also how well they work and why.

After 30 years of trials and errors by many clever researchers, it is tempting to enunciate deep design principles that have guided the progress of networks until today. What is closer to the truth is that, as the needs and technology evolve, different solutions emerge as preferable at various times.

A simple story could be told about why networks should be dumb and all the complex tasks should be left to the terminals. This story would make short shrift of the fabulous success of the telephone network and ignore emerging requirements for sophisticated services that a dumb network may not be able to offer. Instead of limiting itself to this simple story, this text discusses the pros and cons of different designs.

The plan of the book is as follows: Chapter 1, "Introduction to Communication Networks," introduces communication networks and reviews their history. The chapter explains circuit switching and packet switching, as well as the store-and-forward and channel-sharing methods used by networks to transport information. Packet switching, the decomposition of information into packets of bits that are then transported along links of the network before being reassembled at the destination, is the procedure that makes efficient and reliable networks possible. The historical overview is designed to provide a sense of the rapid pace of evolution of network technology.

Chapter 2, "The Way Networks Work," examines the most popular networks and explains their main operating principles. The chapter starts with Ethernet then discusses Internet and ATM networks. We then explore general architecture models of networks to situate the specific examples in a general framework. The chapter concludes with some complementary material that describes some conceptual breakthroughs that have made the information revolution possible.

Chapter 3, "Internet," examines the main protocols and design principles of Internet. The chapter starts with a brief history of this 30-year old network of networks then explains the delivery of packets and the end-to-end operations that supervise these deliveries.

Chapter 4, "Local Area Networks," describes the most widely used local area networks and explains their operating principles and performance characteristics. The chapter focuses on how the networks work and on their main characteristics. The performance of these networks is discussed in complementary sections that can be skipped if desired.

Chapter 5, "Asynchronous Transfer Mode," explains the main features of this transport technology. ATM is a good switching and link technology that is used for backbones of local area networks and for Internet links. This technology may play a role between end terminals that require high-quality connections.

Chapter 6, "Data Link Layer and Retransmission Protocols," explains how reliable transmissions are implemented. This chapter discusses error control mechanisms and congestion control.

Chapter 7, "Physical Layer," explores the digital communication links: optical links, copper lines, and radio waves. The focus is on the characteristics of these links as network components and on the phenomena that limit the bit rate and distance of these links.

Chapter 8, "Security and Compression," explains the main ideas behind these two essential fields. We discuss the major security threats that networked computers face and the protections against those threats. The chapter examines principles behind audio, video, and data compression.

Chapter 9, "Performance Evaluation and Monitoring," discusses the monitoring of networks. The chapter also presents a few simple results on delays in networks and illustrates these results. We conclude with an explanation of the methodology of simulations.

Appendix A, "Probability," is a self-contained introduction to the concepts and methods of probability theory applicable to our topic. For instance, the calculation of average delays of packets subject to retransmission after a timeout is explained. Appendix B, "Queues and Networks of Queues," is an elementary introduction to queueing theory. This appendix explains the analysis of simple queues and networks of queues illustrated in Chapter 9. Appendix C, "Communication Principles," provides some elements of communication theory for readers with little background in that subject.

Each chapter concludes with problems selected to help readers test their understanding of the material in the book. Each problem section includes some more challenging problems, which are marked with an asterisk (*). Selected references are listed at the end of each chapter and a compilation of all references previously cited appears in Appendix D.

Acknowledgments from the first edition

I am grateful to many students and colleagues for their suggestions and their help with this textbook. In particular, N. Bambos (University of California, Los Angeles), D. Browning (Georgia Institute of Technology), A. Fawaz (Teknekron Communication Systems, Berkeley), B. Hajek (University of Illinois, Urbana), T. Hsiao (Purdue University), P. Lapsley (University of California, Berkeley), N. McKeown (University of California, Berkeley),

B. Mukherjee (University of California, Davis), K. J. Pires (University of California, Berkeley), P. Varaiya (University of California, Berkeley), S. Verdu (Princeton University), all have provided valuable comments on the text. P. Konstantopoulos (University of California, Berkeley) formulated the problems in the last four chapters. Part of the text was written when I was enjoying the warm hospitality of F. Baccelli at INRIA in Sophia-Antipolis, France. Marjorie Singer Anderson was an invaluable developmental editor responsible for the final style of the text.

Finally, this book would not have been written without the encouragement and the help of Howard S. Aksen, who is the most graceful, competent, and supportive publisher and editor a technical author can hope for.

Acknowledgments for the second edition

This second edition has benefited from the comments of many kind users of the first edition. I am grateful to all of them and specially to Professor McKeown and his classes at Stanford for trying the manuscript and suggesting improvements. The editing team was most helpful. Without the support of Pat Frederickson, Betsy Jones, and Kelley Butcher, this book would not have been possible.

In addition, special thanks are due to reviewers who provided detailed recommendations, including

Shih-Chun Chang, George Mason University

Costas Courcoubetis, University of Crete

Dr. Richard Edell, University of California at Berkeley

Gale T. Finley, University of District of Columbia

Victor Frost, University of Kansas

Michael Greenwald, Stanford University

Phil Ho, University of California at Berkeley

Herman Hughes, Michigan State University

Doug Jacobson, Iowa State University

Alan G. Konheim, University of California, Santa Barbara

Richard La, University of California at Berkeley

Stephane Lafortune, University of Michigan

Jean-Yves Le Boudec EPFL, Lausanne, Switzerland

Willis Marti, Texas A&M University

Jeonghoon Mo, University of California at Berkeley

W. Melody Moh, San Jose State University

Lawrence J. Osborne, Lamar University

Bart Preneel, University of Leuven, Belgium

Gene Hill Price, Old Dominion University

Matt Siler, University of California at Berkeley

Mark Soper, University of California at Berkeley

Robert L. Stafford, Temple University

William Teter, SUNY, Plattsburgh

K. Wendy Tang, The State University of New York at Stony Brook

Pramod Visnanath, University of California at Berkeley

Laurence Yogman, Stanford University

James M. Westall, Clemson University

Introduction to Communication Networks

1.1 What Are Communication Networks?

Communication networks are arrangements of hardware and software that allow users to exchange information. This very broad definition will help you begin learning about one of the fastest-growing areas in electrical engineering and computer science. Once we examine some common communication networks, we will develop a more precise definition. In this chapter we will elaborate on the importance of this field and review its evolution.

The telephone network is the most familiar and ubiquitous communication network. It is designed for voice transmission. An office computer network is a communication network used by organizations to connect personal computers and workstations so they may share programs and data and to link those computers to printers and, possibly, to some other peripherals (e.g., file servers that provide mass storage or plotters). Computer networks also are used in manufacturing plants to connect machine tools, robots, and sensors. The Internet is a network of computer networks that covers most of the world and allows millions of users to exchange messages and computer files and some limited video and audio signals.

Although all these systems are communication networks, they are quite different in the information that they transmit and in the way they are used. Nevertheless, they operate on similar principles. The unifying characteristics of all networks help us develop a definition of communication networks that describes the arrangements of hardware and software that we study in this text. Each system described is designed to exchange *information,* which may be voice, sounds, graphics, pictures, video, text, or data, among *users.* Most often the users are humans, but they also can be computer programs or devices. Before the information is transmitted, it is converted into bits (zeros or ones). Then the bits are sent to a receiver as electrical or optical signals (electromagnetic waves, to be more precise). Finally, the information is reconstructed from the received bits. This transmission method, called *digital transmission,* reduces the transmission errors.

The transmission from some user A to another user B can take place over a *point-to-point* communication link, that is, over a link that permanently connects A and B. (Figure 1.1). The physical medium that supports this communication may be a cable, copper wires, an optical fiber, a radio link, or a light wave in free space.

A network is a broader arrangement than a single point-to-point link; it connects a large collection of users. A network is almost never built by laying out one point-to-point link between each pair of users (Figure 1.2a) because the cost would be prohibitive and resources wasted. Instead, a network is organized so as to have different information flows *share* communication links (Figure 1.2b). One of the main network design problems is to find efficient ways of sharing communication links. Typically, the sharing of links means that an information flow may have to wait for a link to become free. Quantities that measure that aspect of the behavior of the network are important network design and selection parameters.

From this description we can define a communication network:

A communication network is a set of nodes that are interconnected to permit the exchange of information.

The nodes in the definition can be terminal entities, such as phone sets, computers, printers, file servers, or video monitors. They can also be communication devices, such as telephone exchanges (switches), routers, or repeaters. Thus there are two types of nodes: *terminal nodes* and *communication nodes.* The terminal nodes generate or use the information

FIGURE 1.1

Point-to-point connection between computers A *and* B. *A dedicated link can be used to connect two computers permanently.*

A   B

FIGURE 1.2

Point-to-point links (a) versus shared links (b). Two solutions are possible for connecting terminal devices. In solution a, each pair of devices is connected by a dedicated link. This solution is not feasible for a large number of devices. In solution b, some links are used by different pairs of devices. That solution achieves significant savings in link length.

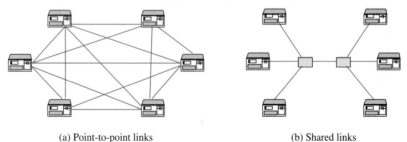

(a) Point-to-point links   (b) Shared links

transmitted over the network. The communication nodes transport the information but do not generate or use it. Note that terminal nodes perform some communication functions: at the very least they have to receive or transmit information.

Our new definition of communication network is still rather general. But it is accurate enough for our purpose now in introducing the subject and our goals in this book.

1.2 Why Should You Learn about Communication Networks?

The design, manufacture, and maintenance of hardware and software for communication networks are among the fastest-growing engineering areas. This growth is fueled by the rapid progress in computer and communication technologies and by the substantial increase in productivity generated by improved communications. This section discusses the impact of communication networks and future applications, emphasizing the expanding opportunities available in this field.

Computers and workstations have had a dramatic impact on the workplace. When these machines are organized as a computer network, their capabilities are extended, increasing even further the range of applications available to users. Computer networks offer immediate cost benefits by allowing many computers to access the same expensive peripherals (a high-quality color printer, for instance). Beyond this benefit, computer networks make efficient communications possible through electronic mail and file transfers. Such communications can improve productivity by reducing the time needed to disseminate information throughout an organization.

New communication networks merge the capabilities of the telephone network and of computer networks, enabling the simultaneous transmission of voice and data. This merging makes new applications possible, such as voice mail, in which a voice message is stored on the disk of the destination workstation. The workstation serves as an answering machine controlled by the caller from his or her phone keypad. Users are able to annotate typed texts or graphics with spoken comments. Speech recognition and speech synthesis systems provide convenient interfaces for visually impaired or for hearing-impaired users. The combination of telephone and computer networks makes existing services more convenient. For instance, the workstation can find and dial phone numbers automatically. The phone–computer combination requires a single hookup and a single *address* instead of a telephone number and a computer address.

Under development are high-speed networks that transmit high-quality, full-motion video. These networks make connections with video and audio channels economical and an effective alternative to many business trips. The entertainment industry will use high-speed networks to distribute digital audio or video programs. Other services can also arise from these networks: instant access to news, to libraries, to help services (medical information, auto and appliance repair, cooking, gardening, quantum mechanics, movie theater schedules, restaurant menus and reservations), and video dating. Some of these high-speed networks combine the technologies of cable television and computer networks.

The possible applications of the future communication networks are so numerous that they will affect most sectors of society. The impact of the future communication networks will be comparable in magnitude to that of the telephone network. All this translates into a range of opportunities for those responsible for working to construct such networks and to apply them to new tasks. The information providers and the telecommunication industry have recognized the vast potential of the new high-speed networks. A number of cable television, entertainment, and telecommunication companies have merged since the early 1990s to combine their expertise and begin to tap into that potential. Cable television companies have announced experimental telephone services and telephone companies have started projects to deploy integrated video, audio, and data networks. Since the breakup of AT&T, local telephone companies have been prohibited from competing in the cable television market. Court rulings are removing those restrictions, and it appears certain that such competition will be allowed in the near future. This development is significant because of the important economy that the industry achieves by combining video, audio, and data services on the same transmission network.

1.3   What Should You Learn about Communication Networks?

A general understanding of the way network operations are organized and of the telecommunication principles on which networks rely helps everyone working with communication networks use them more effectively. This book is intended to provide you with that understanding, which can form the foundation for the more specific knowledge that you may need as a network user, manager, buyer, or designer.

To benefit from a network, the user must understand the capabilities of the available hardware and software as well as the way the information flow should be structured to improve work productivity. Often, a company must reorganize its work procedures to take advantage of the communication facilities created by the network. For instance, for the network to be an efficient tool, department managers must convince all employees to use electronic mail instead of paper mail and to maintain an up-to-date schedule on the network calendar for automatic appointment scheduling.

The network manager must understand the network facilities in order to update the software and hardware and to perform the necessary file backups. The network manager should also monitor the network performance to identify problems quickly and to initiate repairs.

The network buyer faces the complex task of assessing the potential benefits of a network. Before selecting equipment, the buyer should analyze the information flows in the company, determine how new communication channels would affect existing procedures, and assess the value of electronic communications with other company branches or with professional databases. Criteria for selection of hardware and software composing the network are ease of use, ability to be upgraded, and cost.

A network designer must be familiar with the possible services of existing networks and of forthcoming enhancements. The designer cannot construct an adequate system without understanding the needs of the users and the added value provided by new services.

The author of any textbook on communication networks faces a dilemma in presenting details about specific network implementations. As you will see, communication networks are all based on relatively few simple principles. However, an actual implementation has to follow a set of detailed specifications that are, to a large extent, arbitrary. Those detailed specifications are required to guarantee the compatibility of products from different vendors. The details are not necessary to understand how the network operates. For instance, the precise format of the information units being transmitted is important mostly to engineers who design new software or hardware for a given network. To obtain the background, the engineer facing such a task needs to consult official network standards and product documentations. It is not our intention to reproduce that information in this book. However, some details are useful for making the descriptions of networks realistic. For example, knowing the number of address bits used by a network allows you to consolidate your understanding of related size parameters of that network. As a consequence, we have opted for providing enough details to give you a thorough understanding of network operating principles but have avoided the tendency to drown important concepts in a sea of inessential facts.

As background for the current and future technology and applications of communication networks, we now briefly review how networks evolved and describe some of the major trends that affect those systems.

1.4 Evolution of Communication Networks

In this brief historical overview, we examine the main steps in the evolution of the telephone and computer networks.

1.4.1 Telephone Network

The telegraph was developed by Samuel Morse in the 1830s and the telephone by Alexander Graham Bell in 1876. The original point-to-point telephone lines connecting pairs of users in the first telephone systems gave way in the 1880s to lines switched by human operators. Electromagnetic switches appeared in the 1890s. Computerized switches started being deployed in the 1970s.

Digital Transmission and CCS

Two developments in telephone networks paved the way for modern networks: *digital transmission* and *common channel signaling* (CCS). Digital telephone transmission transmits the voice signals as bit streams. Such transmissions have low noise levels. Moreover, digital transmissions facilitate the switching of signals and the simultaneous transmission of many signals on the same line. Common channel signaling transmits connection control information among telephone switching equipment. This control information permits efficient implementations of many services, such as call forwarding, credit card calls, and 800 numbers. It also leads to better utilization of the network lines by providing better control over how the lines are selected to carry the phone calls.

Integration of Services

Digital transmission and common channel signaling form the basis of the *integrated services digital networks* (ISDNs). The basic ISDN connection (called *basic access*) provides a subscriber with three *full-duplex* (i.e., two-way) digital connections: two with a rate of

64,000 bits per second (called the *B* channels) and the third with a rate of 16,000 bits per second (called the *D* channel). The *B* channels can transmit voice or computer data. The *D* channel can transmit alarm, monitoring information, control signals for lights and such appliances as air conditioning and heating units, or network control information as needed in services offered by the ISDN. The common channel signaling system controls the ISDN connections. It is being expanded to carry some of the user data transfers.

The rate of 64,000 bits per second for a telephone call comes from the transmission of 8000 values of the voice signal per second and the encoding of each value with 8 bits. These specific values are selected so that the quality of the transmission is adequate. (See Appendix C for more details.)

Synchronous Network

The telephone networks are implementing a transmission technology called the synchronous optical network (SONET) in the United States, compatible with and similar to the synchronous digital hierarchy (SDH) used in Europe and Japan. In a SONET or SDH network, the transmitters of the telephone company are synchronized to a common clock. Equipment combines a set of synchronized signals from different transmitters by interleaving their bytes (1 byte = 8 bits). Such a straightforward interleaving is not possible with transmitters that are not synchronized. With the SONET or SDH technology, the equipment needed to combine different signals is much simpler, and therefore less expensive.

Cellular Telephones

Cellular telephones enable their users to place and receive telephone calls from a wireless terminal. The 1990s have seen a rapid growth in the market penetration of cellular telephones.

Circuit Switching

Before leaving the subject of telephone networks, let us highlight the main mechanism that these networks use to deliver information. This mechanism is called *circuit switching,* and this term refers to the fact that the computer sets up a circuit between the two telephone terminals for the duration of the conversation and releases the circuit when the call terminates. Here, *circuit* means the resources needed to transmit 64,000 bits per second on a set of links between the two terminals. That is, for the duration of the conversation, these resources are allocated to that conversation and cannot be shared by any other communication.

1.4.2   Computer Networks

Computers were born in the 1940s and began to multiply in the 1960s.

Asynchronous Link

The need for standardized connections of a computer to such peripherals as tape drives, keyboards, printers, disk drives, and terminals led to the publication of the specifications of the connection RS-232-C by the Electronics Industries Association (EIA) in 1969. This connection allows transmission rates of up to 38,400 bits per second over 4 to 12 wires for distances of up to 15 meters. The RS-232-C connection still is widely used and is known as the *serial port.*

Data transmitted over this connection is sent and received character by character. Specifically, one character is sent at a time, and two characters must be separated by a minimum time interval. A character is represented by a group of 7 or 8 bits, depending on the code being used. This type of transmission is called *asynchronous* because successive characters can be transmitted at arbitrary times, other than for the minimum separation. The receiver starts its clock when it detects the start of the character. Using this clock, the receiver reads off the successive bits. The number of bits in a character must be small for this procedure to work: If the number of bits were too large, any small difference between the rates of the transmitter and receiver clocks would lead to errors since the receiver would skip bits or read the same bit twice.

RS-232-C was replaced by the current RS-232-D standard. Standards for faster asynchronous connections (RS-449, RS-422-A, and RS-423-A) followed in the late 1970s.

Modems

In the 1960s, *modems* (*mo*dulator-*dem*odulators) were developed to permit the transmission of bits over telephone lines. The modulator of a modem converts bits into sounds in the frequency range transmitted by telephone lines, and the demodulator converts such sounds back into bits. To implement a long-distance connection between a terminal and a computer, one can connect the terminal to a modem, the modem to a telephone line, and the other end of the line to another modem that is connected to the computer. By dialing different phone numbers that access different modem-computer hookups, a terminal connected to a modem can access different computer services.

Synchronous Link

Computers were first interconnected with point-to-point links in the mid-1960s. The need for fast connections with automatic error control led to the development of a set of procedures called *data link protocols* known under the abbreviations SDLC, LAPA, LAPB, and HDLC. These connections are *synchronous:* They transmit the information in the form of *packets.* A packet is a group of bits, typically from a few hundreds to thousands, that are transmitted at precise times (Figure 1.3). The transmitter encodes the packet bits in a signal that contains the timing information. The receiver extracts that timing information to recover the bits accurately. This synchronization of the receiver through the timing information contained in the signal enables the transmission of long packets. Such long packets are not possible in an asynchronous link.

There is a minimum separation requirement between successive packets instead of between successive characters, as was the case for asynchronous transmissions. This

FIGURE 1.3

Typical packet. *A typical packet consists of a header, user information, and a trailer. The header and the trailer contain control information used by the network to transmit the packet and to verify its correct reception. The header usually contains the addresses of the destination and of the source of the packet. The header may also indicate a sequence number that the destination uses to verify that all the packets were received or to reorder them. The trailer contains error control bits that the nodes use to verify that they received the packet correctly.*

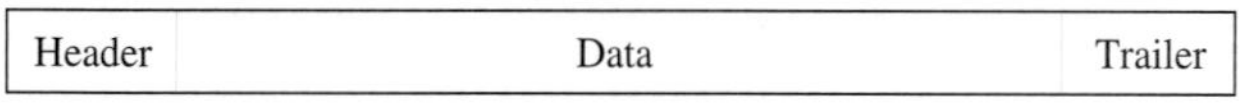

Header	Data	Trailer

difference makes synchronous transmissions faster than asynchronous transmissions: The amount of inserted idle time per transmitted bit is smaller for synchronous transmissions than for asynchronous transmissions. Synchronous connections also can be implemented over telephone lines with a modem at each end. The figure shows that the data bits are framed in a specific packet format that contains additional control information.

Note that the meaning of synchronous in a synchronous versus asynchronous link is not the same as in SONET. In SONET, synchronous refers to the fact that the transmitters are synchronized to a common clock to enable the byte-interleaved multiplexing. In a synchronous link, synchronous refers to the synchronization of the receiver by the timing information contained in the signal that carries the bit stream. There is no master clock in an SDLC link, only a synchronization of the receiver by the signal it receives. Thus, different transmitters on distinct SDLC links use clocks that are not synchronized. In contrast, all the SONET transmitters are locked to a common master clock.

Store-and-Forward Transmission

The development of data link protocols led to the idea of indirect connection of computers. Consider Figure 1.4: If computer B is connected to computers A and C by two point-to-point links, it is possible to send messages from A to C by sending them first from A to B, then from B to C.

This idea is called *store-and-forward transmission.* A store-and-forward transmission from A to C via B is more efficient if the transmission from B to C can start before that from A to B is completed. To do this, one must decompose the messages into relatively small packets. This decomposition allows the transmission from B to C to start as soon as

FIGURE 1.4

Indirect connection of A and C through B. Computer A sends packets to C via B. A packet may be sent from B to C while another is being sent from A to B. If the transmission of a packet from A to B or from B to C takes 1 second, then 60 packets can be sent from A to C in 61 seconds. This method is called store-and-forward packet switching. *If a block of 60 packets was sent in 60 seconds from A to B and then from B to C in another 60 seconds, the full transmission would take 2 minutes instead of 61 seconds when store-and-forward packet switching is used.*

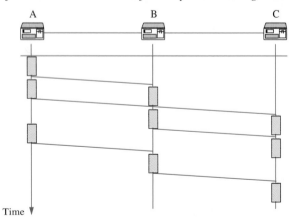

B has received one packet from A instead of having to wait until B has received a complete message from A. To make this idea precise, assume that A wants to send a long message to C and that the transmission of that message on a direct link takes 1 minute. If A sends the complete message to B (which takes 1 minute) before B sends it to C (which takes another minute), then it takes 2 minutes before the message is received by C. However, if the message is decomposed into 60 packets that take 1 second each to be transmitted on a direct link, then during the first second packet 1 is sent from A to B, during the next second packet 2 is sent from A to B and packet 1 is sent from B to C, and so on. After 1 minute and 1 second, the complete message is received by C. The savings achieved by decomposing of the messages into small packets increase with the number of intermediate nodes: if there are N intermediate nodes, the transmission of the undivided message takes $N + 1$ minutes while that of the 60 packets takes only 1 minute and N seconds. The transmission of messages as small packets is called *store-and-forward packet-switching.* The reduction in the delivery time achieved by store-and-forward packet switching is a *pipelining gain.*

Note that packets with different sources and/or destinations can be sent one after the other on a link provided that their source and destination are marked on the packet. When this is done, the link is not reserved by any one communication. Instead, the link is shared by many communications and each uses the link only when it has a packet to send on that link. If a source sends packets only sporadically, then such a mechanism is more economical than reserving the full link for the transmission.

From ARPANET to Internet

When many computers are connected by point-to-point links in a meshlike network, the network designer must resolve a number of questions to ensure effective use of the network resources:

- *Routing:* What paths should the packets follow in the network? In particular, where should packets be duplicated if they are destined to multiple users?
- *Flow control:* How can one regulate the flow so as to avoid having parts of the network become congested?
- *Error control:* How can the network automatically correct errors that occur during packet transmissions?
- *Addressing:* What is a convenient way of specifying the addresses of the terminal nodes?
- *Security:* How can the privacy of the information transmitted over the network and the integrity of the nodes of the network be maintained?
- *Standards:* How can the characteristics of the nodes be described so that vendors can build compatible hardware and software?
- *Presentation:* How can one enable many different types of terminal equipment to communicate?
- *Management:* How can one supervise the operations of a network so as to provide timely and relevant information for detecting and correcting faulty operations and for longer-term network planning?

To study these questions and to develop solutions, the Department of Defense initiated the development of the ARPANET network in the late 1960s. ARPANET was the first large-scale store-and-forward packet-switched network.

The main goal of the Department of Defense in funding the ARPANET project was to develop a robust communication network that could survive the failures of some of its links and nodes. The routing mechanism of ARPANET adapts automatically to changing conditions. If a node or link fails, then the network routes subsequent packets along a different path that avoids the failed link or node.

ARPANET has evolved into the Internet, a collection of interconnected networks that are used today by millions of people. In 1994, the Internet entered the general public's consciousness mostly because of the World Wide Web and its popular interfaces, Mosaic and Netscape. Companies are developing simple computers designed to serve as Web browsers. These devices would have only limited onboard software and they would execute programs downloaded together with the documents from the Web.

Many companies developed store-and-forward packet-switched networks in the mid-1970s. IBM introduced its Systems Networks Architecture (SNA) in 1974, Digital Equipment its DECnet in 1975, Sperry-Univac its Data Communications Architecture in 1976, Siemens its Transdata in 1978, CII-Honeywell-Bull its Distributed Systems Architecture in 1979, etc. These networks differ in how they solve the routing, flow control, addressing, and other problems and also in the way these functions are organized. Standards for *public data networks,* networks that can be accessed by any subscriber, were published in the 1970s by the Consultative Committee for International Telegraphy and Telephony (CCITT). They are known as the X.25 standards. [The CCITT has now become the International Telecommunications Union (ITU)].

The store-and-forward networks that we have discussed are classified as *wide area networks* (WANs). WANs can cover very large areas (hundreds or thousands of miles), and they typically use telephone lines leased from phone companies. The transmission rates range from 56,000 bits per second to 155 million bits per second.

Datagram Packet Switching

The Internet transports packets one at a time. Each packet is marked with its destination address. When a node gets a packet, it reads its destination address and figures out where to send the packet next. If conditions change in the network, successive packets that belong to the same communication between two computers may follow different paths in the network.

ALOHA: The Birth of Multiple Access

In parallel with the evolution of store-and-forward networks that send information over a mesh of point-to-point links, the 1970s saw the development of networks in which nodes share a single communication channel. These networks are called *multiple-access* networks (Figure 1.5). An early example of a multiple-access network is ALOHA, which was developed to connect computers situated on different Hawaiian islands. The computers of the ALOHA network transmit on the same radio channel whenever they have a packet to transmit. When two or more computers try to transmit simultaneously, the information is *garbled.* Thus, the computers transmit their packets, then check whether the transmissions were garbled; if so, they wait for a random time before trying again.

FIGURE 1.5

Multiple-access networks. ALOHA is a packet radio network developed at the University of Hawaii in the early 1970s. A variation of the protocol of ALOHA is used by Ethernet, developed in the mid 1970s at Xerox PARC. The token ring used a token-passing protocol to regulate the access to the channel; it was developed by IBM in the early 1980s.

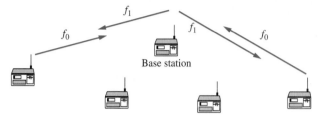

Ethernet

An idea similar to that of ALOHA was used in the early 1970s by researchers from the Xerox Palo Alto Research Center (PARC) to connect computers with a single transmission line. The resulting network—Ethernet—differs from ALOHA in that the transmitters interrupt their transmissions as soon as they detect simultaneous transmissions. This detection may occur long before the packets are transmitted completely (at least, if the cable is not too long). Consequently, Ethernet is more efficient than ALOHA.

Local Area Networks

The development of Ethernet coincided with the explosion in the use of personal computers. It is the most widely used *local area network* (LAN), that is, a network for computers that are geographically close. Another popular LAN is the *token ring,* developed in the early 1980s by IBM. Other LANs that were also introduced in the 1970s and 1980s include the *token bus,* the *slotted ring, AppleTalk, DataKit,* and *StarLan.* The LANs have high transmission rates, typically a few million bits per second; geographical ranges limited to a few miles; and short delays, usually fractions of a second, before the start of transmissions. In the late 1980s, the *fiber distributed data interface* (FDDI) was introduced. This network is similar to a token ring except that it operates at 100 Mbps (one hundred million bits per second), instead of 4 or 16 Mbps for the token ring) and that it controls the transmissions so that no node has ever to wait for more than an agreed-upon time before it can transmit. The Institute of Electrical and Electronics Engineers (IEEE) standardized the 100-Mbps Ethernet in 1994 and is developing a Gigabit Ethernet standard for 1997. The transmission rate of a LAN is shared by all the stations attached to it. To increase the total transmission rate, network managers decompose a LAN into domains that they interconnect with a LAN switch. This device enables simultaneous transmissions in the different domains and it automatically forwards packets that must go from one domain to another.

Asynchronous Transfer Mode

Another important development in the early 1990s is the *Asynchronous Transfer Mode* (ATM). An ATM network transports information in fixed-size packets of 53 bytes (called cells). The transmission rate ranges from about 25 Mbps to 155 Mbps for local ATM networks and from 1.5 Mbps to multiples of 155 Mbps for wide area networks. The ATM networks are flexible in the transmission services that they offer to the users: from best-effort delivery of packets to transmission with guaranteed small delays and high rate. In 1996, equipment for setting up local ATM networks was available from dozens of vendors and wide area ATM links were used by the Internet.

In ATM, asynchronous indicates that the cells are sent at arbitrary times, in contrast with the voice samples sent in a SONET network or a circuit-switched network.

Virtual Circuit Switching and Comparison of Switching Mechanisms

The ATM cells that belong to a given connection all follow the same path in the network. When the connection is set up, the ATM network selects a path for its cells.

We have discussed three different switching mechanisms in this section: circuit switching used by the telephone network, datagram store-and-forward packet switching (e.g., in the Internet), and virtual circuit switching (e.g., ATM).

We come back to these mechanisms when we study specific networks. For now, let us recall their main features. Circuit switching minimizes the delay of the information transfer because all the forwarding decisions are made once and for all when the connection is set up. However, circuit switching reserves fixed communication resources for the duration of a connection even though these may not be needed all the time. Datagram is very flexible because it can adapt to changing conditions during a given connection. Also, in datagram switching, a connection uses a link only when it transmits a packet on the link. Finally, virtual circuits have the advantage of making the routing decision only at the start of a connection and of enabling the sharing of resources by different connections.

The Future of Networks

In late 1997 the future of communication networks looks as follows. The Internet will keep on growing with more users and faster communication links and routers. Internet engineers will keep on improving the router technology and the protocols to offer better quality of service, especially for distribution of video and also for video conferences. The cable TV and telephone companies will deploy video-on-demand networks using a hybrid fiber–coaxial cable technology and using high-speed transmissions over telephone lines. LAN switches will be introduced to increase the throughput of LANs, first for Ethernets but also for token rings. More 100-Mbps Ethernets will be deployed and gigabit Ethernets will be deployed. ATM networks may replace other LANs first as interconnection backbones and then as connections to the desk. Companies and institutions will start using video conferences from the desktop.

Some difficult questions about the future evolution are the following:

- What are the future roles of ATM and of the Internet? Will the Internet protocols and technology progress enough to make migration to ATM unnecessary or will there be coexistence of ATM and Internet networks used by different applications?
- Will the telecommunication companies set up a commercial wide area network that will complement the Internet and provide a better alternative for video conferences or other real-time or interactive services?
- What is the future role of satellite distribution systems for video or information-on-demand networks?

- Will the long-anticipated demand for cellular data networks and devices materialize?

In the book we provide you with the knowledge to appreciate many aspects of these questions.

1.5 Organization of the Book

Our objectives in this text are to explain the principles that guide the design and operations of networks, to describe some of the popular networks, and to identify important trends.

As a rule we have placed material that can be read out of sequence or that some readers may want to skip in "complements." By selecting complements, the instructor can adjust the level of the course. Without the complements, the material is suitable for a junior level course in electrical engineering or computer science.

Chapter 2 explains the main operating principles of networks. The objective of that chapter is to provide a big picture of networks so that you can better appreciate how the various pieces fit together.

Chapter 3 is a walk through the Internet protocols. By studying that short chapter you should become familiar with the main ideas of the Internet and with recent developments.

Chapter 4 discusses some local area networks: Ethernet, token ring, and FDDI. The chapter is streamlined to make the main concepts accessible. More advanced issues are studied in complements.

Chapter 5 is devoted to ATM. This short chapter discusses the architecture of ATM protocols and explores the classes of service and the routing in more details. The chapter also explains two types of internetworking with ATM: LAN emulation and Internet Protocol (IP) over ATM.

Chapter 6 presents the data link layer and retransmission protocols. The mechanisms of this chapter (framing, error control, retransmissions) are used by all networks.

Chapter 7 discusses the physical layer: how to transmit bits over copper, fibers, or radio waves. This chapter has a more "electrical engineering" flavor than the rest of the book, and it can be read independently.

Chapter 8 explains the main mechanisms that security systems use and illustrates them with two widely used systems: Kerberos and PGP. The chapter also discusses audio, video, and file compression. This technology makes multimedia network applications possible.

Chapter 9 is devoted to the evaluation of measures of networks' performance by monitoring analytical methods or simulations.

Appendix A reviews some basic probability theory which is useful for most of the text. Appendix B covers elements of queueing theory which are needed in Chapter 9. Appendix C is a review of communication theory for interested readers with no training in that field.

Summary

- A communication network is a set of nodes interconnected to permit the exchange of information.

- Communication networks are a rapidly growing engineering field which will significantly affect our personal and professional lives.
- The text will provide you with a general understanding of the operating and design principles of networks.
- The evolution of networks began with digital communications, followed by the introduction of store-and-forward and of multiple-access packetized communications.
- Communication networks are getting faster and are used frequently to transmit audio, video, and data.

CHAPTER 2

The Way Networks Work

In this chapter we describe how some widely used networks work. We keep the discussion at an elementary level by avoiding most considerations of performance and details of packet formats. This chapter gives you a basic understanding of what networks do and provides you with the background necessary to appreciate the more detailed discussions of the rest of the book.

This chapter is a quick survey of different network technologies. Our objective is to introduce the operating principles of these technologies and to provide a "big-picture" look at networks. We revisit the networks discussed in this chapter later in the book when we explore their operation in more details. The chapter should help you situate the trees in the forest.

We start our tour of networks in Section 2.1 by looking at the most popular local area network: Ethernet. A local network is designed to connect up to a few hundred computers typically situated in the same building. We study Ethernet in more detail in Chapter 4. In Section 2.2 we explore the Internet. The Internet is a worldwide network of networks and computers. You can connect your home computer to the Internet through an Internet service provider via the telephone line. Your office computer is probably attached to an Ethernet connected to the Internet by a router. Chapter 3 is devoted to a discussion of the protocols of Internet. Section 2.3 discusses asynchronous transfer mode (ATM) networks that are being deployed as a complementary technology to Ethernet and the Internet. We revisit ATM in more depth in Chapter 5. In Section 2.4 we explain architectural models that constitute frameworks that you can use to consolidate your understanding of networks. Section 2.5 is complementary material that describes some conceptual breakthroughs that have made the information revolution possible. These breakthroughs are the move from analog to digital representation of information, the fundamental understanding of the limitations of such representation and of the transmission of bits through noisy channels, and packet switching.

2.1 Ethernet

Ethernet was developed at the Xerox Palo Alto Research Center in the mid-1970s and was further refined in a joint project between DEC, Intel, and Xerox in 1980. IEEE published the first versions of the specifications in 1981, 1982, and 1985. Extensions to different media and higher transmission rates were published later. The specifications based on the DEC, Intel, Xerox (DIX) design are slightly different from the IEEE 802.3 standards.

We first explain how the computers exchange packets. We then explore how the computers discover the addresses they need for the packets. Finally, we examine the interconnection of Ethernets.

2.1.1   Shared Ethernet: Hubs and Collisions

Layout

Figure 2.1 shows the Ethernet to which my office computer is attached. Twisted wire pairs connect my computer to a box situated about 80 meters away in a wiring closet. That box is attached to other boxes in the same closet and about 50 computers on the same floor are attached to one of those interconnected boxes.

Such a box is called an Ethernet hub, an appropriate name if one views all the computers attached to the hub by wires that look like the spokes of a bicycle wheel. My computer comes with an Ethernet connection, as do most personal computers and workstations. The bulk of the cost of my connection to this Ethernet is the wiring cost; the rest is the fraction of the cost of the hub and of the salary of our network administrator. The computers attached to the same set of interconnected hubs are said to be in the same "collision domain." This term refers to the fact that packets sent by these computers may collide when they arrive at the hub at the same time.

When a computer sends a packet—that is, a group of bits—onto the Ethernet, its electronics encodes the bits into electrical signals. These signals propagate at a fraction

FIGURE 2.1

A shared Ethernet. Each computer has a name, a network address, and an Ethernet address. The computers are attached to a "hub" which repeats incoming packets on its other ports. The hub also informs the computers of "collisions" that occur when two or more packets arrive at the hub at the same time.

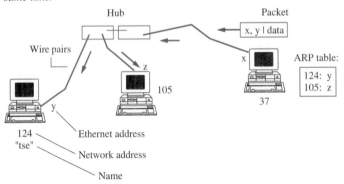

(about 60 percent) of the speed of light from the computer attachment to the hub. The hub repeats, or reproduces, the signals onto all its other connections, or ports, toward the other computers attached to the same Ethernet.

Thus, all the computers on the Ethernet "see" the packet. To identify the source and the destination of a packet, the sending computer marks the packet with its Ethernet address and with that of the destination. The Ethernet address is a unique 48-bit string that comes wired into the electronics of the Ethernet attachment. When a packet arrives at an Ethernet attachment, the electronics compares the destination address with its own and discards the packet if the addresses differ, unless the address is a special "broadcast address" which signals that the packet is meant for all the computers.

In this way, the computers on a common Ethernet can send packets to one another, provided they know each other's addresses. The transmission rate on an Ethernet is either 10 million bits per second (Mbps) or 100 Mbps. Soon, probably in 1998, we will be able to buy Ethernets with a transmission rate of 1 billion bits per second (Gbps). At 100 Mbps the network is called Fast Ethernet and at 1 Gbps it is Gigabit Ethernet. These versions differ in how they transmit the bits.

Collisions

The Ethernet just described is called a "shared Ethernet" because all the computers share the same wires: all the packets get transmitted on all the wires. The computers take turns transmitting their packets. What if two computers start transmitting at the same time? When this happens, the hub sees two packets arriving at the same time on two different ports. The hub then knows that it cannot transmit these two different signals at the same time on its other ports because they would clobber each other. Instead, the hub sends out a signal to all the computers to warn them that a "collision" is occurring. The computers that are transmitting stop when they get the collision warning. They then wait for some random time before they try again to transmit their packets. Moreover, a computer knows that it should not start transmitting if there is another transmission in progress. With these simple rules—called a media access protocol—all the computers eventually get their turn to transmit. The media access protocol that Ethernet uses is called *carrier sense multiple access with collision detection* (CSMA/CD).

2.1.2   *Discovering Addresses: ARP and RARP*

When sending a packet to another computer, a computer refers to the destination by its "network address." The network address is based on the location of the computer and is used by the network to forward the packet to its destination. This network address is independent of the 48-bit Ethernet address of the destination, which has nothing to do with the computer location. However, to send the packet the sending computer must attach the destination's Ethernet address. How do the computers find each other's Ethernet address? We first describe an elementary method and we then explain how to improve it.

Naive Strategy

In this elementary method, the first step is to allocate identification numbers to the computers on our Ethernet. We call these numbers "network addresses" to differentiate them from the Ethernet addresses. We register the network addresses with the network manager to make sure that they are distinct. In the second step, the network manager goes to each computer and runs a little program that reads the internal Ethernet address of the computer. The network manager then writes a configuration file "My-Address" into each computer that tells the computer its network address. In the third step, the network manager goes back to each computer and completes a configuration file "List" which lists the network address and Ethernet address of every computer attached to the Ethernet. To send a packet to computer "124" on the Ethernet, my computer "37" checks its file "List" and attaches the 48-bit Ethernet address of "124" as the destination address of the packet and its own Ethernet address as the source address.

ARP

We can improve the strategy by automating the third step as follows. Each computer has a file "My-Address" with its own network address and a file "List" that is initially empty. When I ask my computer to send a message to "124," my computer looks into its file "List" to see if it contains an entry for "124." Since there is no such entry, my computer sends a broadcast message on the Ethernet asking "If you are 124, please reply to me." A broadcast message is one that is destined to all the computers on the network. Such messages have a special destination address, called a broadcast address. The computers see this message, compare "124" with their own network address stored in their file "My-Address." The computer "124" replies to "37" which finds out the Ethernet address of "124," say x, by reading the Ethernet source address of that reply. My computer then writes an entry into its file "List." That entry says "124 = x" and the computer can finally send the packet to 124. The computer keeps that entry in the file for future use.

This strategy is called the address resolution protocol (ARP) and is the method used in by the protocols of the Internet suite. Other protocol suites use variations of this strategy.

RARP

Some computers do not have a disk of their own, and they boot from a network server. In that case, the list of names is on the server. When the computer starts, it sends a special broadcast message that asks "Server, please tell me my network address." The server can give the appropriate network address that corresponds to the Ethernet address of the asking computer. This strategy, called the reverse ARP (RARP), is used by the protocols of the Internet suite.

2.1.3   *Interconnecting Ethernets: Switches and Routers*

The number of nodes on an Ethernet is limited for performance reasons. Moreover, the length of wires is also limited to about 100 meters.

The performance issue is easy to appreciate. We explained in Section 2.1.1 that the computers on the Ethernet share a set of wires on which they can transmit bits. If 10 computers attached to one shared 10-Mbps Ethernet are trying to transmit files at the same time, each sees an average transmission rate of 1 Mbps. In this situation, transmitting a 1-MByte file takes at least 8 seconds. (In actuality, this transfer will take longer because computers waste time when their packets collide.)

For these reasons, one is often led to interconnect multiple Ethernets. We interconnect Ethernets with switches or with routers. (Switches used to be called bridges. We no longer use this terminology here although you will still find it in the literature and some authors are attached to it because of the differences in operations of some switches.) We examine these two situations next.

Switches

Figure 2.2 shows three Ethernets attached together with switches. A switch is a device attached to two or more Ethernets. In particular, one can replace the hub of an Ethernet with a switch. When we do this, the "shared Ethernet" becomes a "switched Ethernet." A switch is capable of transmitting multiple packets at the same time, provided that the packets have different input and output ports. This method multiplies the throughput (the rate of transmission) of the networks by a factor which depends on the number of distinct input/output port pairs of packets sent at the same time. Thus, if most of the transmissions take place between disjoint sources and destinations, this switched Ethernet has a significantly larger throughput than a shared Ethernet. However, if all the packets are sent to or from the same file server, then the switched Ethernet has the same throughput as a shared Ethernet.

When computer S wants to send a packet to computer D, it sends an Ethernet packet on the Ethernet E_1 with the Ethernet source address s of computer S and Ethernet destination address d of computer D. The switch B sees all the packets and looks at the destination address d. The switch maintains a list of pairs (Ethernet address, switch port). The list specifies which link each Ethernet address is attached to. By consulting this list, the switch finds out that d is not yet in the list; it then copies the packet to all its ports, except the incoming port of the packet. In the example of the figure, switch B copies the packet on Ethernet E_2.

How does the switch B maintain its list? It reads the source address of all the packets that other nodes transmit and adds them to the list. Thus, if computer D sends a packet that

FIGURE 2.2

Switched Ethernets.

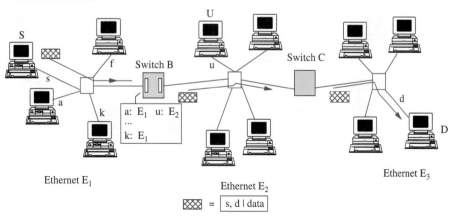

switch C copies onto E_2, then switch B adds d to its list as being attached to E_2. Each entry has a time to live. That is, if a packet with a given source address has not been seen for some time T, then the entry is deleted from the list.

You will note that if the switch copies a packet from E_1 to E_2 even though the destination is on E_1 (but is not yet on the list), then no harm is done: the packet is ignored by all the nodes other than the switches and eventually disappears.

An interesting twist to this procedure occurs when we attach Ethernet with switches in a way that creates loops in the network. Loops are desirable for reliability: if one switch fails, there is another path for the packets. Figure 2.3 illustrates an interconnection with a loop.

A packet sent from S to D is copied by B from E_1 to E_2. When it appears on E_2, the other switch F copies it back to E_1. The switch E_1 then copies it again onto E_2, and so on. How do we prevent this behavior? One possibility would be for B to keep track of which packets it has copied previously so that it does not copy them again. This is clearly not a good solution because of the complex bookkeeping and comparisons required. The standard solution is the "spanning tree" algorithm where the switches execute an algorithm that identifies a tree in the network. A tree is a graph without loop that reaches all the Ethernets. Such a tree is shown in Figure 2.4. The procedure, called the *spanning tree routing,* is then that only the switches that are on the tree copy packets.

Router

Figure 2.5 shows Ethernets attached together with a box that has multiple ports. The box is a router: it decides where to send a packet next.

Recall that a switch makes the decision to copy a packet from the packet's Ethernet address and copies a packet without modification. A router operates differently. The router makes the decision of whether and where to forward a packet based on the network address of the destination, not the Ethernet address. Moreover, the router modifies the Ethernet addresses of the packet.

FIGURE 2.3

Loop in switched Ethernets.

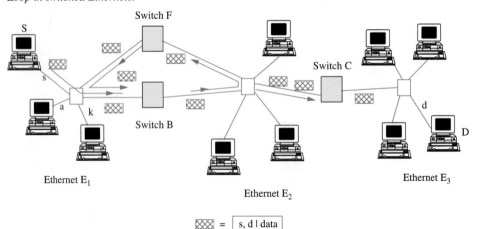

FIGURE 2.4

Spanning tree. To prevent loops, the switches run an algorithm to select a tree. The tree is shown with thick lines.

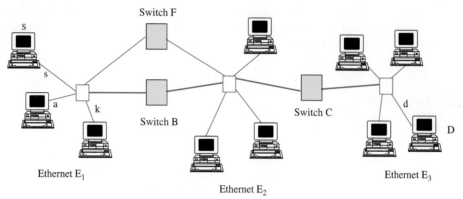

FIGURE 2.5

Ethernets interconnected by routers.

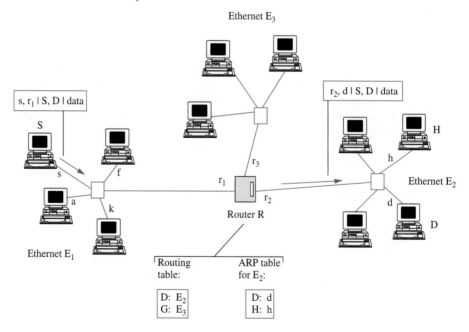

Assume that computer S in Figure 2.5 wants to send a packet to computer D. Here, S, D, R are network addresses of computers. The addresses s, d, r_1, and r_2 are Ethernet addresses. Computer S sends the packet [s, r_1|S, D|*data*] with Ethernet source address s, Ethernet destination address r_1, network source address S, and network destination address D. The router reads that packet and looks at its network destination address D. The router then searches a routing table that indicates the next link for destination D. In this example, the next link is Ethernet E_2. The router then sends the packet [r_2, d|S, D|*data*] onto E_2 and eventually, computer D gets the packet.

The routers need to maintain routing tables to guide the packets to their destination. A great deal of the cleverness of the Internet is in how its routers maintain their routing tables. We sketch the main ideas in the next section.

2.2 Internet

The Internet is a global network of networks that use a specific set of protocols. The Internet exploded into the public consciousness in the mid-1990s with the World Wide Web, a hyperlinking of documents located in computers around the world. The documents, described in an easy-to-write language called HTML, can include text, graphics, video clips, audio, and links to other documents. Users navigate in this worldwide collection of documents by simple mouse clicks on links or locate them by using a search engine. The World Wide Web, with its popular interfaces Netscape Navigator and Microsoft Explorer, breaks down distances, integrates multimedia, enables users to find information, and, equally importantly, makes every user a potential publisher. The Web merges computers and communication and transforms every personal computer into a personal communication device.

The eventual impact of the Internet on society is hard to fathom. Many sociologists predict that it will be larger than that of the printing press, the telephone, television, and radio put together. As many computer and communication experts have been insisting to an incredulous audience since the early 1970s, we are in the middle of the Information Revolution, the third wave. The value of connectivity is much larger than its cost, and it increases with the number and quality of information available on the Web and with the number of people you can reach with it. As a result, the number of users of the Internet is growing exponentially fast. In 1996, about 40 percent of U.S. homes have a personal computer (PC) and approximately 25 percent of them are using the Internet.

The technology of the Internet, although remarkable, it still primitive. It connects millions of heterogeneous computers with best-effort services, with no guarantee on the delays or even the loss rate of the transmissions. The network is not reliable and offers no security. In the large view, these are insignificant temporary shortcomings. The first telephone connections and the first television transmissions were of low quality and not reliable. In spite of its temporary shortcomings, the Internet is a revolutionary medium.

The Internet has grown from a little experiment in 1969 with a few small nodes to a worldwide collection of a few million interconnected computers. One should marvel at the insight of the original designers that enabled such a remarkable growth. The most astonishing aspect of the Internet is its self-organizing mechanisms. Nobody knows how many nodes are attached or where they are, yet packets find their way to their destination. Moreover, if a link or a router breaks down, packets automatically transit along other paths.

In this section we look at how all this works. We first examine a simple example and we explore how packets get to their destination. We then look at how names and addresses are organized. We outline the routing algorithms and we discuss the transmission of packets and the design of applications. We study the Internet and its protocols in more detail in Chapter 3. This section examines the main steps in sending packets across an internet.

2.2.1   An Example

Figure 2.6 shows a small internet. This is a network that uses the same rules of operation as the Internet. We follow accepted conventions and reserve the capitalized Internet for the (global) Internet.

The nodes all have unique names and network addresses. More precisely, every network attachment of a computer has a unique network address. In addition, we know that every Ethernet attachment has a unique Ethernet address.

We explained in Section 2.1.3 how a packet would get from S to D. We review these steps briefly. Computer S knows its router, possibly after having asked, and also knows its own network address, possibly through RARP. Computer S then sends the packet [s, r_1|S, D|*data*]. Router R checks its routing table and finds that the next link is r_2. This link is not an Ethernet but is instead a dedicated link. Sending packets on this link may require that R format them in a particular way and possibly fragment the packet into smaller pieces. For simplicity, we ignore these steps and simply consider that R sends [S, D|*data*] on the link r_2 to the next router V. Eventually, the packet reaches the router W of the Ethernet E_2. To send the packet on that Ethernet, W needs the Ethernet address d of D. If that Ethernet address is not in the list "List" of the router W, that router uses ARP. It then gets d and sends the packet [r, d|S, D|*data*].

A global directory system enables a host to find the Internet protocol (IP) address that corresponds to a given name. Using this system, called the *Domain Name System* (DNS), an application that wants to send a packet to a computer with name *name* first finds the IP address D of that computer. The transmission of the packet then proceeds as we explained above. We describe the organization and the operations of DNS in Chapter 3.

We examine the routing next.

FIGURE 2.6

A small internet.

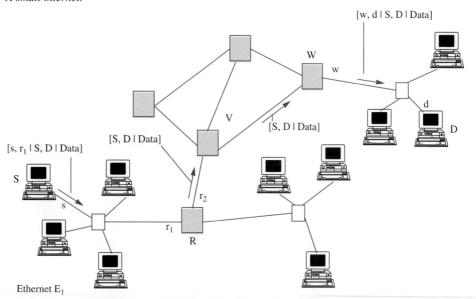

2.2.2   *Routing: OSPF and BGP*

How can the routers maintain tables that indicate the next link a packet should follow to reach a given destination address? We examine four increasingly sophisticated procedures.

Procedure 1: Manual

The first procedure is to have a network manager keep track of the network map, compute routing tables, and send the tables to all the routers.

This procedure would be fine for a small network that changes rarely. However, a link or node failure would cause packets to not find their destination, and correcting the failure would be a slow procedure. It is precisely to avoid such consequences of failures that the original ARPANET experiment started.

Although we will quickly move to better strategies, let us reflect an instant to see how the manager could get the routers to report information that would be sufficient to construct the network map. Imagine that each router can determine the addresses of its neighbors and somehow report the list of such addresses to the manager. The manager can then put those lists together and construct a map. How do the routers send the lists to the manager? One possibility is to flood the network: send a copy along each outgoing link and copy each flooding packet onto all links but its incoming link.

How does a router learn about its neighbors? It sends a message on the link that attaches to another node and asks "Router, give me your network address."

Procedure 2: Maps

A simple modification of procedure 1 becomes feasible for relatively small networks. By using flooding, a mechanism that sends each packet to every node, all the routers get lists that enable them to construct a full map of the network and to identify the paths to be followed. If all the nodes use the same algorithm to construct the paths, they build consistent routing tables.

The routers can be a bit more sophisticated and characterize their outgoing links not only with the address of their neighbor but also with a collection of link parameters. Possible link parameters are the maximum size of packets on that link [called the maximum transfer unit (MTU) of the link], the typical delay of a packet on that link (as estimated from the number of packets waiting to be sent, for instance), the maximum rate of that link, the level of security, and the reliability of the link.

When the routers get these lists of parameters, they can construct the best paths from themselves to all the other nodes. Observe that the routers need only know how to reach the router of an Ethernet, not each node on that Ethernet. This procedure is called open shortest path first (OSPF).

Procedure 3: Hierarchical Routing

The difficulty with procedure 2 is that the routing tables grow with the number of computers or Ethernets. Internet uses two ideas for improving the scalability of the network.

The first idea is to arrange the addresses in a two-level hierarchy. That is, an address has to form network.host and the routing is based on the *network* part of the address instead of on the full address. With this mechanism, the routers need only one entry per *network* instead of an entry per node.

The second idea is a further decomposition of the network similar to what we do when we travel to a distant city. To go from my house in Berkeley to a friend's house in Palo Alto,

I use a local map of Berkeley, a local map of Palo Alto, and a freeway map. I find how to get from my house to the freeway toward Palo Alto, look at the freeway map to make sure I head off along the correct freeway, and I look at the Palo Alto map to get to my friend's house. The routers of the Internet are grouped in clusters called *autonomous systems* (ASs). An AS is supervised by one organization. Thus, the Berkeley campus is one AS. So are the Stanford campus and the Internet routers (the freeway) that connect them together. To send a packet from Berkeley to Stanford, a local algorithm (OSPF) tells the Berkeley routers how to reach the Internet router. Another algorithm tells the Berkeley AS how to reach the Stanford AS, and finally OSPF tells the routers inside Stanford how to reach my friend's computer.

The algorithm between ASs could be a shortest path algorithm such as OSPF. However, since 1989 the Internet has been using an algorithm that is better at detecting loops. Moreover, that algorithm, the border gateway protocol (BGP), makes it easier to base the routing on more complex considerations of security and reliability. The key idea of BGP is that an AS advertises the path it prefers to reach a given destination by listing the ASs that the path goes through and by providing a set of characteristics of the path such as its bandwidth and its average delay. For instance, in Figure 2.7, the AS **Y** advertises that its preferred path to reach the destination **D** is (**Y, X, Z, D**). The AS **Z** advertises that its preferred path to **D** is (**Z, D**). Additionally, the ASs **Y** and **Z** provide the estimated delays 15 and 17 along their preferred paths. You note that these delay estimates are not consistent with the preferred paths: the delay along the path (**X, Z, D**) should be larger than along the path (**Z, D**). This lack of consistency may occur because the delay estimates are based on information that may be somewhat out of date. If it were to use the delay estimates alone, the AS **X** would send a packet to **Y** and **Y** would send the packet back to **X**, thus creating a loop. By looking at the preferred paths, **X** knows that it should send its packet directly to **Z**.

2.2.3   Transmission Control Protocol

The Internet endeavors to deliver packets to the correct destination, but makes no guarantees. Packets may get lost on their way because of congestion in the links and routers, or a link

FIGURE 2.7

Illustration of the border gateway protocol.

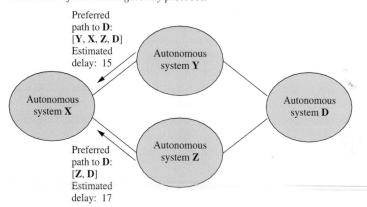

may fail completely. Also, transmission errors may corrupt a packet. To make the Internet usable, applications need a technique for ensuring that packets are retransmitted until they are correctly delivered. This is the job of the transmission control protocol. In addition, the transmission control protocol regulates the rate at which computers send packets to prevent some links from becoming excessively congested.

The transmission control protocol (TCP) is a control mechanism that runs in the source and destination computers. Although the details of TCP are rather involved, the key ideas are elementary and we explain them next.

Consider a transmission of a long sequence of packets from a source computer S to a destination computer D. To supervise the transmissions, D acknowledges every correct packet that it receives by sending back an acknowledgment for that packet to S. Computer S maintains a counter and timers. The counter makes sure that S stops sending packets when it has sent N packets for which it has not yet received the acknowledgment. For instance, S could send N packets, then wait for the acknowledgment of the first packet before it sends the $(N + 1)$st packet, and so on. If N is small and if it takes T seconds from the time S sends a packet until S receives the acknowledgment, then you see that S sends about N packets every T seconds. Thus, by modifying N the computer S can adjust the rate at which it transmits and therefore the load that the network routers face. The idea of TCP is then for S to reduce N when S believes that some congestion is occurring and to increase N otherwise. S can detect congestion by measuring how long it takes for acknowledgments to come back, that is, by estimating the value of T. An increase in T is likely to be a sign of congestion and signals that S should reduce N. Of course, a late acknowledgment may also be caused by a transmission error, but such errors are unlikely if most links are high-quality copper or optical links. If the connection goes over noisy wireless links, then such links should implement their own local retransmissions to avoid burdening the end-to-end connection with retransmissions.

2.2.4   Client/Server Applications

The generic way to design network applications is using the client/server model. In this model, a server is a process designed to service requests sent by clients.

Typically, the client sends messages using TCP to the server. The server checks the TCP connection to wait for requests. A server is designed either to service the requests one at a time or to replicate itself for each new request.

Since TCP is another process that runs inside the computer, to write a client or a server you need to know how to send messages between processes. This interprocess communication depends on the operating system that your computer uses. In UNIX BSD, interprocess communication is implemented trough queues. To send a message to a process, you write into a queue destined for that process. This write operation is called a socket write. Similarly, to read messages the process performs a socket read.

At any given time, a number of network applications may be running in your computer. They correspond to different processes that are differentiated by distinct TCP port numbers.

Generic scripts for such network clients and servers are widely available in C, C++, or Java.

2.3 Asynchronous Transfer Mode

Asynchronous transfer mode, or ATM, is a networking technology designed to complement the Ethernet and the Internet technologies.

It is useful to examine the operating mechanisms of ATM because they differ significantly from those of the Internet. By comparing these mechanisms you will appreciate better the features of both and you will realize that there is more than one way to design networks.

We first look at a few key features of ATM networks, then we examine the routing and the control of quality of service. Chapter 5 is devoted to ATM. In that chapter, we examine the technology and protocols of ATM in more detail and we study how ATM networks can be used to interconnect IP subnetworks and to emulate LANs.

2.3.1 Main Features

ATM transports information in 53-byte cells along virtual circuits. That is, before transmitting information to another computer, the source asks the network to set up a connection. The network identifies a path along which it reserves the resources that are needed for that connection and it then tells the source to start transmitting. The resources include transmission rate and buffer space in the switches. The source packages the information into 53-byte cells that the network transports along the path that it has selected. As it sets up the connection, the network updates routing tables into its switches. The network may have to block a connection request if it is unable to locate a path with sufficient resources.

The cells have a 5-byte header that contains an identification number of the connection. As a switch gets a cell, it checks the connection number, looks into its routing table, and finds out how to handle the cell: its output links, its level of priority, and so on. When the connection terminates, the network informs the switches to update their routing tables and to release the resources it had set aside for that connection.

ATM is designed to accommodate a wide range of applications ranging from video conferences to file transfers and electronic mail. Some of these applications, such as video conferences, generate a constant bit rate traffic and require a small end-to-end delay. Other applications, such as email, can tolerate large delays. The network will take advantage of these different requirements to handle the cells differently. For instance, a switch may serve its video conference cells before email cells.

2.3.2 Routing

Figure 2.8 shows three virtual circuits that are set up in a small ATM network and the routing tables inside the nodes. Note the numbering of the virtual circuits. That numbering need only be unique on each link instead of end to end. Thus, the virtual circuit from B to D uses the number 1 from A to S_2 and the number 2 from S_2 to S_3 and from S_3 to D.

One possible rule for numbering the virtual circuits is to use the smallest number not already used on a link. You can verify that the numbering shown in the figure would arise from applying this rule if the virtual circuits were set up in the order A to D, G, then E to F, then B to D. Note that the cells from A are multicast to D, G by being copied in S_2.

FIGURE 2.8

Virtual circuits and routing tables.

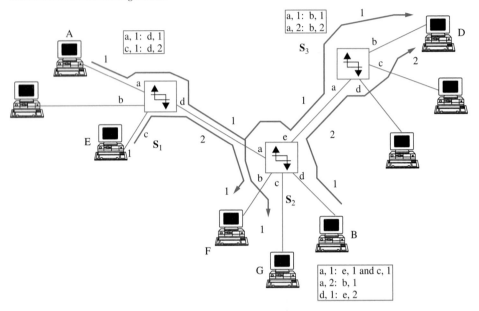

For each incoming link and incoming virtual circuit number, the routing table specifies the outgoing link, the outgoing virtual circuit number, and some indication of how the cell should be handled. For instance, the entry that corresponds to the cells from *B* specifies the identity of the multicast program. The switches use that information to add other destinations to the multicast upon request.

To limit the amount of bookkeeping, virtual circuits are grouped into virtual paths that are routed together. If the cells of a given virtual path are handled in the same way (e.g., same priority and multicast group), then the switch need only maintain one entry per virtual path.

Figure 2.9 shows the most common design of ATM switches. The cells arrive into small input registers that can store one cell each. The cells are then broadcast in turn on the switch bus. Output buffers are attached to the bus and are equipped with filters that check which cells they should copy into the buffer. Thus, the routing tables are implemented in the filters. In addition, each output buffer can be decomposed into a number of parallel buffers that correspond to different priorities of the cells. Note that multicast cells are copied simultaneously into the appropriate output buffers. This switch is limited by the speed of the memories that must be able to handle the sum of the line rates.

2.3.3   Control of QoS: Leaky Buckets

The original motivation for designing ATM networks was to provide a better control of the quality of service (QoS) than that possible in a datagram network such as Internet. Since an ATM network keeps track of individual connections, it can—at least in principle—make sure that it allocates sufficient resources to the connections so that they are delivered without exceeding the promised delay bound and low loss rate.

We start our discussion of quality of service with a simple example that illustrates how the network can guarantee delay bounds and low losses. Figure 2.10 shows 100 virtual circuits that go through a buffer with transmission rate of 155 Mbps.

Each virtual circuit transports a video conference signal. The bit stream of a video conference is constrained at the source not to transport more than $0.1 + 1.5t$ megabits (Mbit) in any period of t seconds. Thus, for large t the average transmission rate does not exceed $(0.1 + 1.5t)/t \approx 1.5$ Mbps. Also, for a short time $t \ll 1$, the source cannot produce a burst of cells that exceeds $0.1 + 1.5t \approx 0.1$ Mbit. A scheme that enforces these bounds is called the leaky bucket controller: the bucket has a capacity of 0.1 Mbit and leaks at the rate of 1.5 Mbps. The source can transmit whenever the bucket—implemented as a counter—is not full, and the bucket fills up by as many bits as the source sends.

FIGURE 2.9

An output buffer ATM switch.

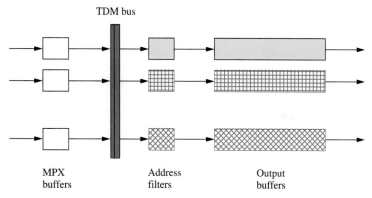

FIGURE 2.10

Sources controlled by a leaky bucket.

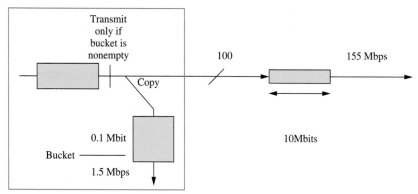

A moment of reflection should convince you that if the buffer capacity B is equal to 10 Mbits, then the buffer will never lose any single bit. Moreover, since the buffer never accumulates more than 10 Mbits and is being served at rate 155 Mbps, no bit will ever face a delay larger than $10/155 \approx 65$ millisecond (ms) through the buffer.

The point of this example is that if the sources are controlled to meet some bounds on the traffic that they produce and if the network sets aside sufficient resources to carry the connections, then it is possible to guarantee delays on bounds and losses.

After this bird's eye view of networks work, we mark a pause in the next section to reflect on different aspects of networks and to begin developing a more structured understanding of the field.

2.4 Network Architecture

In the previous sections we looked at a few network examples: Ethernet, internets, and ATM. These examples are only a few of the many network technologies that are widely used, and many more are being developed.

At the level of implementation details, networks appear to be in a rapid state of flux. Fortunately, at the conceptual level the situation is simpler and changes occur more slowly. Although networks seem very varied, they are all based on a small set of operating principles and are organized according to a similar architecture.

In this section we start sketching out the architecture and operating principles of networks. These principles will remain with you long after the concrete implementations become obsolete.

How should we think about a network? A network is put together by interconnecting various components. Consequently, it is quite natural to view a network as a set of components and their interconnections. Thus, we can look at computers, routers, links, servers. This is the *physical* view of a network.

We can also see a network as a set of information delivery services built on top of one another. This is the *layered* view of a network. In particular, the application designer sees the network as providing end-to-end information delivery services. The characteristics of these end-to-end services depend on their implementation and on the network components. We examine this layered view of networks in Section 2.4.1.

As a first approximation, you can work on network applications without worrying about how the network actually implements the end-to-end services. However, you must have some idea of the characteristics of these services. We discuss representative characteristics in Section 2.4.2.

To design many "demanding" applications, you need a more detailed understanding of the end-to-end services and how they may behave in different environments. Such detailed understanding requires some familiarity with the underlying technology. We start exploring such technology in Section 2.4.3.

2.4.1 Layered Architecture

Network functions are organized in a layered architecture. In such an architecture, the services of one layer are implemented on top of the services provided by the layer immediately below.

Two important benefits result from the layer decomposition. The different layers can be designed more or less independently, which greatly simplifies network design. Another advantage is the *compatibility* derived from the independence of the layers. For instance, the same applications can run on very different networks. And when different networks are interconnected, a computer on one network can access computers on all the networks, independently of the specific implementations of the different networks. Since connectivity is a primary objective of communication networks, compatibility is most valuable.

Networks use different architectures. At one level of description, the similarities outnumber the differences. We present the most prevalent architectures.

OSI

In the late 1970s, to promote the compatibility of network designs, the *International Organization for Standardization* (ISO) proposed an architecture model called the *open systems interconnection reference model* (OSI model). The OSI model is a layered architecture with seven layers. Figure 2.11 summarizes the functions that the layers perform.

The physical layer implements a digital communication link that delivers bits. A communication link is always unreliable. The link may be point-to-point from one transmitter to a receiver, or it may be shared by a number of transmitters and receivers.

The data link layer implements a packet delivery service between nodes that are attached to the same physical link. At the transmitter, the data link layer frames packets so that the receiver can recover them in the bit stream. The data link layer may also arrange for the erroneous packets to be retransmitted.

The network layer guides the packets from their source to their destination, along a path that may comprise a number of links. A typical method is store-and-forward transmission, either as datagrams or along virtual circuits.

The transport layer of the OSI model supervises the end-to-end transmission of packets. This layer may arrange for retransmissions of erroneous packets. This layer also controls the rate of transfer of packets to avoid congesting parts of the network.

The session layer uses the transmission layer services to set up and supervise connections between end systems.

The presentation layer in OSI takes care of data compression, security, and format conversions so that nodes that use different representations of information can communicate efficiently and securely.

Finally, the application layer implements commonly used communication services including file transfer, directory services, virtual terminal.

IEEE 802

The IEEE standards for LANs are known as IEEE 802. The IEEE standard activities for LANs are summarized in Figure 2.12. We study these layers in Chapter 4.

Internet

The Internet uses the model illustrated in Figure 2.13. It is a four-layer model. The layers, from bottom to top, provide the following services:

- *LAN or link layer:* Transmitting packets on one link or across interconnected LANs.
- *Network layer:* Transmitting messages on a sequence of links from one terminal node to another (from end to end).

FIGURE 2.11

The functions of the seven OSI layers.

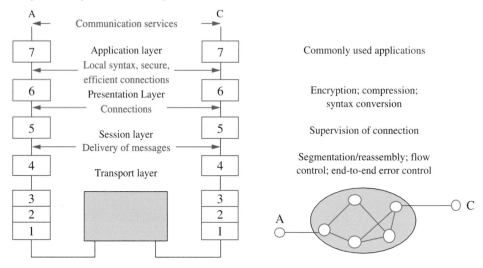

FIGURE 2.12

The IEEE 802 standards. The standards developed by the IEEE 802 working groups are indicated in this figure. These standards specify the physical layer and the data link layer of LANs. The data link layer is decomposed into the media access control (MAC) layer and the logical link control (LLC) layer.

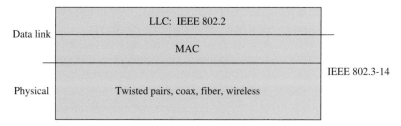

- *Transport layer:* Supervising the transmission of messages from end to end.
- *Application layer:* Providing a number of communication services for the users, including file transfer, remote terminal, electronic mail, and directory services. These applications consist of services that are offered to any application as well as some network-specific applications like email.

User applications are implemented on top of these layers: gopher, World Wide Web, (WWW), and many others. We study these layers in Chapter 3.

ODN

The *Open Data Network* is a model proposed by the telecommunications study group of the National Research Council in 1994. This model, shown in Figure 2.14, has four layers.

FIGURE 2.13

Internet protocols. The functions of the Internet are decomposed into four layers.

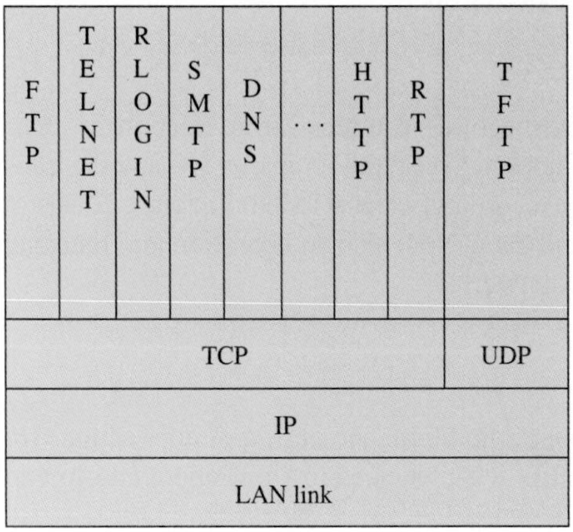

F T P	T E L N E T	R L O G I N	S M T P	D N S	...	H T T P	R T P	T F T P
TCP								UDP
IP								
LAN link								

FIGURE 2.14

Open Data Network model. This model decomposes the network functions into four layers.

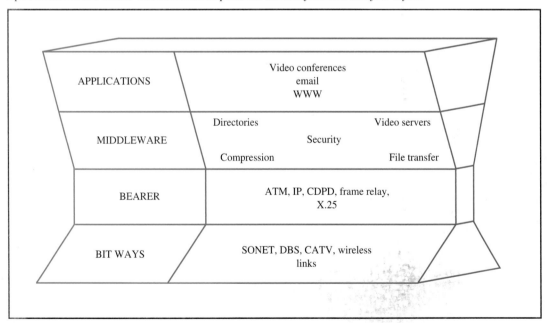

The bottom layer consists of "bit ways" that transport bits along links. The second layer implements bearer services that transport a bit stream or packets from end to end between terminal devices. Thus, the bottom three layers of Internet and of OSI implement bearer services as do the layers up to the ATM layer in ATM networks. Layer 3 implements services that complement the transport of information, such as security and privacy, storage, and directory services. Finally, layer 4 implements applications such as file and compute servers. This model is designed to encompass all data networks in addition to telephone and cable television networks.

ATM

Figure 2.15 shows the ATM architecture. You recall from Section 2.3 that ATM networks use virtual circuits (VCs). In such a network, the setup of a VC and the transport of the data involve different sets of tasks. Accordingly, the ATM architecture consists of two separate sets of layers for the control and data. Note also an operation and management plane that supervises the functions of the network.

2.4.2   End-to-End Services

We use an example to guide our exploration: the design of applications for network computers. Our objective is to identify what we need to know about end-to-end services.

The Application

Our network computer (NC) uses network servers as sources of data and software and also as computing engines. NC is able to perform elementary operations, but it relies on network servers (NS) to perform more complicated tasks. We want to determine the applications that can run on the NC.

One potential advantage of an NC/NS system is that it can be cheaper than full-fledged PCs with comparable capabilities. Another advantage is that the capabilities of the NC/NS system grow as new providers make new services available on the network. Thus, your NC expands automatically. With suitable standardization, the NCs can become universal communicators.

If communications were free, secure, instantaneous, and perfectly available and reliable, the NC/NS architecture would be desirable. We may therefore argue that an NC/NS architecture is a good idea because communication costs and delays are bound to decrease over time and communications can be made secure.

End-to-End Service Model

Before we start designing applications for the NC/NS we need to identify the main characteristics of the end-to-end services that the network provides. To make some headway in

FIGURE 2.15

ATM architecture. The control and data planes are separated.

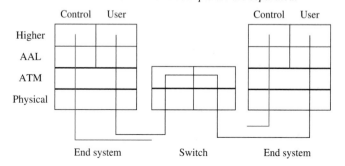

this case study, we model the network delivery services as follows:

- *Service 1:* Deliver message of M bytes with delay T and error probability ϵ.
- *Service 2:* Set up bit link with rate R bps, delay T, and bit error rate ρ.

In these models we must refine the meaning of the parameters. Although we do not want to go into details at this point, let us see how we could be somewhat more precise, just to indicate what is involved in defining the parameters.

In service 1, the delay of a message cannot be predicted fully; it depends on M and on the distance to the NS. We ask our network engineers and they tell us that the network can provide service 1 with $\epsilon \leq 10^{-4}$ for $M \leq 10^6$ bytes and with specific delays for $M = 1$ megabyte. For service 2, our network engineer tells us that $R = 150$ kbps, 0.01 s $\leq T \leq 1.2$ s with probability at least 99.9 percent, $\rho \leq 10^{-8}$ erroneous bit per delivered bit.

How do we use this information to decide the applications that will work on the NC? We examine two hypothetical applications for illustrative purpose. These two applications do not correspond to standards.

As a first example, imagine a videophone application. We will connect our NC to another NC and set up a videophone call. We use a camera and software that converts the video into a 64-kbps bitstream. Since the delay can be as high as 1.2 s, we don't have much choice but to equip the destination NC with a buffer that stores video bits and plays them back at 64 kbps. This buffer absorbs the variable delay of the packets that carry the video bits. No matter what clever algorithm the destination NC uses, the resulting delay from one NC camera to the other NC display is then at least 1.2 s. Unfortunately, such a delay is not acceptable for a conversation application.

The second application that we consider is watching video broadcasts on the NC. That is, the network is carrying some video streams that we can subscribe to. If the video streams have a rate less than 150 kbps, then our network will be able to deliver us these broadcasts and we will not be affected by the 1.2-s delay.

A third example is Web browsing. We click on a link of a Web page and expect to get a new Web page with some pictures. Let us assume that a typical page takes less than 1 megabyte (Mbyte) and that the server is at most 3000 km away. Simple calculations show that a request to the Web server may take up to 0.5 s and that the Web page comes back within at most 5 s. We conclude that this application is feasible.

We will not multiply such examples, but they make the following point. Knowing basic characteristics of end-to-end services enables us to identify which applications are feasible and the quality that the users may expect to perceive. These basic characteristics depend on how the network implements the services. In particular, they depend on the network technology: the components that make up the network. We explore such components next.

2.4.3 *Physical View*

A network consists of links, routers or switches, servers, and computers. The user computers and servers are called end systems or hosts. The routers and switches are transfer systems. In this section, we take a quick look at these basic components and summarize their characteristics.

Links

For our purpose, a link is the implementation of a bit pipe. The link has a rate R, a delay D, and a bit error rate BER. The delay is the propagation time of a signal across the physical medium of the link. This delay is equal to 5 μs/km in an optical fiber, about 4 μs/km in a coaxial cable and twisted pairs, and about 3.3 μs/km in free space.

Imagine that we use this link to transmit a packet of P bits. It takes P/R seconds for the transmitter to send the bits of the packets one after the other. It takes D seconds for the last bit of the packet to reach the other end of the link. Thus, if we start transmitting the packet at time 0, the last bit of the packet reaches the destination at time

$$T = D + P/R. \tag{2.1}$$

If we assume that the bits are corrupted independently of one another by the transmission, then the probability that the packet arrives correctly is given by

$$1 - \text{PER} = (1 - \text{BER})^P. \tag{2.2}$$

Indeed, each bit arrives correctly with probability $1 - $ BER. Consequently, the probability that the P bits arrive correctly is the product of the probabilities that the P bits arrive correctly (by independence; see Appendix A).

We call PER—the probability that the packet is corrupted by the transmission—the *packet error rate*. (To understand the above calculation, you need to know that the probability that independent events all occur is the product of the probabilities of the individual events.)

Note that this independence assumption is simplistic, since errors may occur in bursts that corrupt a number of successive packets. The PER that corresponds to bursty errors is less than the value predicted when assuming independent bit errors. For instance if a burst always corrupts all the bits in a packet, then PER = BER, which is much smaller than the value predicted by Formula (2.2).

Switches and Routers

At our level of discussion, a switch or router is a box with input and output links. As a packet or ATM cell arrives, the router or switch figures out the output link. After some delay, the router or switch sends the packet or cell on the output link.

The main characteristics of the switch or router are the rates of the links, the number of links, the maximum total rate of packets or cells through the switch, and the delay across the switch. Another important characteristic for ATM switches is the time to set up a virtual circuit.

Servers

Servers are computer systems that store files or provide other services on the network. Examples include network file servers, Web search engines, network caches, and network compute servers.

The main characteristics are the storage size, the transfer rates, and the number of transactions per second. We hesitate to indicate numbers here because they are changing as we speak.

2.5   Complement 1: Insights behind the Information Revolution

In this section we reflect on the insights that have made the information revolution possible. Progress in semiconductor manufacturing and in computer-aided design tools for building increasingly complex integrated circuits, improvements in hard disks, and advances in optoelectronics and in optical fiber technology were all essential ingredients in this revolution.

In addition, a number of conceptual breakthroughs—different ways of looking at familiar questions—were equally essential. It is these conceptual shifts, made possible only by profound insights, that we explore in this section. Appreciating these insights reveals the importance of extracting structure and patterns behind apparently messy and complex situations. The insights were accessible to researchers with remarkable capacities for abstraction, for building simple models of the physical reality, and with a deep desire to really understand the limits of what can be done.

We would like to share our enthusiasm for this aspect of engineering—the reduction to basic principles, the modeling and analysis for the sheer joy of understanding. We believe that in this age of rapid obsolescence of technology it is all the more essential to study the fundamentals and to appreciate the continuity from fundamentals to designs to devices to systems to commercial products. These fundamentals and, more importantly, the process of discovering the basic principles and the key mechanisms, will stay with us long after specific standards have become obsolete.

In this section we focus on three insights, three deep ideas that changed information technology. The examples are not exhaustive but we selected them because they meet two criteria: (1) they have a fundamental impact and (2) they truly require a conceptual breakthrough and not simply a continuous improvement.

2.5.1   The Digital Revolution

Today's computers, audio discs, some audio and some video cassettes, and most telephone lines are digital. Tomorrow's televisions and possibly radios will also be digital.

Edison's phonograph encodes a sound by tracing the displacement of a diaphragm in a metallic or wax cylinder. To play back the recording, the reverse process takes place: a needle follows the groove in the cylinder and mechanically transfers its motion to a diaphragm that recreates the air vibration that initially created the recording. The shape of the groove is analogous to the variation in the sound pressure that produces it. The acceleration of the needle is proportional to that pressure. Vinyl recordings work in the same way. To play back a record on a loudspeaker instead of through a direct mechanical device, the motion of the needle is converted into a varying electrical current by a piezoelectric or electromagnetic or other transducer. That current is then amplified by an analog amplifier that multiplies it by some approximately fixed factor and the amplified current is injected into an electromagnetic coil that moves in the magnetic field of a permanent magnet. The coil is attached to a cone that communicates its motion to the air. This playback process can be reversed to replace the direct mechanical recording by an electromechanical system; the loudspeaker then becomes a microphone, and it can be made very small. The microphone and loudspeaker can be attached together via a pair of wires to construct a basic telephone. Mechanical recording on vinyl disks can be replaced by recording on ferromagnetic tapes where the local magnetization of the tape is proportional—analogous—to the sound pressure. A similar process takes place for video: A two-dimensional image is converted into a sequence of scanning lines whose varying brightness can be changed into an analogous varying current that can then be transmitted or recorded and that a reverse process converts back into a picture in a cathode-ray tube.

Considerable effort has been devoted to perfecting the above procedures. When I was in college, in the 1970s, many classmates were fanatics of high-fidelity audio. They would compare notes and specifications of their heavy-platter turntables on granite blocks, precision phonograph cells, and delicate arms. Most of these devices were equipped with careful brushes that would clean up the vinyl groove just before the needle would trace it. Despite meticulous care, the result was often marred by clicks and pops caused by dust particles or minute fractures of an aging record. Some invested in record cleaning machines that improved only marginally the quality of the playback.

A similar story can be told about the attempts to protect the amplification circuitry from electromagnetic interference, the rumble induced by power supplies, the parasitic electromagnetic noise of appliances, and the thermal noise in the wires and resistors of the circuitry. Fancy shielded connection cables were fashionable, preferably with gold-plated connectors. We would compare notes on the relative merits of preamplifiers built with field-effect transistors or with vacuum tubes. Clever designs were invented to combat distortions in the amplification process. The transmission of telephone signals was a similar battle against noise and distortions, and comparable struggles took place in the recording trenches.

The Achilles' heel of all these systems was the same. Since the signals are proportional to the sound pressure or the image brightness, the slightest perturbation of the signal results in a commensurate degradation of the audio or video quality.

Compare this situation with the robustness of the handwriting process. We can read even very poor handwriting. I have a dear friend whose letters are as challenging to decipher as they are welcome. However, remarkably, it almost never happens that I cannot recognize a word. There are only 26 letters in our alphabet. Even a very distorted *b* is difficult to mistake for another letter. Moreover, even if one or more letters cannot be recognized, chances are that the word they are part of can be recognized from the other letters. In addition, even if a word cannot be deciphered, it can usually be identified from the rest of the text.

Thus, two mechanisms help us read a poor handwriting. The first mechanism is that there are only finitely many symbols. A distorted symbol is unlikely to look more like another symbol than the original symbol. The second mechanism is that there is redundancy in the words and sentences so that we can fill in gaps of recognition and correct errors.

The first mechanism is exploited in digital encoding of information. Instead of representing a sound pressure or a brightness by an analog signal proportional to that physical quantity, we represent it by a string of bits, a binary word. We then represent each bit by one of two different signals that encode a 1 and a 0, respectively. The two signals are chosen so that it is unlikely that the signal that represents a 0 gets distorted so that it looks more like the signal that represents a 1, and conversely.

The second mechanism, using the redundancy of information, is exploited in error correction codes. These codes deliberately insert some redundancy in the bit stream that is recorded or transmitted. The redundancy is introduced in a systematic way so that the receiver or playback device can "fill in the gaps."

Since the preamplifier, transducers, and recording system now only need to handle 0's and 1's, they become less delicate to manufacture. Consequently, the cost of digital devices is considerably less than that of analog devices of comparable quality. A significant additional benefit is that the information is now encoded into bits that can be processed with the same technology as numbers. Thus, all sources of information become represented by bits and an integrated communication infrastructure is possible. A bit is a bit is a bit.

2.5.2   *Source and Channel Coding*

In the late 1940s, Claude Shannon was trying to understand the meaning of "information" and whether information can be quantified. The idea of quantifying, of measuring, a nonphysical quantity such as information was, you will admit, quite radical.

On the other hand, the potential payoff is clear. If we can measure the amount of information in a picture, a video clip, or a book, then we can begin to study how much silicon or ferromagnetic material we need to store that information and maybe even how long it might take to transmit that piece of information over a telephone line, say. When Shannon was worrying about this basic question, his contemporaries at Bell Laboratories were busy designing and building transmission lines and switching systems and we may suspect that his investigations appeared far-fetched and gratuitous to them.

Source Coding

Say that we suspect that it should be possible to quantify information. The next step is to find the simplest possible piece of information and to try define a quantity for it. Here, the breakthrough is to realize that there is information only in learning something that was not known before. Shannon modeled the output of a source as a sequence of random variables, a useful quantitative model of uncertainty.

Shannon showed that there is a minimum average number of bits per symbol required to encode the output of a source. This number of bits per symbol is called the *entropy rate* of the source. This entropy rate determines the limit of any compression algorithm.

Compression algorithms have been designed that attempt to approach the entropy rate per symbol. We study such algorithms in Chapter 8. Shannon's theory enables us to evaluate how well practical algorithms perform compared to the absolute limit and to find out if further improvements are possible.

Assume that the entropy rate of a source is very large and that we cannot afford to store or to transmit all those bits. What happens if we encode a source output with fewer bits per symbol than the entropy rate? Not surprisingly, in that case the best reconstruction of the source output from the encoding bits is not perfect. We can define a measure for the *distortion* of the reconstructed source output by computing some distance between the original source output and the reconstruction. Shannon's theory tells us that if we accept some distortion ϵ, then there is a minimum average number of bits $H(\epsilon)$ required per output symbol. If ϵ is very small, then $H(\epsilon)$ is close to the entropy rate of the source. For instance, imagine that we want to compress a picture and are willing to tolerate some distortion. It is in principle possible to calculate the average number of bits that we need to encode the picture. We study practical algorithms for this type of *lossy* compression in Chapter 8. We explore the concept of entropy of a source in Section 8.10.

Channel Coding

How fast can we transmit bits over a telephone line or over an optical fiber or between two microwave antennas? A better question is how fast can we transmit the bits reliably. Indeed, it is plausible that if we are willing to receive incorrectly a fraction of the bits, then we should be able to send them faster, a little like trying to throw balls into a basket.

As soon as we try to make the question precise we run into an apparent contradiction. Indeed, no matter what communication equipment we use, it is impossible to guarantee that any single bit will be received correctly. Thermal noise, electromagnetic interference, or other sources of noise have some positive probability of corrupting any given bit. More-over, we can also argue that any finite set of bits also has some positive probability of being entirely corrupted. Consequently, it seems that the only possible answer to our question is that the maximum rate at which we can transmit bits reliably is 0: We can not transmit bits reliably at any rate.

Once again, Shannon asked the precise question and answered it. The precise question is what is the maximum rate at which one can transmit bits with an arbitrarily small proba-bility of errors. Say that that rate is C bps. The meaning of C is that, for any given ϵ with $0 < \epsilon \leq 1$ and any $R < C$, one can transmit bits at rate R bps with the fraction of the bits that are not received correctly at most ϵ. The specific procedure that we use depends on ϵ, and this is why we cannot transmit the bits fully reliably.

Since Shannon's work in the late 1940s, many extensions have been developed and practical coding methods with efficient decoding algorithms are now available and are implemented in modems and other communication link components.

We study channel coding further in Section A.5.

Source and Channel Coding

We can combine the above two ideas to try to find out how fast we can send pictures over a telephone line.

We first use the source coding result of Shannon to find out how many bits we need per picture, on average. Say that we need H bits per picture. We then use the channel coding result to find out the maximum rate C at which we can send bits reliably through the telephone line. We then conclude that we can send C/H pictures per second through the telephone line.

These results do not provide recipes for how to send those pictures. We address practical aspects of source coding in Chapter 8.

2.5.3   *Packet Switching*

Packet switching consists in sending information by first packaging it into groups of bits with suitable control fields and then sending the packets one at a time. This procedure is by now so familiar that it seems to be obviously a good way to send information. How-ever, in the early 1960s this mechanism—developed by Paul Baran, then with the Rand Corporation—was radically different from how the telephone network and every other electrical communication system transported information.

Packet switching is essentially how the post office transports mail: one letter or package at a time. The telephone network uses circuit switching: When Alice and Bob want to talk to each other, the telephone network sets up a connection between them; Alice and Bob then communicate; the connection is released when the conversation is over. To set up a connection, the network reserves capacity on links between Alice and Bob and connects those links together. The capacity is allocated to Alice and Bob for the duration of their conversation, whether they use it or not.

Two features of circuit switching make it wasteful in many situations when compared to packet switching. The first feature is that Alice and Bob may not need the reserved capacity all the time during their communication. For instance, in the case of a telephone call, Alice and Bob do not talk all the time. The second feature—which is somewhat more difficult to quantify—is that a circuit-switched network requires considerably more bookkeeping than a packet-switched network to supervise the connections.

Statistical Multiplexing

In 1995, the average bit rate per Internet user was about 40 bps. For instance, 40,000 users on the Berkeley campus shared a 3-Mbps connection to the Internet with capacity to spare. Although users can transmit on the Internet at any time, they do so only occasionally. When users transmit or receive messages, they do so with a large bit rate, typically a few tens of kilobits per second.

Let us say that users transmit at 40 kbps and that their average transmission rate is only 40 bps. These numbers imply that a user transmits only 0.1 percent of the time. Transmissions are usually short email messages of about 10^3 bytes and infrequently files with a fraction of 10^6 bytes; only a small fraction of transmissions are much larger. Consequently, most transmissions take between a fraction of a second and some take up to a few minutes.

Consider how a circuit-switched network transmits short email messages. To transmit a message from a computer A to another computer B, the network sets up a connection from A to B; computer A sends its message to B; the network then closes down the connection. To estimate the efficiency of such a system, we assume that the connection rate is 64 kbps, that the connection setup and tear-down take 10 s and that the message is 1 kbyte long. Once the connection is set up, the message transfer takes 8000 bits/64 kbps = 0.125 s. Thus the network transfers bits for 0.125 s and is busy for a total of 10.125 s to connect the computers. That is, the transmission lines are used only about 1.2 percent of time to send bits and are used the rest of the time to set up the connection and to release it.

A packet-switched network such as the Internet transfers messages very differently from a circuit-switched network. If the computers are connected to the Internet, computer A sends a short email message in an IP packet, as we saw in Section 2.2. That packet then travels from router to router until it gets to computer B. The IP packet uses a transmission line between two routers only when it is being transmitted. The network does not waste resources to set up or tear down a connection from A to B. The network wastes some time transporting the packet headers that specify the source and destination addresses plus some additional control information. In addition, the network routers exchange some other information to update their routing tables and to supervise the transmissions. This control traffic represents only a few percent of the total Internet traffic. Consequently, the lines of the network are used almost 100 percent of the time to transmit user bits. This example shows that packet switching is about 100 times more efficient than circuit switching to deliver electronic mail.

This increase in efficiency exploits the *statistical multiplexing* gain. This gain comes from the ratio of the "peak rate" over the average rate of a connection. In our example, a typical IP connection on the Berkeley campus sees a rate of about 40 kbps whereas the average rate per user is only 40 bps (because that user transmits only a very small fraction of the time).

As you may have noticed, this efficiency superiority of packet switching is less dramatic for longer transmissions. To demonstrate that fact, let us examine the case of Internet phone—telephone calls placed via the Internet instead of the telephone network. In an application, Alice's computer converts the voice signal into bits, compresses the bit stream to reduce the amount of bits that have to be transported, introduces some redundancy for error correction, then packages the bits into IP packets. The Internet delivers the packets to Bob's computer which converts back the bits into Alice's voice. The savings in bit rate that compression achieves can range from a factor 2 to a factor 10, depending on the sophistication of the algorithms and on the desired voice quality. Another source of savings is that the silent periods do not have to be transmitted: Alice's computer can append a short symbol to a packet to represent one or more packets of "zeros." Since telephone users typically speak only 40 percent of the time, this method divides the average bit rate by about 2. The necessary headers in IP packets eat up some of these savings (about 15 percent to 30 percent). Putting these numbers together, we find that a good-quality transmission might use 30 percent of the 64 kbps that the telephone network would use. Thus, for a typical implementation of Internet phone, the gain that packet switching achieves over circuit switching is only a factor of 3 and not a factor of 100 as for email transmissions.

Summary

- We have explored ways of arranging computers into networks. We started with a shared Ethernet where all the computers share a common transmission medium. Source and destinations of packets are identified by Ethernet (48-bit) addresses wired into the interfaces.

- We then moved on to interconnected Ethernets, with a brief allusion to switched Ethernets. We saw that Ethernet switches operate on the basis of Ethernet addresses but that routers use network addresses. This led us to discuss procedures to learn addresses (ARP and RARP).

- Our next topic was internets. We discussed the routing and explained that it is hierarchical in two levels: autonomous systems that use OSPF internally and the routing between autonomous systems that uses the BGP. We indicated that flows and errors should be controlled and we sketched how this is done in TCP.

- We then turned our attention to ATM networks. In these networks, routing is based on virtual circuits. The main objective of ATM designers is to better control the quality of the service.

- Finally, we examined the architecture of networks. We explained the layered architectures. We then discussed end-to-end services and the physical view of networks.

Problems

These problems are designed to help you consolidate your understanding of the material in this chapter. The more difficult problems are indicated by *.

1. Consider the network of Figure 2.16. In this network, B is a switch and C is a router. Indicate the format of packets going from S to D on the various links of the network.
2. Repeat Problem 2.1 assuming that B and C both are switches.
3. For the network of Figure 2.16 where B is a switch and C a router, show the entries of the routing tables at B and C that are needed to send a packet from D to S.
4. Figure 2.17 shows an Ethernet switch. Assume that 16 computers are attached to the switch and that the transmission rates on all the links are 100 Mbps.
   a. Calculate the maximum total transmission rate over this network and describe the assumptions under which this rate can be achieved.
   b. Describe a situation where all the computers have a lot of traffic to send but the average transmission rate is much smaller than in part a.

FIGURE 2.16

Network for Problems 1, 2, and 3.

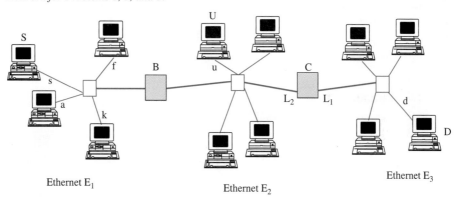

FIGURE 2.17

Ethernet switch for Problem 4.

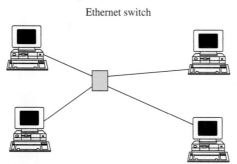

5. Figure 2.18 shows an Ethernet switch that connects 16 computers at 10 Mbps and a file server at 100 Mbps. Assume that all the traffic goes from the server to the computers.
   a. Estimate the total transmission rate achievable under this arrangement.
   b. Compare with the case where the server is attached with a 10-Mbps link.
   c. Discuss the transmission of a packet from a computer to the server. In particular, explain whether the switch can start forwarding the packet before it has fully received it.

6. This problem explores the hierarchical routing described in Section 2.2.2. Figure 2.19 shows three networks (a, b, c) interconnected with routers. The IP addresses have the structure network.host. For instance, the computer with name m on network a has IP

FIGURE 2.18

Ethernet switch for Problem 5.

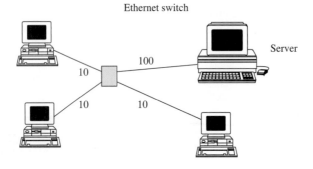

FIGURE 2.19

Network for Problems 6 and 7.

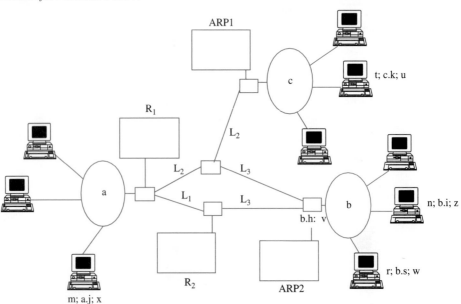

address a.j. The entries of the routing tables R1 and R2 correspond to the network part of the destination addresses. Provide the entries in those tables needed to route a packet from computer m to computer n.

7. Consider once again the network of Figure 2.19. The address resolution tables ARP1 and ARP2 are specific to the subnetworks c and b, respectively. Provide the entries of these tables that are needed to send packets to the computers t, n, and r.

8. Figure 2.20 shows an ATM network. Assume that virtual circuits are set up in the following order: A to E, E to D, F to D, A to E. Indicate the entries in the routing tables needed to set up these circuits.

*9. This problem is somewhat more complex than the others. Figure 2.21 shows two computers m and n that are attached to two networks: an IP network with routers R1,

FIGURE 2.20

ATM network for Problem 8.

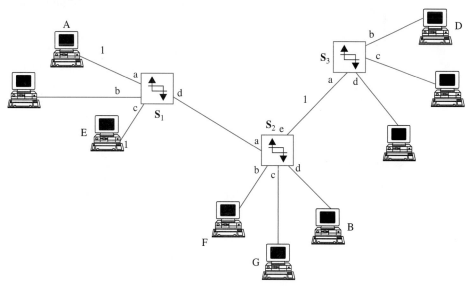

FIGURE 2.21

Network for Problem 9.

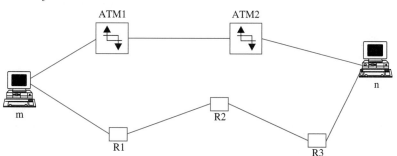

R2, R3 and an ATM network with switches ATM1 and ATM2. Since setting up a virtual circuit through an ATM network requires some time, it should be sensible to send small groups of packets via the IP network and long groups via the ATM network. Exhibit a model of the set up times and of the communication delays that confirms this basic intuition.

References

Since this chapter is a general overview of ideas that we explore further later, the references that you might want to consult contain general information about the Internet, Ethernet, and ATM. We recommend that you browse the following Web pages to become familiar with the standardization bodies of networks.

For the Internet, you should consult the home page of the Internet Engineering Technical Forum that supervises the development of the Internet protocols: http://www.ietf.cnri.reston. va.us/.

For Ethernet, I recommend Charles Spurgeon's Ethernet page at the University of Texas: http://www.ots.utexas.edu/ethernet/ethernet.html. You should consult the home page of the ATM Forum that organizes the recommendations that those networks should follow: http://www.atmforum.com/.

There are many excellent texts on networking. For the general description of networks, the best reference is Tanenbaum (1996). Peterson (1996) is also highly recommended. Partridge (1994) is an accessible discussion of high-performance networks. The publications of the IEEE, specially the *Communications Magazine,* are a good source of information on recent developments. The ODN model was proposed in Kleinrock (1994) an excellent discussion of the issues raised by modern communication networks.

We give more specific references in subsequent chapters.

Bibliography

Kleinrock, L., editor, *Realizing the Future; The Internet and Beyond,* National Academy Press, 1994.
Partridge, C., *Gigabit Networking,* Addison-Wesley, 1994.
Peterson, Larry L., and Davie, Bruce S., *Computer Networks: A Systems Approach,* Morgan Kaufmann, 1996.
Tanenbaum, A. S., *Computer Networks,* 3rd ed., Prentice-Hall, 1996.

CHAPTER 3 — Internet

The Internet is growing rapidly in the number of computers and networks it connects, in the volume of traffic it carries, and in the range of applications it enables. We sketch the history of the Internet in Section 3.1.

The main protocols of the Internet are the Internet protocol (IP) and the transmission control protocol (TCP). Many companies build isolated networks also using TCP/IP. These networks are called intranets. Such networks may be attached to the Internet, usually through a "firewall" that protects the security of the intranet. We call a network that uses the TCP/IP protocols an internet. We discussed the basic operations of an internet in Chapter 2. In this chapter we examine those operations in more detail.

We take for granted that computers attached to a common link can exchange packets. In Chapter 2, we sketched how computers exchange packets on an Ethernet. Computers attached by a LAN other than an Ethernet or by a point-to-point link or another packet delivery service such as an ATM virtual circuit can also exchange packets. Usually, such packet transfers are not reliable, but have a small packet error rate.

We review the architecture of TCP/IP networks in Section 3.2. We then explain the addressing in Section 3.3 and routing in Section 3.4. We discuss the transport layer in Section 3.5. After this basic discussion of the major Internet protocols, we present a few complements on some aspects of these protocols. All these complements can be skipped without harm to the understanding the rest of the book.

3.1 A Brief History

A few of the major milestones in the history of the Internet are the following.

- 1962: Paul Baran, of the Rand Corporation, proposes packet switching as a robust networking mechanism.

- 1969: The Department of Defense Advanced Research Projects Agency funds a project on packet switched networks. The first four nodes of ARPANET are connected.
- 1974: Vint Cerf and Bob Kahn publish the basic mechanisms of TCP.
- 1982: The protocol suite TCP/IP is defined for ARPANET.
- 1984: Domain name system is introduced.
- 1986: NSFNET, the backbone (at 56 kbps) of Internet is created by the National Science Foundation (NSF).
- 1992: The World Wide Web (WWW), designed by Tim Berners-Lee, is released by CERN, the European Organization for Nuclear Research.

The telephone networks were designed in a centralized way by major corporations. In contrast, the evolution of the Internet has been decentralized, almost chaotic. No one can estimate reliably the number of hosts on the networks or the number of users. Such a distributed growth is possible because of the specific protocols that control the operations of Internet, as we explain in this chapter.

Figure 3.1 illustrates the Internet in the United States around 1990. The highest part of the figure shows NSFNET, the backbone of the network. Regional networks are attached to the nodes of the backbone. This arrangement is sketched in Figure 3.2.

FIGURE 3.1

The Internet around 1990.

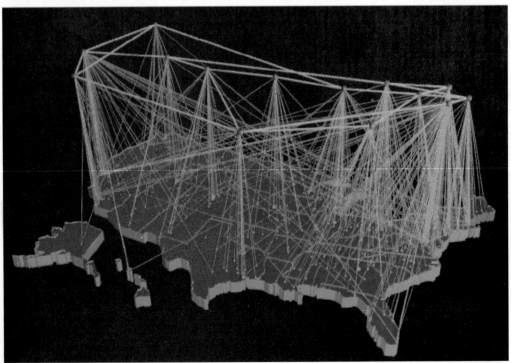

In 1997, the Internet structure was as shown in Figure 3.3. As the figure shows, Internet has three backbones that are maintained by network service providers. Users are connected to the Internet through a large number of Internet service providers.

The Internet is a collection of interconnected subnetworks. The terminology is that the Internet consists of *autonomous systems* that are connected together by routers called *border gateways*. In Figure 3.3, the autonomous systems are the backbone networks,

FIGURE 3.2

Sketch of the Internet around 1990.

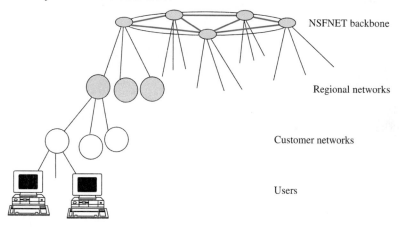

FIGURE 3.3

Sketch of the Internet in 1997.

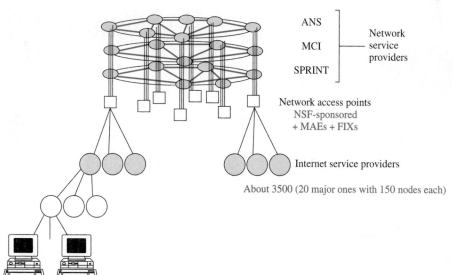

the regional networks, and the customer networks. Roughly, an autonomous system is a network that is managed by an independent authority. Thus, the Berkeley campus network is an autonomous system. To go from one host to another, a packet may have to go across a number of autonomous systems and there may be a few possible alternative paths.

FIGURE 3.4

Locations of network access points.

FIGURE 3.5

Pacific Bell NAP.

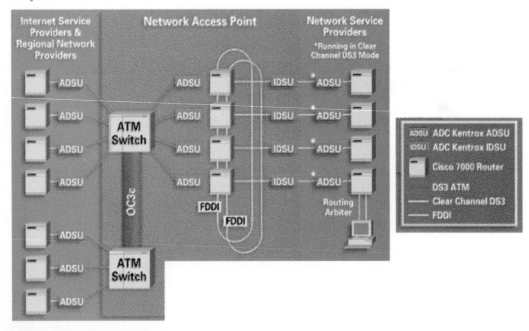

Figure 3.4 shows the location of the network access points (NAPs), the nodes of the backbone. One such network access point, the Pacific Bell NAP, is shown in Figure 3.5.

3.2   Architecture

The architecture of TCP/IP networks is shown in Figure 3.6. We comment briefly on each of the layers.

3.2.1   LAN-Link Layer

The LAN-link layer transmits packets between a specific pair of nodes. In most cases, these two nodes are attached to a common physical channel. For instance, if you use your home computer to communicate with your office computer, you may be using a link protocol such as SLIP or PPP. These protocols—explained in Section 3.6—transmit packets between the modem of your home computer and that of your office computer. As another example, computers attached to the same Ethernet exchange packets.

To understand some networks, it helps to extend the notion of a link to a packet delivery system that may involve more than one physical link. For instance, consider a network of IP routers interconnected by an ATM (permanent) virtual circuit. In looking at this network as an internet, focusing on the TCP/IP protocols, it is convenient to view the ATM network as implementing links between pairs of IP routers. Thus, although each ATM link may consist of several physical links interconnected with ATM switches that perform some routing and flow control functions, the IP layer sees this link as a *simple* link.

The LAN-link layer delivers packets, possibly with occasional transmission errors. The main characteristics of a link are its maximum transfer unit (MTU), its packet error rate (PER), and its transmission rate. The MTU is the maximum size of packets that the link can transmit. For instance, the MTU of Ethernet is 1500 bytes. That of SLIP is 512 bytes. Typical transmission rates range from 9600 bps for a slow modem to 622 Mbps for a fast ATM LAN. This rate is 1 Gbps for the gigabit Ethernet. Some specialized links and

FIGURE 3.6

TCP/IP architecture.

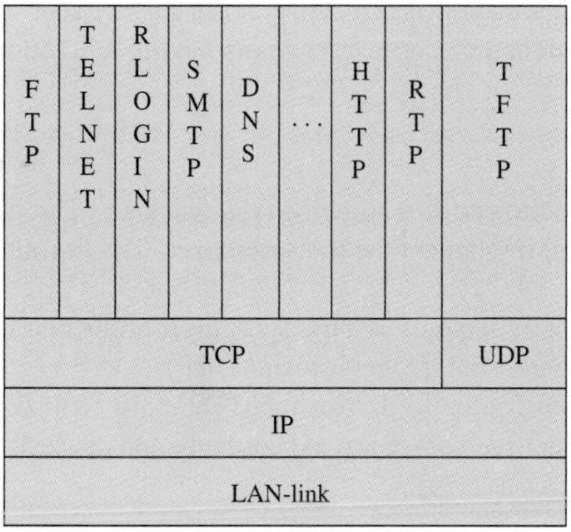

high-speed LANs have larger transmission rates. The PER is very small for most wired links (a small fraction of 1 percent). For wireless links, this PER can be larger and losses are often bursty. That is, errors tend to occur in groups.

3.2.2 Network Layer

The network layer of an internet, the *Internet protocol* (IP), implements the end-to-end delivery of packets of up to 64 kbytes. IP supervises the addressing of nodes and the routing of packets. The nodes of an internet use ICMP (Internet control message protocol) to supervise the delivery of packets.

The routers route the packets one at a time. Each packet carries its full source and destination addresses. The routers maintain routing tables, as we explained in Chapter 2. The router checks its routing table to find the next link for the packet based on its destination address. This form of routing is called *datagram* routing. When an IP packet is larger than the MTU of a link, IP fragments the packet and transmits the fragments one at a time. The destination host reassembles the packet from the fragments.

One major design objective of ARPANET, which evolved into the Internet, was that the network should be robust. Thus, router failures should have minimal consequences. Accordingly, the routers should remember as little information as possible, for that information is lost in case of failure. Since a packet carries its destination address, a router can determine where it should send that packet next by consulting a map of the network, without requiring a memory of previous packets.

Recall from Chapter 2 that ATM routing uses virtual circuits. In a virtual circuit network, after a connection is set up, the packets carry only a virtual circuit number. The router must remember how it must handle packets with a given virtual circuit number. We say that an ATM switch has a state and that an IP router is *stateless*. Note that the terminology is somewhat oversimplified, since the IP routers remember a routing table. The point of this distinction is that the IP router does not keep track of individual connections as an ATM switch does.

Thus, one design rule of TCP/IP networks is to minimize the information that the network elements maintain to make the network robust.

As an analogy, if a mailman discovers a misplaced bag of envelopes, he can resume the delivery of the envelopes without having to check any records. All the information he needs is on the envelopes. Similarly, if an envelope is mistakenly sent to Paris, France, instead of Paris, Texas, the error can be corrected without having to trace a paper trail.

3.2.3 Transport

The transport layer supervises the end-to-end delivery of packets. The two end systems (the two ends of a transmission) implement the transport layer. The IP routers are unaware of this layer.

The Internet protocol suite implements two transport layer protocols: the transmission control protocol (TCP) and the user datagram protocol (UDP).

UDP delivers packets from source to destination. The only error control that UDP provides is that it checks whether the packet arrived correctly and discards it if it did not.

TCP is a more sophisticated transport protocol than UDP. TCP performs two main tasks: controlling errors and controlling the flow of packets. TCP controls errors by arranging for erroneous packets to be retransmitted. It controls the flow by slowing down the transmissions when it detects congestion. Thus, the end systems perform the important tasks of error control and congestion/flow control, not the IP routers. This choice reflects a design principle: "Do not ask the network to do what you can do yourself."

This principle is consistent with what we learned in the network layer: keep the network as simple as possible and let the end systems perform most of the tasks. This approach makes the network as scalable as possible. Since the internal network elements are unaware of connections, the number of connections can grow without bounds: As more end systems are attached, they can manage the additional connections.

This "keep it simple" principle is prevalent in the design of internets. One interesting question in network engineering is whether this principle might not ultimately limit the capabilities of TCP/IP networks in an unacceptable way.

3.2.4   *Applications*

The applications layer implements information delivery services and accessory services that user applications need. We briefly comment on a few representative applications.

FTP

Three basic operations are supported by file transfer protocol (FTP): A user on a machine can send a file to another computer, get a file from that computer, and transfer files between two remote machines. FTP is usually used interactively. It provides a large number of options for creating, changing, or consulting a remote directory, deleting and retrieving a remote file, choosing the transfer mode (stream, block, or compressed), and sending files.

The *stream* mode is used by default. The file is sent without modification. The *block* mode partitions the file to be transferred into blocks. This mode is used to simplify data recovery in case of an error. Finally, the *compressed* mode is used to avoid sending long strings of repeated characters (e.g., spaces). This mode uses the Lempel-Ziv algorithm.

FTP uses two TCP connections: one for the commands/responses and the other for the data transfers/acknowledgments. A host has an FTP process constantly running and ready to process commands. These commands reach the machine on a TCP connection, using a special port number (21). An FTP request from another machine may require the user to be authenticated by a password.

SMTP

The *simple mail-transfer protocol* (SMTP) is used to transfer electronic mail messages between hosts. A mail server process is always running, ready to handle messages.

SMTP accepts a message from the user along with a list of destinations. A copy is then sent to each destination, except when different users are on the same host. In that case, the message is sent only once, together with the list of destinations on the corresponding host. When the delivery of a message is not successful, SMTP will attempt delivery a number of times on successive days before giving up and indicating failure to deliver it to the user.

TELNET

TELNET is the virtual terminal protocol of Internet. It enables a user to simulate a direct connection from a terminal to a remote host.

This function is implemented by defining a standard character code for the network. In TELNET, this is ASCII (the usual symbols plus a set of control codes). This standard code corresponds to a *virtual terminal* that all the hosts are able, it is assumed, to interact with. This interaction takes place by converting the characters sent by the actual terminal into the network standard. Similarly, the virtual terminal input is translated into the input expected by the host.

The transmission is done by means of TCP. Special commands can be sent as expedited data so as to bypass queued data.

The TELNET virtual terminal is a basic scroll-mode terminal. That is, when the end of a line is reached, a new line starts and the others move up, as in most video displays. Many screen commands such as "home" and "clear screen" are not supported. The selection of these basic terminal actions guarantees that most existing terminals can support the features of the virtual terminal. The disadvantage is that it limits the capabilities of the terminal. In order to make the connection of more sophisticated terminals possible, TELNET has an *option negotiation* phase that allows users to agree on a set of options to be supported by the virtual terminal. For instance, the *echo* (i.e., displaying the sender's commands on the sender's screen) can be handled either locally or by the remote machine. In the latter case, each character is sent in a separate TCP packet to the remote machine and is then sent back for echoing. In the case of local echoing, the characters can be sent together in one packet. The *sizes* of the output (line length, page size) can also be negotiated.

rcp

The *remote file copy* (rcp) command is used to copy files between machines. It can also be used to copy files from one remote machine to another remote machine. An option allows the user to copy all the files of a subdirectory.

rsh

The *remote shell* (rsh) command is used to execute a command on a remote machine and to see the results on the local output. Thus, *rsh* first connects to the remote machine, then sends the command to the machine, returns the result of the command, and finally terminates the connection when the command is executed.

rlogin

rlogin is the remote log in command. It is used to connect the local terminal to a remote host. The remote host will verify that the user is authorized to log in and will then execute the log in without asking for a password.

The echoing is done by the remote machine. Special commands to control the flow of data (the "stop" $\hat{S}$ and the "resume" $\hat{Q}$ commands) are sent as expedited data.

TFTP

The *trivial file transfer protocol* enables the transfer of files between two processes over UDP. A file to be transferred is decomposed into blocks of up to 512 bytes. Each block is sent as a UDP packet together with a block number to enable the receiver to reassemble the file. The blocks are acknowledged by the receiver. The sender retransmits blocks that are not acknowledged before a timeout.

HTML

HTML is a *hypertext marking language.* Figure 3.7 shows the a simple HTML source code of a page and, to its right, a screen capture of that page. The source code is almost self-explanatory. Note that the link "http://www.eecs.berkeley.edu/~wlr" specifies the protocol (http), the server (www.eecs.berkeley.edu), and a document (here indicated by the link wlr).

HTTP

The *hypertext transfer protocol* HTTP uses TCP to get documents from a server. You use HTTP when you click on a hypertext link on a web page using a web browser.

The sequence of operations is as we explained in our discussion of TCP: Your web browser is the client and the remote machine indicated by the link is the server. When you click on the link, HTTP opens a connection with the server (with a three-way handshake), the server checks for authorizations and asks for your password if necessary, then the server sends you the requested document and closes its connection to the client, and then the client closes the connection to the server.

Future versions of HTTP will enable a user to start a few connections at a time and will also enable the same client/server to use the same TCP connection for multiple documents. Such modifications will improve the utilization of the network links.

RTP

RTP, a *real-time transfer protocol,* is designed to transmit audio and video over the Internet with little latency. The source encodes the video or audio signal and compresses it. The source then places that signal into packets whose header specifies the compression algorithm

FIGURE 3.7

Example of Web page in HTML (left) and as displayed.

```
<HTML>
<HEAD>
 <META NAME="GENERATOR" CONTENT="Adobe
PageMill 2.0 Mac">
 <TITLE>HTML Example</TITLE>
</HEAD>
<BODY BGCOLOR="#fedede">
<P>This is a simple example of an HTML page.
It has a figure: </P>
<P ALIGN=CENTER><IMG
SRC="/MacintoshHD/Documents/CNAFC/cn.gif"
WIDTH="72" HEIGHT="72" ALIGN="BOTTOM"
NATURALSIZEFLAG="3"></P>
<P>and a link:
<A HREF="http://www.eecs.berkeley.edu/~wlr">
Click here to go to Walrand's home page</A>.
</BODY>
</HTML>
```

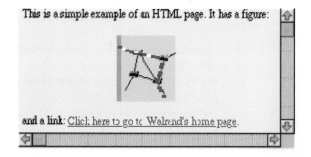

and the time when the packet was produced. The receiver buffers the packets to absorb the delay jitter in the network, decompresses them, and plays them back at the appropriate time.

Note on Java

Java is a programming language that was designed at Sun Microsystems. The initial motivation was that Java should be a lightweight programming language that can run on small controllers embedded in consumer appliances as well as in powerful desktop computers. To be universal, Java runs on a virtual machine that is simulated by the actual computer. In this way, the same code can be executed on all computers, independently of their operating system.

The importance of Java in network applications is that it could potentially resolve the compatibility problem that new applications face. A document that requires a specific viewer can be sent with its viewer written in Java so that all receivers can view it.

3.3 Names and Addresses

To ensure that packets arrive at the correct destination, the network must know the location of a destination before selecting a path leading to it. As we explained in Chapter 2, each node has a unique network address and a unique name. In the Internet, the name has a hierarchical structure based on that of the name-granting authorities; the network addresses have a hierarchical structure that is geographical.

3.3.1   Names

The network—more precisely, a server in the network—must translate the destination name into a network location. To find the network address of a destination, the sending computer consults a directory, as you do to find a telephone number. The organization of the automatic directory service for the Internet is similar to that of telephone directory assistance. Telephone operators are responsible for the information in one area. To learn a telephone number anywhere in the United States, you need only know the area code, say XYZ, of that number and call the directory assistance of that area (1-XYZ-555-1212). Thus, the directory database is partitioned by area codes. This partition localizes the assistance calls and simplifies the updating of the database.

Similarly, the host names in Internet are grouped into *domains*. Note that the division into domains is not geographical. Instead, the division is based on the hierarchical organization structure of the authorities that supervise the naming of nodes. When a host needs the address of some other host, it places a request to a local *name server* process that runs on a computer whose address must be known. This process checks whether the node name belongs to the domain for which it is responsible. If so, the server replies with the requested network address. Otherwise, the name server determines which other name server is responsible for the name's domain, and it forwards the request to that other server. When it gets the address, the host caches it for future use.

This decomposition of the set of nodes into domains can be extended further by decomposing the domains into subdomains and subdomains into subsubdomains, and so on.

Such a decomposition is called *hierarchical naming*. The structure of the Internet names is shown in the left part of Figure 3.8. The name space is divided into domains and each domain into subdomains. The domains are .com (private companies), .edu (educational institutions), .gov (governmental agencies), .int (international organizations), .mil (military), .net (network service providers), .org (nonprofit organizations), and countries such as .be (Belgium), .ca (Canada), .ch (Switzerland), .fr (France), .in (India), .jp (Japan), and .uk (United Kingdom). For instance. berkeley.edu is the subdomain "Berkeley campus" of the domain "educational institutions." The subdomains can be further divided. For instance, eecs.berkeley.edu is the collection of names in the Electrical Engineering and Computer Sciences Department of berkeley.edu. The names are managed by the Internet Assigned Number Authority (IANA), which distributes the allocation authority to three organizations called Internet registries (IRs): RIPE for Europe (http://www.ripe.net), APNIC for Asia and the Pacific (http://www.apnic.net), and InterNIC for the United States and the rest of the world (http://www.internic.net). To connect to the Internet, a company asks its service provider for a collection of names. The provider in turn gets the names from its IR. Each domain is equipped with a *domain name server* (DNS) which maintains the list of Internet addresses of its names. In late 1996, the Internet Engineering Technical Forum was considering adding new domain names.

As an illustration, let us examine how an application such as email or WWW running in a computer A in France finds the Internet address of computer B with name diva.eecs.berkeley.edu. (See the right part of Figure 3.8.) Computer A first searches a cache for diva.eecs.berkeley.edu. If that cache contains the IP address of that computer, the procedure is terminated. Otherwise, A asks its local name server, say C, for the address. Computer C asks the root name server (duplicated in France) for the address of the name server, say D, of berkeley.edu. Computer C then asks D for the address of the name server, say E of eecs.berkeley.edu. Finally, C asks E for the address of B and it gives that address to A.

FIGURE 3.8

Name structure in IP (left) and messages exchanged to find the IP address of diva.eecs.berkeley.edu.

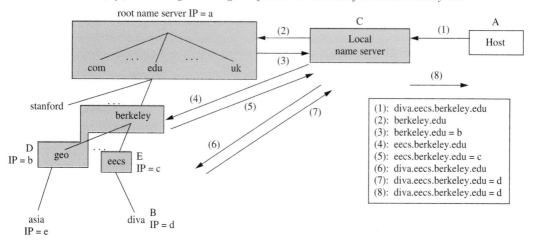

A network resource (a file to be accessed with a specific protocol) is identified by a resource identifier. The most common resource identifier is the URL (uniform resource locator). The URL specifies the protocol (e.g., FTP or HTTP) and the location of the resource. The location is the network address the computer and the path name of the file in the directory of the computer. For instance, http://www.eecs.berkeley.edu/~wlr is the URL of my home page. In this URL, http specifies that the protocol to be used is the hypertext transfer protocol, www.eecs.berkeley.edu is the name of the WWW server of the EECS Department, and ~wlr is a link to my homepage.

3.3.2 Addresses

Addressing in Internet has gone through three steps: class-based addresses, subnetting, and classless addressing. We explain and justify these three steps.

Class-Based Addressing

The class-based Internet addresses have the form "network.host." For instance, the address of the network U.C. Berkeley is 128.32, which corresponds to two successive 8-bit words with decimal values 128 and 32 (thus, 1000'0000'0010'0000). Similarly, the address of U.C. Riverside is 192.31.146. The rule is that small networks have large addresses, since fewer bits are needed to discriminate the few hosts on such networks, while large networks have small addresses; the balance of the address bits are used for the host address.

There are three classes of networks shown in Figure 3.9: large, or class A (network address = 8 bits, host address on network = 24 bits); medium, or class B (16, 16); and small, or class C (24, 8). A fourth class of addresses, D, is reserved for multicast groups.

One problem with these address classes is that they correspond to network sizes that are not well matched to what users need: class A networks are huge and class C networks are very small; there are almost no more free class B addresses. One partial solution to this problem is subnetting, which we explain next.

Subnetting

Subnetting enables a large network, say class B, to be split into subnetworks. With subnetting, five 2000-node subnetworks can share to the same class B address instead of using five different class B addresses. For instance, the U.C. Berkeley network partitions its hosts into subnetworks; the subnetwork is indicated by the third byte and the host on that network is

FIGURE 3.9

Class-based addresses. The figure shows the format of the addresses and the number of networks and hosts of the different classes.

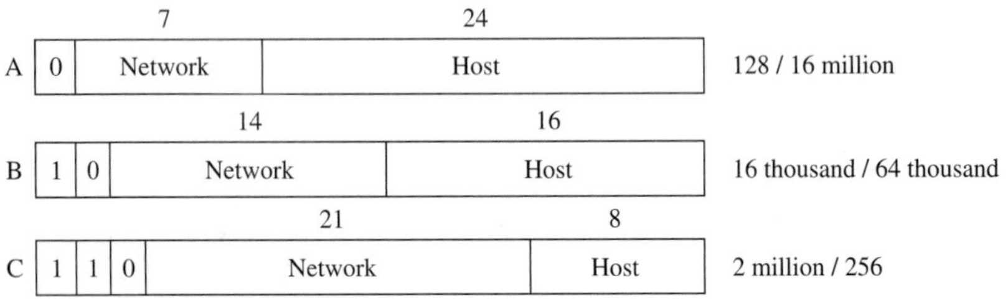

identified by the last byte. Thus, 128.32.134 is a particular subnetwork on the U.C. Berkeley network and 128.32.134.56 is a computer on that subnetwork.

When subnetting is used, a host is given an IP address and a subnet mask, as shown in Figure 3.10. For instance, the computer 128.32.134.56 also has the mask 255.255.255.0, that is, 24 ones followed by 8 zeros. By looking at the first bits of the address (100..), one finds that this address is of class B and, consequently, that the network part of the address is 128.32. In addition, by using the mask, one finds that the subnetwork of that computer is 128.32.134.

Figure 3.10 shows a packet that goes from a host with IP address A and mask N to a host with IP address D and mask M. To send that packet, host A compares $A \otimes N$ and $D \otimes M$. If these values are different, then A is not on the same subnetwork as D. Accordingly, A must send the packet first to the router of the subnetwork $A \otimes N$. The router finds the route to reach D by checking the entry of the routing table that corresponds to the network part of the address D. Eventually, the packet reaches the router R. Since $R \otimes M = D \otimes M$, the router knows that D is on the same subnet. Router R then consults its address resolution protocol (ARP) table, or uses ARP, to find the media access control (MAC) address of D and sends the packet to D over the LAN.

For instance, assume that computer A with address 128.32.152.26 and mask 255.255.255.0 needs to send a packet to computer B with address 128.32.134.56. Computer A finds out that $128.32.152.26 \otimes 255.255.255.0 = 128.32.152.0$ is not the same as $128.32.134.56 \otimes 255.255.255.0 = 128.32.134$. Thus, computer A knows that it is not on the same subnetwork as computer B. Computer A then sends the packet to the router of its subnetwork.

This subnetting procedure uses addresses more efficiently. However, an even better scheme is explained next.

FIGURE 3.10

Subnetting. Each host has an IP address and a subnet mask. The subnetwork of the host is determined by a bit-by-bit "and" of the IP address and of the mask.

Replace Network.Host by Network.Subnet.Host

Host has IP address (associated with subnet)

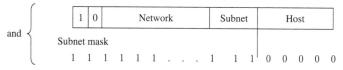

Note: Subnet address: IP⊗Mask (⊗ = AND)

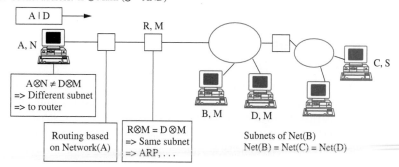

FIGURE 3.11

Supernetting in CIDR. In this addressing scheme, the router checks the longest prefix matching entry in its routing table.

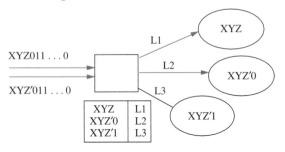

CIDR

To make the addressing more flexible, the Internet designers defined more general prefix-based addresses called CIDR (for *classless interdomain routing*). The general idea of CIDR is that the addresses are organized as a prefix-free code. That is, the initial segment of an address can define a domain if shortening it does not define another domain. Thus, the initial prefix, say the first 3 bits, could specify the continent, the next 7 bits the country. If the country is France, then the next 7 bits could be the department. Depending on the department, a number of bits then define the town, and so on. If the country is Belgium, then 4 bits suffice to define the province, then a few for the town, and so on. The routing can then be based on the structure of the address.

Figure 3.11 illustrates CIDR. The figure shows a supernet which consists of all the hosts with addresses that share a prefix XYZ. Another supernet corresponds to the prefix XYZ′0, and a third supernet corresponds to the prefix XYZ′1. The entries of the routing table correspond to the longest matching prefixes.

The 32-bit addresses are in principle sufficient to accommodate about 4 billion hosts. With class-based addressing, the utilization of the addresses was very inefficient and we would have run out of addresses by this time. Internet experts predict that with an efficient use of CIDR (which requires renumbering of hosts), the 32-bit addresses may be sufficient for the foreseeable future.

The next version of IP (called IPv6; see Section 3.9) uses 128-bit addresses. Its main motivation is to overcome the limitations of the 32-bit addresses. It is possible that CIDR will make the 128-bit addresses unnecessary, thus eliminating the principal justification for IPv6.

3.4   Internet Protocol

The Internet protocol is the main network layer protocol of an internet. This protocol supervises the routing of packets to their destination. The network nodes exchange control packets using ICMP (Internet control message protocol) to implement IP. We explain the services that IP provides in Section 3.4.1. In particular we discuss the format of IP packets.

The routing in Internet is hierarchical and divides nodes into subnetworks called *autonomous systems* (ASs) that are interconnected by *border gateways*. An AS is a subnetwork controlled by a single organization, such as a university campus or a company network. Within the AS, the routing algorithm is OSPF (open shortest path first). We explain that algorithm in Section 3.4.2. A number of routers still use RIP (routing information protocol), a routing protocol based on the Bellman-Ford algorithm, which we explain in Section 3.8.1. Between ASs, the routers use a routing algorithm called the border gateway protocol (BGP). This algorithm, which we explain in Section 3.4.3, is implemented by a router in each AS. Sections 3.6 to 3.8 contain additional material on routing algorithms.

3.4.1   IP Datagrams and ICMP

IP is a datagram protocol. It delivers datagrams of up to 64 kbytes. That is, the layer above the network layer—the transport layer—gives IP a datagram of up to 64 kbytes together with a network destination IP address. If all goes well, IP in the destination node eventually gives the datagram to the transport layer. A few things may go wrong. For instance, the destination may be unreachable or a packet may loop in the network. In such a case, the Internet control message protocol (ICMP) informs the source host. ICMP specifies the format of control messages and when routers should send them. ICMP messages are delivered by IP. Transmission errors may also corrupt the packet or a packet may arrive at a router that is full and get dropped. In these situations, IP cannot inform the source of the packets it lost and it is up to the transport layer to take corrective actions.

Next, we explain the format of the header of an IP packet.

IP Header

The packets contain an *IP header*. The basic header, without options, is illustrated in Figure 3.12, where each tick indicates a bit position. The *version* field allows new versions of the IP protocol to be installed while the network is operational. The *Internet header length* (IHL) indicates how long the header is. The Service type field specifies the quality of service desired (e.g., low delay, high throughput, high reliability); few routers offer this choice. Identification, Flag, and Fragment offset allow reassembly of fragmented datagrams (see below). The Time to live field indicates how long the packet can still remain in the network. Each router decrements the value in that field by one when it gets the packet, and discards the packet if the value reaches zero (ICMP then informs the source). The Protocol field indicates what higher level protocol is contained in the data portion of the IP packet (e.g., TCP, UDP, or ICMP). The Header checksum is a checksum calculated on the bytes in the IP header. Higher-level protocols must control the errors in their data. The Source and Destination network addresses indicate the sending host and the intended recipient host for this datagram.

Fragmentation

Links have a maximum size of packets they can transmit. This maximum size is the link's *maximum transfer unit* (MTU).

If a packet that arrives at a router is larger than the MTU of the outgoing link, the router fragments the packet. The destination host reassembles the packet. Thus, a packet may be fragmented a few times between the source and destination. The specific numbering scheme of the fragments makes the reassembly procedure straightforward and independent of the number of fragmentations. Looking at the IP header in Figure 3.12 we see that the

FIGURE 3.12

IP header.

<table>
<tr><td colspan="4">0</td><td>31</td></tr>
<tr><td>VER</td><td>IHL</td><td>Service type</td><td colspan="2">Total length</td></tr>
<tr><td colspan="3">Identification</td><td>Flag</td><td>Fragment offset</td></tr>
<tr><td colspan="2">Time to live</td><td>Protocol</td><td colspan="2">Header checksum</td></tr>
<tr><td colspan="5">Source network address</td></tr>
<tr><td colspan="5">Destination network address</td></tr>
<tr><td colspan="4">Options</td><td>Padding</td></tr>
</table>

fragments are marked with the identification number of the original datagram and with the offset of the fragment relative to the first byte of the datagram.

We illustrate the procedure in an example. Consider a datagram with identification number X and length L. Assume that a router first fragments this datagram into four pieces with lengths K, K, K, K', respectively. Here, K is the MTU of the outgoing link of the router that fragments the packet and $K' = L - 3K \leq K$. The first fragment is marked with $(X, 0)$, the second with (X, K), the third with $(X, 2K)$, and the fourth with $(X, 3K)$ to indicate the identity X of the original datagram and the offsets of the fragments. Now assume that these fragments reach a router with an outgoing link that has an MTU equal to M where $M < K$. Let us assume that $2M < K = 2M + M' < 3M$. The third fragment—that marked with $(X, 2K)$ after the first fragmentation—gets decomposed into three fragments. These fragments have lengths M, M, M' and are marked with $(X, 2K)$, $(X, 2K + M)$, $(X, 2K + 2M)$, respectively. Two points should be noted: (1) the router can figure out the offsets of the new fragments from the original offset and the new MTU and (2) the destination has no problem reassembling these fragments.

3.4.2 OSPF

The open shortest path first algorithm is based on the shortest path algorithm of Dijkstra. This algorithm constructs the shortest path from one node of a graph to all the other nodes of the graph. The length of a path is defined as the sum of the lengths of the links along the path.

A Physical Model

There is an easy way to visualize the operations of the algorithm. Imagine a collection of N balls that are attached to each other by strings. The strings have different lengths. We select one ball and we call it "ball 1." We want to find the shortest path from ball 1 to each of the other balls. To find these paths we put down all the balls on the floor and we start lifting ball 1 slowly. We call the next ball to rise from the floor "ball 2." When ball 2 rises from the floor, we have found the shortest path from ball 1 to ball 2: it is the string that attaches balls 1 and 2 together. As we continue to lift ball 1, another ball, ball 3, rises from the floor. The shortest path from ball 1 to ball 3 is either a link from 1 to 3 or is made up of a link from 1 to 2 and another from 2 to 3. Continuing in this way, eventually we have lifted n balls from the floor. When next ball, ball $n + 1$, rises, we have found the shortest path from 1 to $n + 1$. Eventually, we find the shortest paths from 1 to all the other balls.

Dijkstra's Shortest Path Algorithm

The algorithmic translation of the physical process that we just described is called *Dijkstra's shortest path algorithm.* The algorithm is based on the following observation: The distance from ball 1 to ball n cannot decrease after ball n rises from the floor. Consequently, if we keep track of an estimated distance from ball 1 to ball n, we know that this estimated distance eventually settles to the shortest distance, and that this happens when the ball rises from the floor.

How do we know which ball rises next? Say that ball k rises and let d be the distance from ball 1 to ball k. We update the estimates of the distances from ball 1 to the balls attached to ball k as follows. If some ball j is not up yet and is attached to ball k with a string of length s, we replace the current estimated distance x from 1 to j by the minimum of x and $d + s$. When we have updated all the neighbors of ball k, we must find the ball that will rise next. That ball is the ball on the floor, say ball p, with the current smallest estimated distance away from ball 1. We can then continue the process with ball p.

We first explain the algorithm on the simple network shown in the left part of Figure 3.13. The objective is to find the shortest paths from A to all the other nodes. The lengths of the individual links are marked next to them. On such a small network, a simple inspection shows that the shortest path from A to the bottom node has length 5 and goes through the right middle node. Dijkstra's algorithm is a systematic procedure for discovering such a shortest path even in a large network.

Next to each node, we mark the current estimate of the length of the shortest path from A to that node. The symbol ∞ means that no path to that node has been found yet. The algorithm starts by considering the node with the smallest label, in this case A. The algorithm explores the links going out of that node. The left link leads to the left middle node which can then be reached from A with a path of length 3. Accordingly, we reduce the label of that node from ∞ to 3. Since the label is reduced by using that left link out of A, we mark the link as being the current candidate for the shortest path to the left middle node.

FIGURE 3.13

Steps in Dijkstra's algorithm.

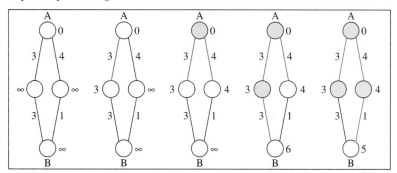

A similar step is performed for the right link out of A. We then shade node A to indicate that we have examined all its outgoing links. At the next step, we examine the unshaded node with the current smallest label. That node is the left middle node. We examine the outgoing link of that node and find a candidate shortest path to the bottom node with length $3+3 = 6$. We shade that node and then examine the unshaded node with the smallest label: the right middle node. With the link out of that node, we find a shorter path to the bottom node. Accordingly, we "unmark" the link from the left middle node to the bottom node because we know that the link is not on the shortest path to the bottom node. In addition, we mark the link from the right middle node to the bottom node. Since the bottom node has no outgoing link, the algorithm terminates. The marked links define the shortest paths from A to the other nodes and the node labels are the lengths of the shortest paths from A to the nodes.

Next we illustrate Dijkstra's shortest path algorithm with the example shown in Figure 3.14. We start with node A as the source and we mark each node with an estimated distance from A to the node. The initial estimates are infinite, except that of node A which is 0. The first node to "rise from the floor" is node A. We shade that node to remember that it is off the floor and we update the estimate of all the neighbors B, C, D of A. For instance, we replace the current infinite estimate of B by the sum of the estimate of node A (equal to 0) plus the length 4 of the link from A to B. The new estimates (4, 3, 2) are underlined in the second part of the figure. At that point, we determine the unshaded node with the smallest estimate. That node is node D and is therefore the next node to rise from the floor; we shade that node D. Note that all the links from node A to its neighbors are colored blue

FIGURE 3.14

Illustration of Dijkstra's algorithm.

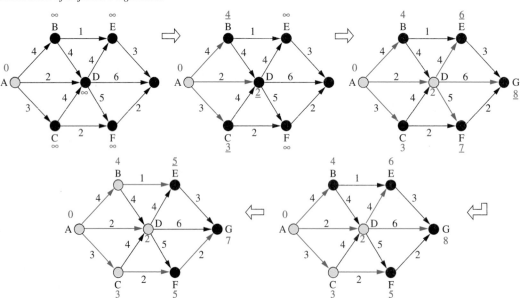

to remember that these are candidates for shortest paths from A to other nodes since these links have provided us with the smallest estimates of the distances to the nodes B, C, D so far. In the third part of the figure, we explore the neighbors of node D and we update their estimates and color the links from D to these neighbors blue. In the fourth part of the figure, we have located node C as the unshaded node with the smallest estimate; we shade C and we explore its neighbors and update their estimates. Note that the link CF produces a smaller estimate (6) for node F. Accordingly, we change the link DF from blue to black and we color the link CF as blue. In the last part of the figure, we locate the unshaded node B with the smallest estimate; we shade node B and update the estimates of its neighbors. Note that the link from B to D does not reduce the estimate of node D because $4 + 4 > 2$. However, the link from B to E reduces the estimate of E because $4 + 1 < 6$. Consequently, to reflect this reduced estimate, we color the link from B to E blue and we change the link from D to E from blue to black. Indeed, the link from D to E cannot be on a shortest path from A to E since it yields a longer path than a path that reaches E via link BE. The left-most part of the figure shows the shortest paths.

We analyze Dijkstra's shortest path algorithm in Section 3.5.

Implementation of OSPF

Each router executes Dijkstra's shortest path algorithm. To execute the algorithm each router needs a map of the network with the link lengths. Each router knows its own network address, which is configured when the router is set up. Each router talks to its neighbors to find their network addresses. As the network is in operation, the router calculates a metric for each output link. In the simplest case, the metric is 1 when the link is operating and is infinite when it is not. A more accurate metric takes into account the link transmission rate and the average delay to go through its buffer. The metric is selected by the network administrator and all the routers (in a given AS) use the same metric.

The routers of an AS can also execute Dijkstra's algorithm for a few different metrics. For instance, the metrics can be measures of reliability, transmission rate, and delay. After running Dijkstra's algorithm for each of these three metrics, each router knows the most reliable path, the path with the largest throughput, and the path with the smallest delay to each of the other routers in the AS. A packet can then specify which criterion the routers should use to select its path. We continue our discussion assuming a single metric, but it can be adapted easily to the case of multiple metrics.

Router i prepares a message that lists the metrics of its outgoing links. The router appends a sequence number s to the message M. Thus, M might look like this (we do not explain the actual message formats in bits and bytes):

$$M = [i|s|k_1, d_1|k_2, d_2| \ldots |k_m, d_m].$$

In this notation, k_1 identifies a neighbor of router i and d_1 is the estimated metric of the link from i to k_1, the meaning of $k_2, d_2, \ldots, k_m, d_m$ is similar. Router i sends message M on each of its outgoing links.

When a router j gets such a message, it checks the original source i of that message. If $i = j$, then router j discards the packet. Otherwise, it checks the largest sequence number of routing messages from i that it has received to date. If that sequence number is smaller than s, indicating that M is a new message from i, router j updates that largest sequence number, stores the message for its own use, and sends a copy along each link other than the incoming link of the message. Eventually, the message is sent once to all the reachable nodes. This distribution method to all the other routers is called *flooding*.

As the router gets messages, it can construct a map of the network and execute Dijkstra's algorithm. A router knows when it gets an updated message by checking the sequence number and it then knows that it should run the algorithm again. A router can send a new message when a link metric has changed significantly enough.

3.4.3 BGP

The routing between ASs is based on the *border gateway protocol* (BGP). In contrast with OSPF, BGP is a distributed protocol. That is, the routers that execute the protocol have different information about the network. Another key difference with OSPF is that BGP is a *preferred path* algorithm instead of being a *metric* algorithm, as we explain below.

A First Look

Let us start with a rough sketch of how BGP works. We then refine that sketch. Say that the routers want to construct a path to some destination host D. At some step of the algorithm, some of the autonomous systems have identified preferred paths to D.

Consider an AS X that is attached to the AS Y and the AS U. (See Figure 3.15.) The AS Y advertises to X the path that Y prefers to reach D by sending the message, [Y, X, U; 17]. In this message, (Y, X, U) is the path that Y prefers to reach D and 17 is the estimated metric of that path, say the expected delay (in milliseconds). Similarly, the AS U advertises its preferred path to D by sending the message [U; 18].

When it gets these messages, X must select the path is prefers to reach D. Assume that X chooses to send a packet destined to D to Y because the advertised metric of that path is smaller than that advertised by U. When Y gets the packet, it sends it along its preferred path. Accordingly, Y sends the packet first to X. Consequently, the packet will loop between X and Y. However, if X examines the path advertised by Y, it notices that this path goes through X, so that X should not choose that path. Thus X will select the path going through U as its preferred path and it will advertise it as [X, U; 23], where the metric

FIGURE 3.15

Illustration of BGP.

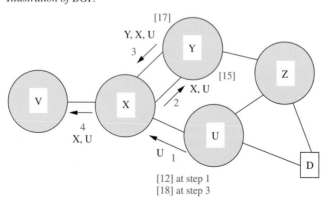

23 is obtained by adding to 18 the metric of the connection from the entry of V into X to the exit of X toward U (which we assume to be 5 here). This metric (5) is internal to the AS X and is derived by the routers of X running OSPF. The metrics that the different ASs use do not need to be identical. However, the metrics through an AS used to update the BGP path metrics should be consistent: a longer path inside X should have a larger metric.

Note that the metrics that Y and U advertise to X are inconsistent with the topology of the paths. Indeed, since the path from Y to D goes first through X then through U, its metric should be larger than the metric of the path from U to D. Such inconsistencies may result from dated estimates. The figure illustrates a sequence of updates that leads to inconsistent estimates. At step 1, U advertises a metric equal to 12 to reach D. At step 2, X propagates that estimate to Y. At step 3, Y uses that estimate to calculate the metric 17 that it advertises to X. At the same time, step 3, U updates its estimate of the metric to D and advertises the new estimate (18) to X. The increase in the metric is due to a buildup of congestion inside U that Y is not aware of yet at step 3.

Implementation

An AS may be connected to other ASs with more than one border router. One router inside the AS is selected by the network manager to execute BGP. That router is called the *BGP speaker*. As in OSPF, BGP can use a number of different metrics. In addition, each AS can maintain lists of ASs that it does not want to send its packets through. For instance, some ASs may belong to a commercial competitor and security or reliability might be doubted.

Note the economy of information exchange in BGP. Instead of flooding the network, each AS sends only short messages to its neighbors. Distributed algorithms run the risk of creating loops. We saw in the previous section that if X bases its path selection only on the advertised metrics, then the packets it sends enter a loop. Such a loop is created when the nodes have inconsistent estimates of metrics. BGP prevents loops by advertising the paths in addition to their metrics.

3.4.4   *Plug and Play: DHCP*

A number of study rooms and lecture rooms on our campus are equipped with Ethernet outlets. A student can plug a laptop computer into the Ethernet outlet and start using electronic mail, Web browsers, and file transfer applications. The protocol that makes such connections possible is called the dynamic host configuration protocol (DHCP) and is known under the descriptive name of *plug and play*.

Remember that to send a packet on the Internet, a computer must first have a network address. This address is associated with the location of the computer, specifically with the network to which the computer is attached. In our example, the laptop computer must find the Ethernet router and ask for a network address. The laptop computer sends a special broadcast (DHCP) message asking "Router, give me a network address." The router maintains a pool of free network addresses and gives one to the laptop with a specified time to live (say 20 minutes). The laptop can then use the standard Internet applications. As the time to live gets close to zero, the laptop asks the router for a new 20 minutes, which is normally granted. If the student unplugs the laptop, the network address returns to the pool of free addresses when its time to live expires because the laptop does not ask for an extension. The same mechanism can be used with a wireless Ethernet.

FIGURE 3.16

Illustration of mobile IP.

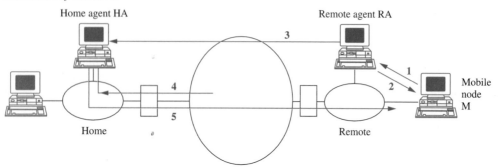

1. M tells RA: I am visiting, my home agent is HA
2. RA tells M: your temporary IP address is IP.temp
3. RA tells HA: M is now at IP.temp
4. Packet destined to M is intercepted by HA
5. HA sends packet to M (at IP.temp)

3.4.5   Mobile IP

The DHCP that we just explained can be combined with another protocol so that the student's email is automatically forwarded to the laptop. This protocol, mobile IP, works as follows. The student is assumed to have a home router where the computer is normally attached. When the student moves to another location, the computer automatically registers with an agent on the network the student is visiting. That agent gives the student the temporary network address that we saw in DHCP. Additionally, the agent contacts an agent at the student's home router to inform that agent of the student's new location. That agent at the home router intercepts the packets intended for the student and forwards them to the remote agent which delivers them to the temporary address. When the time to live of the temporary address expires, the foreign agent informs the home agent to stop forwarding the packets and to hold them until the student returns or until a new location is announced. This mechanism is illustrated in Figure 3.16.

3.5   End-to-End Transmission

We learned that the network layer implements the end-to-end delivery of packets. The transport layer supervises two aspects of this delivery: error and congestion/flow control.

3.5.1   Overview

Errors

Errors occur in the delivery of packets for two reasons. First, transmission errors corrupt packets. Second, a router discards a packet that arrives when the buffer is full.

Two versions of the transport layer are implemented in the Internet: TCP, the *transmission control protocol,* and UDP, the *user datagram protocol.*

TCP establishes a full-duplex (i.e., bidirectional) virtual byte link between two computers that we call *end computers* to recall that they are at both ends of the connection. That virtual link is perfect, except for the delays of the bytes. That is, TCP delivers the bytes without errors and in the correct order. To implement this service, the end computers use a retransmission protocol called the *selective repeat protocol* (SRP) that we explain below.

UDP delivers datagrams and makes no attempt to retransmit erroneous packets.

Congestion and Flow Control

Flow control is the procedure a source uses to adjust its transmission rate so as not to overwhelm the receiver. In TCP, the receiver tells the source the number of packets it is willing to receive.

Congestion control is the mechanism that sources use to limit the congestion in nodes of the network. In TCP, the sources use the delays of acknowledgments as indicators of congestion and adjust the window size of the retransmission protocol accordingly, as we explain below.

3.5.2   *Retransmission Protocol*

The selective repeat protocol attempts to retransmit only the packets that do not get correctly to the destination, either because of transmission errors or because they are dropped by a full router.

The source and destination computers implement SRP. When the source sends a packet, it starts a timer. The destination sends back an acknowledgment for every correct packet that it receives. The source assumes that a packet fails to reach the destination correctly when the acknowledgment does not come back within some time called the *timeout.* The source then retransmits a copy of that packet.

More precisely, the source uses a window size W for SRP. The source sends the packets $1, 2, \ldots, W$ and waits for the acknowledgment of packet 1 before it sends packet $W + 1$. If everything goes well, the source gets back the acknowledgments of packets before a timeout and it can keep on transmitting. If the acknowledgment of some packet i fails to arrive within a timeout, the source retransmits that packet.

When it gets a correct packet, the destination sends an acknowledgment with a sequence number equal to $K + 1$ if the destination has received the packets $1, 2, \ldots, K$ but has not received $K+1$. Thus, if the destination receives the packets $1, 2, \ldots, K, K+2, K+3, K+4$, then it sends the acknowledgments with sequence numbers $2, 3, \ldots, K + 1, K + 1, K + 1, K + 1$. In such a situation we say that the last three acknowledgments are *duplicate acknowledgments.* The destination delivers the packets $1, 2, \ldots, K$ that it got in order and stores the packets $K + 2, K + 3, K + 4$ to be able to deliver them after packet $K + 1$ that it is waiting for. When the sender sees three duplicate acknowledgments, it assumes that packet $K + 1$ has been lost and it retransmits that packet, without waiting for a timeout of the acknowledgment $K + 1$. This mechanism is called *fast retransmit*. Section 6.11 explains the protocols in more detail.

3.5.3   TCP

TCP implements a full-duplex byte link between two end computers. The virtual link transports streams of bytes. The application that uses TCP injects a stream of bytes and listens to the stream coming toward it. The stream is error-free, at least as long as the connection stays up. A computer may have a number of connections open at the same time. These TCP connections are differentiated by a *port number*. Some applications are always listening to the same port number. For instance, the electronic mail process listens to port number 25. Other applications use a temporary port number that is allocated by a "port server" when the application is invoked. Figure 3.17 shows how an application sees TCP.

We view TCP as setting up two connections: one from the client to the server and one from the server to the client. Here are typical steps in a TCP connection.

1. The client starts up a duplex connection with the server and requests a document.
2. The server sends the document to the client.
3. The server closes the connection to the client.
4. The client closes the connection to the server.

This sequence of events is shown in Figure 3.18.

The sequence in the figure shows the acknowledgments that are needed because the network is not fully reliable. When the server sends data to the client, the server sends a stream of bytes to TCP. These bytes enter a buffer and TCP puts a group of these bytes into a *segment* that it gives to IP when the buffer is full. (See Figure 3.17.) The size of the buffer is MSS, for maximum segment size. MSS can be specified as an option at the start of a connection. The default value of MSS is 536 bytes for most links; it is 1460 bytes for an

FIGURE 3.17

TCP viewed by a client and a server.

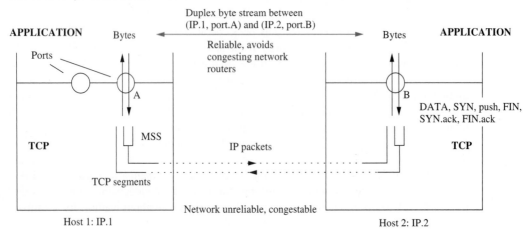

FIGURE 3.18

Typical sequence of transfers by TCP.

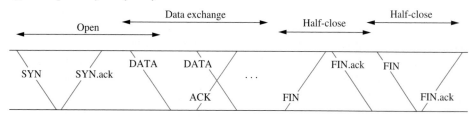

Ethernet and 256 bytes for SLIP. The source can issue a *PUSH* command to TCP to avoid having to wait until the buffer is full. When it gets such a command, TCP puts whatever bytes it has in its buffer into a segment it gives to IP.

Note the control messages (SYN, SYN.ack, FIN, FIN.ack) that the two end computers exchange. These control messages are contained in the TCP header of the packets. This header has the following components:

$$[S.port \mid D.port \mid Seq \mid Ack \mid FLAG \mid Window \mid CKS \mid URG]$$

In this header, $S.port$ and $D.port$ are the source and destination port numbers. Seq and Ack are the packet and acknowledgment sequence numbers. Specifically, the source computer numbers the bytes it sends (in the byte stream) consecutively, starting with an initial sequence number that is determined by a clock-driven counter in the source computer. The objective of this initial sequence number is to make it impossible for a delayed packet in transit in the network to arrive at the destination and be mistaken for a packet of the currently open connection. The clock-driven sequence number wraps around after a time that exceeds the maximum lifetime of a packet in the network, thus preventing such confusions. Seq is the sequence number of the first byte in the packet and Ack is the sequence number of the next byte that the destination expects. That is, the destination has received all the bytes with sequence numbers up to $Ack - 1$ but has not received the next byte (although it may have received bytes with sequence numbers larger than $Ack + 1$). *Window* is the window size that the destination is willing to accept. At any time, the source must use SRP with a window size equal to the minimum of *Window* and the window size W calculated by the slow start and congestion avoidance algorithm. CKS is a checksum that the source computes on the full TCP packet, including the data. URG is a pointer to the last urgent byte in the stream. The *FLAG* indicates whether the packet is a start of connection (*FLAG* = SYN), the acknowledgment of a start of connection (*FLAG* = SYN.ack), an urgent packet (*FLAG* = *Urgent*), a push command (*FLAG* = *PUSH*), an end of connection (*FLAG* = FIN), the acknowledgment of an end of connection (*FLAG* = FIN.ack), or a *reset* to force the termination of a connection (*FLAG* = *reset*).

The protocol specifies what happens when an acknowledgment fails to arrive. Without going into all the details, here are the main points:

- The start of the connection is based on a *three-way handshake* and proceeds as follows. The client asks for a connection to the server by specifying the client's IP address and port number, the server's IP address and port number, and the initial sequence number of the connection based on a clock register in the client. The server sends back an acknowledgment with the same addresses and port numbers and sequence number. The client then sends the first data packet with the initial sequence number. This three-way handshake prevents the client and server from being confused by unrelated delayed packets that might reach them. Such delayed packets would not have the same sequence numbers.

* If the start is not acknowledged in time, then the client tries again after 6 seconds. If this fails, the client tries three more times, every 24 seconds, and then gives up.

* The data transfer is supervised by SRP.

* Each half of the connection is closed by a two-way handshake: "I am closing my connection to you" followed by an acknowledgment.

* After the server sends the acknowledgment of the close of connection from the client to the server, the server stays alive for some time to make sure the client gets the acknowledgment and does not send again its close message, expecting another acknowledgment.

If you want to write client/server network applications, you need to learn how to use TCP and IP. Specifically, you need to know how to start a connection, how to exchange data, and how to close a connection. You can find the implementation details in many books (e.g., Stevens, *UNIX Network Programming,* Prentice-Hall, 1990, and Stevens, *TCP/IP Illustrated,* Addison-Wesley, 1994).

3.5.4   UDP

The *user datagram protocol* is a datagram communication service built on top of IP. It adds multiplexing and error detection to the IP capabilities. As TCP, UDP uses 16-bit port numbers to distinguish source and destination processes.

The UDP header contains 16 error-detection bits which are set to zero when they are not used. The UDP header also specifies the IP source and destination addresses that UDP obtains from IP (which selects the path and, in particular, the network interfaces which determine the source and destination IP addresses). In contrast to TCP, UDP does not use acknowledgments; it does not retransmit erroneous packets or control the flow.

3.6   Complement 1: Link Protocols

In this section we explain two protocols that are commonly used to connect a computer to the Internet with a point-to-point link. Typically, the link is a telephone line and the transmission takes place with modems at each end of the line.

3.6.1   SLIP

SLIP, the *serial line IP,* is a protocol to transport IP packets between computers attached to a common point-to-point link. Many people use SLIP to access the Internet from their home computer either via an Internet service provider or via an office computer.

SLIP starts and ends every IP packet with a special ASCII character $0xC0$ called END. If a byte of the IP packet is the END character, SLIP replaces it by the two characters ESC END where ESC $= 0xdb$. If a byte of the IP packet is the SLIP character ESC, SLIP replaces it with the two characters $0xdb, 0xdd$.

When the receiver gets the bytes, it detects the start of the IP packet when it sees the first END. When it sees ESC, the receiver knows that this byte should be skipped and it looks at the next byte; if that next byte is END, then the receiver knows that this END is a byte inside the IP packet and does not confuse it with the end of packet; if the next byte is $0xdd$, then the receiver knows that a byte $0xdb$ was inside the IP packet.

For the two computers to be able to talk to each other using SLIP, they must know each other's IP address. Because the SLIP frames have no field for the frame type, the link can only be used for SLIP. Moreover, SLIP frames have no error detection field. Errors must be controlled by higher layers or by the line framing that modern modems introduce.

Most SLIP implementations support a variation called *compressed SLIP* (CSLIP). CSLIP recognizes that many packets on the link belong to the same TCP connection and avoids replicating the headers. To avoid repeating headers, CSLIP maintains the state of up to 16 TCP connections and transmits only the changes in the fields of the headers that change in subsequent packets of a given TCP connection.

3.6.2   PPP

PPP, the *point-to-point protocol,* is becoming increasingly popular and is replacing SLIP in many point-to-point links.

PPP, like SLIP, allows IP packets to be transported over a point-to-point link. The link can be either an asynchronous link—a link that transmits one ASCII character at a time—or a synchronous link that transmits large packets. PPP enables the two devices to negotiate options such as deciding which protocol is used on the link, agreeing on the compression of headers as in CSLIP, and agreeing to skip addresses and control fields of IP packets. In addition, PPP packets contain an error detection field (a 2-byte CRC).

3.7   Complement 2: Analysis of Dijkstra's Shortest Path Algorithm

In this section we give a formal definition of Dijkstra's shortest path algorithm and we analyze its properties. Specifically, we prove that it produces the shortest paths and we analyze its numerical complexity.

3.7.1   Definition

We are given a set $\mathbf{S}$ of N nodes and a set of link lengths between these nodes: $\{L(i, j), i, j \in \mathbf{S}\}$. By convention, $L(i, j) = \infty$ if there is no link from i to j. We want to find the shortest path from some node $a \in \mathbf{S}$ to each of the other nodes.

We initialize the algorithm by letting $n = 1$ and defining

$$b_1 = a, d_1(a) = 0, \mathbf{U}(1) = \{a\}, \mathbf{F}(1) = \mathbf{S} - \mathbf{U}(1), \text{ and } d_1(i) = \infty, P_1(i) = \emptyset \text{ for } i \in \mathbf{F}(1).$$

We then determine

$$\mathbf{N}(n) = \{j \in \mathbf{F}(n) | L(b_n, j) < \infty\} \tag{3.1}$$

and

$$d_{n+1}(i) = \min\{d_n(i), d_n(b_n) + L(b_n, i)\} \text{ for } i \in N(n). \tag{3.2}$$

Moreover,

$$P_{n+1}(i) = \left\{ \begin{array}{l} P_n(i) \text{ if } d_n < d_n(b_n) + L(b_n, i), \\ \{b_n\} \text{ if } d_n \geq d_n(b_n) + L(b_n, i). \end{array} \right. \tag{3.3}$$

We set

$$n = n + 1 \tag{3.4}$$

and

$$b_{n+1} = \arg \min\{d_{n+1}(i), i \in \mathbf{F}(n)\} \tag{3.5}$$

and we define

$$\mathbf{U}(n + 1) = U(n) \cup \{b_{n+1}\} \text{ and } \mathbf{F}(n + 1) = \mathbf{S} - \mathbf{U}(n + 1). \tag{3.6}$$

Finally, we set

$$n = n + 1 \tag{3.7}$$

and we repeat steps (3.1) to (3.7) until $n = N + 1$.

When the algorithm has executed, $d_N(i)$ is the length of the shortest path from node a to i, for $i \neq a$. Also, $(P_N(i), i)$ is the link that reaches i along the shortest path from a to i.

We now describe the connection between this formal definition and the physical description of Section 3.4.2. The set $\mathbf{U}(n)$ is the set of balls that are "up" at step n whereas $\mathbf{F}(n)$ is the set of balls that are still on the floor. The ball b_n rises from the floor at step n, and $\mathbf{N}(n)$ is the set of its neighbors that are still on the floor. In step (3.2) we update the estimates of the nodes in $\mathbf{N}(n)$ and, if the estimate of node i is reduced, we note in step (3.2) that (b_n, i) is on the shortest path from a to i. Indeed, the ball b_n that just lifted from the floor yields a shorter path to i than the previously known ones. In step (3.5) we find b_{n+1}, the next ball to lift up.

Next we prove that the algorithm produces the shortest paths.

3.7.2 Shortest Paths

We prove that the algorithm produces the shortest paths by induction on n. That is, we show that at each n the algorithm has found the shortest paths to the nodes in $\mathbf{U}(n)$ and that these paths have the lengths $d_n(i)$ for $i \in \mathbf{U}(n)$.

This statement is certainly true at time $n = 1$ since the shortest path from node a to a has length 0. Assume that the statement is true at time n. We claim that it must then be true at time $n + 1$. To prove this claim, we argue by contradiction. That is, we assume that there is a path from a to $f := b_{n+1}$ that has length strictly less than $d_{n+1}(f)$. We show that this cannot be by considering the last node in $\mathbf{U}(n)$ along that path. Call this node j and let $m \leq n$ be such that $b_m = j$. Let also k be the next node along that path to b_{n+1}. We have

$$d_m(j) + L(j, k) = d_n(j) + L(j, k) = d_{n+1}(k) \geq d_{n+1}(f).$$

Indeed, if this inequality did not hold, then k would have been selected before f by steps (3.2) and (3.6). This contradicts the existence of a shorter path to f and concludes the proof.

3.8 Complement 3: Other Routing Algorithms

3.8.1 Bellman-Ford

The Bellman-Ford algorithm is a distributed algorithm that routers can use to discover the shortest paths to a destination. This algorithm was implemented until 1983 in the Internet under the name of the routing information protocol (RIP). It was then replaced by OSPF because of a few undesirable properties of RIP that we discuss after having described the algorithm and analyzed its main properties.

Description
The problem formulation is identical to that of Dijkstra's shortest path algorithm. One is given a set $\mathbf{S}$ of N nodes and a set of link lengths between these nodes: $\{L(i, j), i, j \in \mathbf{S}\}$. By convention, $L(i, j) = \infty$ if there is no link from i to j. The nodes, which represent routers, want to discover the shortest paths to a destination $D \in \mathbf{S}$.

The Bellman-Ford algorithm is distributed. That is, the routers have different and incomplete information about the graph. At each step of the algorithm, a node computes an estimate of its shortest distance to the destination and advertises that estimate to its neighbors.

To simplify the description of the algorithm, we assume that all the steps are synchronized. We also assume that the links are duplex. That is, we assume that if $L(i, j) < \infty$, then $L(j, i) < \infty$ for any i and j. Thus, if there is a link from i to j, then there is a link from j to i. The algorithm performs correctly without the synchronization and the duplex link assumptions, as you can verify.

At step 0, node i has the estimate $x_0(i) = \infty$ for all $i \in \mathbf{S}$ with $i \neq D$ and that $x_0(D) = 0$. The nodes advertise these estimates to their neighbors.

At step $n \geq 1$, node j updates its estimate of its distance to D to some value $x_n(j)$ and advertises that value to its neighbors. At step $n + 1$, node k examines the messages that it just received from its neighbors at the previous step n. For instance, j might be a neighbor of k—that is, $L(k, j) < \infty$—and j informs k at step n that the shortest estimated distance from j to D is equal to $x_n(j)$. Node k gets similar messages from all its neighbors. Node k then knows that if it chooses to send a message to D by first sending it to its neighbor j, then the distance that the message will travel to reach D is estimated to be $L(k, j) + x_n(j)$. Indeed, this estimate is equal to the distance $L(k, j)$ from k to j plus the estimated shortest distance $x_n(j)$ from j to D as advertised by j at the previous step. Since node k can choose the neighbor to which it sends the message destined to *D,* node *k* estimates the shortest distance to *D* as

$$x_{n+1}(k) = \min\{L(i, j) + x_n(j), j \in \mathbf{S}\}, k \in \mathbf{S}, k \neq D. \tag{3.8}$$

This set of $N - 1$ equations, computed in parallel by $N - 1$ nodes, produces the next set of estimates $\{x_{n+1}(i), i \in \mathbf{S}, i \neq D\}$. These equations are the Bellman-Ford algorithm.

An Example

Figure 3.19 illustrates the steps of the Bellman-Ford algorithm. The network on the left shows the link lengths and the current estimated lengths of the shortest paths from all the nodes to the bottom node.

 The objective is to discover the shortest path from all the nodes to the bottom node. The algorithm proceeds from the destination (the bottom node) to the other nodes. In the first step, shown in the second drawing from the left in Figure 3.19, the bottom node informs its neighbors that its distance to the destination is 0. When they get that information, the middle nodes can update their estimates of their shortest distance to the destination. These estimates correspond to paths that go through the links that we mark thick. At the next step (the right-most drawing), the middle nodes inform their upstream neighbor (the top node) of their estimated distance to the destination. The top node then updates its estimate and the shortest path uses the right outgoing link from the top node. We mark that link thick. The algorithm stops at that step and the marked links are the shortest paths to the destination from all the nodes, with their lengths indicated by the node labels.

Distributed versus Centralized

The implementation of the Bellman-Ford algorithm that we described is *distributed.* This means that the different nodes have incomplete information about the state of the network. It is possible to implement the Bellman-Ford algorithm centrally. To do this, all the nodes broadcast, by flooding, the lengths of their outgoing links to all the other nodes. All the nodes can then execute the Bellman-Ford algorithm on the basis of the full information.

FIGURE 3.19

Illustration of the Bellman-Ford algorithm.

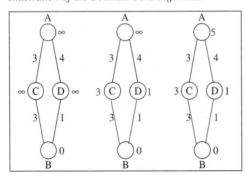

Analysis

We show that this algorithm converges in finite time to the correct estimates. That is, there is some $m < \infty$ such that for all $n \geq m$, $x_n(i)$ is the shortest distance from i to D for all $i \in \mathbf{S}$.

Consider some arbitrary $i \neq D$ and assume that there is a path π from i to D with $n \geq 1$ links. A moment of reflection shows that at step n of the algorithm, node i calculates an estimate that cannot be larger than the length of the path π. Indeed, at step 1, the last node before D along π gets the message from D and calculates an estimate equal to the length of the last link of π and sends that estimate to the previous node of π. At step 2, that previous node calculates an estimate that cannot be larger than the sum of the lengths of the last two links of π, and so on. Thus, at step n, the estimate of node i is the minimum of the lengths of paths of up to n links from i to D. Since the shortest path from i to D is loop-free, the estimate of i must reach the shortest distance from i to D after $m(i)$ steps, where $m(i)$ denotes the maximum length of a loop-free path from i to D. Consequently, the algorithm converges in at most m steps, where $m = \max\{m(i), i \in \mathbf{S}\}$.

Adaptation

One can show that this algorithm converges for arbitrary nonnegative initial estimates $x_0(i), i \in \mathbf{S}$ with $x_0(D) = 0$. To see this, consider the algorithm started with the initial estimates $x_0(i) = 0$ for all $i \in \mathbf{S}$ and denote by $\{y_n(i), i \in \mathbf{S}\}$ the estimates at time n that result from these initial values. You can verify by induction over n that

$$y_n(i) \leq y_{n+1}(i), i \in \mathbf{S}, n \geq 0 \tag{3.9}$$

and

$$y_n(i) \leq x_n(i), i \in \mathbf{S}, n \geq 0. \tag{3.10}$$

Since the estimates $x_n(i)$ converge in finite time to the shortest distance $L(i)$ from i to D, the inequalities (3.9) to (3.10) imply that estimates $y_n(i)$ must also converge. Denote by $V(i)$ the limit of $y_n(i)$ as $n \to \infty$ for $i \in \mathbf{S}$. We claim that $V(i) = L(i)$ for $i \in \mathbf{S}$. To prove this claim, note that the algorithm (3.8) implies that, for all $i \in \mathbf{S}$,

$$V(i) = \min\{L(i, j) + V(j), j \in \mathbf{S}\}. \tag{3.11}$$

Since $V(D) = 0$, these equalities imply that $V(i)$ is the shortest distance from i to D for all $i \in \mathbf{S}$, so that $V(i) = L(i)$.

Finally, consider arbitrary but nonnegative estimates $x_0(i)$ with $x_0(D) = 0$ and denote by $z_n(i)$ the estimate of node i at time n that results from these initial estimates. You can verify by induction on n that

$$y_n(i) \leq z_n(i) \leq x_n(i), i \in \mathbf{S}, n \geq 0. \tag{3.12}$$

Since the estimates $x_n(i)$ and $y_n(i)$ converge to the same value $L(i)$, it follows that the estimates $z_n(i)$ also converge to $L(i)$.

The practical importance of this remark is as follows. Assume that the nodes execute the algorithm and that their estimates have reached some values. Imagine that the situation in the network changes and that this change results in different link lengths. The nodes keep on executing the algorithm and we know from the result of this remark that the estimates will converge to the new shortest distances.

Slow Convergence

Although the Bellman-Ford algorithm converges to the correct values of the shortest distances to the destination, the convergence may be very slow, specially in the case of a link failure. Figure 3.20 illustrates this slow convergence. In the example shown in the figure, the destination is D and the nodes maintain estimates of the shortest distance to D by running the Bellman-Ford algorithm.

The left part of the figure shows the estimates after the algorithm has converged when $L(A, B) = 1 = L(B, D)$ and $L(A, D) = 100$. In the right part of the figure, it is assumed that the link $(A < D)$ fails and has now an infinite length. The numbers below the nodes A and B show the estimates that these nodes compute at the successive steps of the algorithm. As you see, it takes 98 steps for the algorithm to converge to the correct values.

Oscillations

Another undesirable property of the Bellman-Ford algorithm is that it may result in oscillations. The example in Figure 3.21 shows what we mean by oscillations and explains the phenomenon. The figure shows a node A attached to two routers R_1 and R_2 connected to some destination D with two links. Node A has a large amount of traffic to send to D, say B bps. We assume that the delay on link (R_1, D) and on link (R_2, D) is equal to 1 when almost

FIGURE 3.20

Slow convergence of Bellman-Ford algorithm.

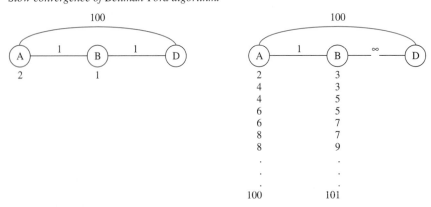

FIGURE 3.21

Oscillations of distributed algorithm.

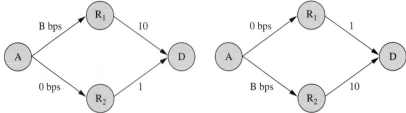

no traffic is sent along that link. Thus, 1 is the propagation time of a signal along the link. We also assume that the delay is equal to 10 when the link carries B bps. For simplicity, we neglect the delays on the links (A, R_1) and (A, R_2).

Assume that initially node A sends all its traffic to D via node R_1. The situation is then as shown in the left part of Figure 3.21: The delay along link (R_1, D) is equal to 10 and the delay along link (R_2, D) is equal to 1. These routers advertise these delays to node A which then finds out that the preferred path to D starts with R_2 and not with R_1. Node A then sends all its traffic to R_2 and the new situation is that shown in the right part of the figure. At that point, R_1 and R_2 advertise the delays 1 and 10, respectively. This leads node A to send all its traffic to node R_1. As a result of this algorithm, the situation alternates between the left and right parts of the figure.

We say that the routing algorithm oscillates between two states. Note that the two states are not desirable. A better routing strategy would consist in node A sending half of its traffic to R_1 and the other half to R_2. This desirable routing can be approached if the Bellman-Ford algorithm is modified so that node A can shift only a small fraction of its traffic in each iteration of the algorithm. However, this modification makes the algorithm even slower to reach to link failures.

3.8.2 *Spanning Tree*

A number of applications require finding a spanning tree in a network. For instance, the transparent routing in bridged LANs that we explained in Section 2.1.3 uses a spanning tree to avoid loops. In this section we explain an algorithm, developed by R. C. Prim, for constructing a spanning tree whose sum of link lengths is minimum.

Description

One is given a set **S** of N nodes and a set of link lengths between these nodes: **L** := $\{L(i, j), i, j \in \mathbf{S}\}$. By convention, $L(i, j) = \infty$ if there is no link from i to j. We assume that $L(i, j) = L(j, i)$ for all i, j. In a typical application, the links are full-duplex and $L(i, j) = L(j, i) = 1$ if there is a link between i and j whereas $L(i, j) = L(j, i) = \infty$ otherwise.

The nodes want to discover a *minimum spanning tree*. Recall that a spanning tree in a graph $\{\mathbf{S}, \mathbf{L}\}$ is a subgraph that spans all the nodes in **S** and has no loop. A minimum spanning tree is a spanning tree with the smallest sum of link lengths. In general a minimum spanning tree is not unique. Prim's algorithm finds one of the minimum spanning trees.

The algorithm starts by selecting an arbitrary node i_0. At each subsequent step, the algorithm attaches the closest node to the partial tree that it has built so far. Figure 3.22 illustrates the algorithm.

Analysis

We prove that Prim's algorithm constructs a minimum spanning tree. We first show that a spanning tree **T** is minimum if and only if it contains the shortest link of every cutset. A cutset is a collection of links whose removal separates the set of nodes into two disconnected subsets. Since, by construction, Prim's tree satisfies that condition, it will follow that it is a minimum spanning tree.

FIGURE 3.22

Prim's minimum spanning tree algorithm.

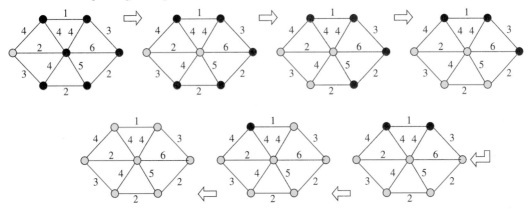

To see why the condition is necessary, note that, because it is a tree, **T** contains at least one link in every cutset. If it does not contain the shortest link in some cutset, then one can make the tree shorter by replacing a link of **T** in that cutset by the shortest one and obtain a sorter spanning tree.

To see why the condition is sufficient, we argue that it should not be possible for another tree **T′** to be the minimum spanning tree when **T** satisfies that condition. To see this, consider a link L in **T′** that is not in **T**. By removing that link we separate the set **S** of nodes into two subsets **A** and **B** such that all the nodes in **A** are still connected by **T′** after having deleted that link, and so are all the nodes in **B**. All the links that join a node of **A** and a node of **B** form a cutset. By assumption, **T** contains the shortest link $L′$ in that cutset. If $L′$ is shorter than $L,$ then replacing L by $L′$ in **T′** forms a shorter tree, which is not possible if **T′** is a minimum spanning tree.

3.9 Complement 4: IPv6

IPv6, version 6 of the Internet Protocol, is being designed to correct the following short-comings of IPv4:

- We are running out of 32-bit addresses.
- IPv4 has poor control of the quality of service it provides.
- IPv4 has no built-in security.

IPv6 uses 128-bit addresses, which is plenty.

The header of IPv6 packets consists of a basic 40-byte header that may be followed by extension headers. The basic header has the following format:

[*Version | Priority | Flow | Total Length | Next Header | Hop Limit | SA | DA*].

The *Flow* field indicates the quality of service that the packet expects and, possibly, characteristics of the connection that generates the packet. The router can use that indication to manage resources for that connection and to schedule the packet for transmission. The *Hop Limit* plays the same role as the time to live in IPv4. The *Next Header* field points to the transport layer protocol (e.g., TCP or UDP) if there is no extension header, or to the next extension header. So far, six extension headers have been defined:

- *Hop-by-hop:* used by routers to indicate information about packet delivery
- *Destination:* indicates options that the end-stations agree upon
- *Routing:* specifies a preferred path when source routing is used
- *Fragmentation:* for fragmenting big datagrams
- *Authentication:* specifies the authentication rules
- *Encrypted payload:* contains encrypted payload

Each extension header has a *Next Header* field that points to the transport layer protocol if there is no other extension header or to another extension header.

Normally, the source computer—not intermediate routers—fragments the packets in IPv6. To perform this fragmentation, the sending host must discover the maximum transfer unit of the path to the destination. IPv6 routers implement an algorithm to calculate the path MTU. A sending host can always force routing through a specific router or provider P either by a *tunnel* (encapsulating into IP packets to P) or with a routing extension header. For routing, IPv6 uses IDRP (interdomain routing protocol), a preferred path routing algorithm more powerful than BGP that can work with multiple families of addresses such as IPv4 and IPv6 addresses.

3.10 Complement 5: Multicast Routing

A multicast routing algorithm sends packets from a source S to a set **G** of destinations. We explain a few multicast routing algorithms.

Flooding

Flooding means sending a packet to all the nodes in the network. USENET news messages are flooded to all the news servers in a distribution group. We also explained that the routers that implement OSPF use flooding to distribute their link state information.

To flood the network, each router transmits every packet it receives once on all its links other than the incoming link of the packet. To prevent multiple transmissions of the same packet, either the router maintains a list of identification numbers of packets it has received previously (as done by OSPF), or the packet itself contains a list of the routers it has already visited (in USENET news).

Spanning Tree Routing

As we explained in Section 2.1.3, the switches in interconnected LANs use a spanning tree to prevent the transmission of multiple copies of packets. Routers in a network can use this method to transmit one copy of a packet to all the hosts.

Reverse-Path Forwarding

This algorithm first constructs the spanning tree of shortest paths to the source S from all the nodes with the reversed link metrics, that is, with $L(i, j)$ replaced by $L(j, i)$. The algorithm then prunes back the tree so that it reaches only the nodes in the multicast group **G**. The routers can use the Bellman-Ford (BF) algorithm (see Section 3.8.1) to construct the tree of shortest paths to the source. The MBONE (multicast backbone) routers use this combination of BF tree with reverse metrics and pruning in the "distance vector multicast routing protocol." In the Internet, only a subset of the routers can implement multicast routing and duplicate the packets. These multicast routers send packets to each other through nonmulticast routers along paths that network administrators set up explicitly (a procedure called *tunneling*).

The multicast routers along the tree duplicate the packets as needed. Note that this algorithm does not minimize the total length traveled by the copies of a packet. However, this algorithm is easy to implement.

One method for pruning back the tree is to send the first packet from the source to all the routers. If a router does not have any group member in its domain, it sends back a "prune" message to the router upstream in the tree. A router that gets a prune message on each downstream tree link sends one "prune" message on the upstream tree link. Each router maintains a table with one entry for each source and each multicast group. The prune messages have a time to live, and packets are again forwarded when the prune messages have expired, possibly triggering new prune messages.

Core-Based Trees

The core-based tree routing was designed to avoid having to send packets periodically to all the nodes and also having the routers maintain potentially large tables. In this algorithm, a router (called the core) in the network is associated with a multicast group. The users who want to receive the multicast send a "join" request toward the core. A router that receives a "join" request marks the incoming link of the packet as belonging to the tree of that multicast group. If the router is not yet part of the tree, it forwards the "join" request toward the core. Eventually, a tree gets built by this procedure and the tree reaches only the hosts that want to be part of the group. Unfortunately, the tree that this algorithm constructs is not minimum, unlike that of the reverse-path forwarding algorithm.

MOSPF

Multicast OSPF is designed by an IETF working group to provide efficient multicasting within an autonomous system. MOSPF uses the network map that the routers maintain to run the OSPF algorithm (Dijkstra's) and calculates the tree of shortest paths from the source to all the nodes. MOSPF then prunes back the tree to the group members. By convention, all border routers are considered to be part of all multicast groups for the backbone routing. Note that MOSPF requires running OSPF for every multicast source and every multicast group.

PIM

The *protocol-independent multicast* (PIM) routing algorithm is an alternative algorithm proposed by the IETF working group. There are two versions of PIM. The first version is to be used when the group members are "dense" and the other version when they are "sparse."

The dense version of PIM assumes that the links are symmetric and that the routers use some unicast routing algorithm. A router R that receives a message from source S

Loan Receipt
Liverpool John Moores University
Learning and Information Services

Borrower ID: 21111057300113
Loan Date: 06/03/2007
Loan Time: 10:56 am

Communication networks: a first course /
1111009773068

Due Date: 27/03/2007 23:59

Please keep your receipt
in case of dispute

for the multicast group G checks that the packet arrived on the link from R to S that the routing algorithm prefers. If that is not the case, R drops the packet and sends "prune (S, G)" message on the incoming link. If that is the case, R forwards a copy of the message on all the links from which R has not yet received a prune (S, G) message. If R has already received such a message from all its links, R drops the message and sends a prune (S, G) message on the incoming link of the message. The algorithm resolves ties to determine the "preferred" path to S by preferring the neighbor with the largest IP address. If router R is attached to a broadcast network E_1 (e.g., an Ethernet) to routers R_1 and R_2 and sends a message from S to G on that network, then it may be that R_1 is attached to a member of G and that R_2 is not. In that case, R_2 will send back a prune (S, G) message on the broadcast network to R. If R believes this message, it will stop transmitting these messages to R_1. The solution implemented by PIM is that R should defer acting on the prune (S, G) message it gets on the broadcast network and that R_2, when it sees this message, should send a "join (S, G)" message back to R. That message voids the prune message. If another router, say R_3, is attached to the same broadcast network and has group members, that router cancels its join messages when it sees one already sent on the broadcast network. PIM must deal with one last difficulty. Imagine that the two routers R_1 and R_2 are both attached to E_1 as before and also to some other broadcast network E_2 to which some member M of G is connected. When R sends a message (S, G) on E_1, both R_1 and R_2 copy that message on E_2 and M gets two copies. To prevent these duplications, PIM proposes that R_1 and R_2 should notice the duplication and that they should send an "assert (S, G)" message on E_2 which indicates their distance to the source S. The router with the largest distance should remove its interface on E_1 as a preferred path to S. The routers cache the prune messages with a time to live of, say, 1 minute.

The sparse mode of PIM, designed for use when group members are few avoids broadcasting a packet from S every minute when the cached entries expire. Instead, a receiver joins a group G by sending a "join G" message to a rendezvous point RP associated with G, using PIM as the routing algorithm. The routers that carry the join message cache that information in a routing table for messages for G. The source S sends its first message for G to RP and when it gets that message, RP sends a join message to S and appends that link to S to the routing tree for G. Using this procedure, a tree is constructed for G that goes through RP. Once they get messages from S, members of G can send a join G message directly to S together with a prune (S, G) message along the link of their original join G message to RP. The tree is then modified partially. If all the group members were to send join messages to the sources, the routing would become RPF.

Summary

- The Internet is expanding rapidly and is having a profound impact on many aspects of society. In this chapter we studied the main protocols of Internet.
- To find the Internet protocol (IP) address of a destination, a sending computer uses the domain name system. This system is a hierachical arrangement of name servers.

- The IP calculates a route to reach every network in the Internet. The routing is organized in a two-level hierarchy: BGP is used to find a path across different autonomous systems and OSPF to find a path inside an autonomous system.
- The transmission control protocol (TCP) supervises end-to-end transmission by controlling errors and the flow. TCP uses a retransmission protocol whose window is adjusted to use the links' efficiency without congesting them.
- The complements provide details on the transmission of IP packets on a point-to-point link and on the routing algorithms. We defer the detailed discussion of retransmission protocols and flow control until Chapter 6.

Problems

Problems with a * are somewhat more challenging. Those with a ^{c} are based on material in a complement. Problems with a ^{1} are borrowed from the first edition.

1. Hierarchical naming reduces the size of the name server tables. This exercise explores that reduction. Imagine that there are 10^9 names in the Internet. Propose a decomposition into domains and subdomains. How many entries will there be in each name server? Next, propose a three-level decomposition: domains, subdomains, subsubdomains. How many entries in each name server? What happens as the number of levels in the hierarchy increases?

*2. To find the address that corresponds to a name, the host invokes the local server, then the root server, then the domain server, then the subdomain server, and so on. As the number of levels in this hierarchy increases, so does the number of steps. Assume that each step takes one unit of time for the communication plus $a \times \log(N)$ units of time, where N is the size of the table in the name server consulted at that step. Compare the costs of two, three, and four levels in the hierarchy. Comment on the effect of caching and of server mirroring on the actual costs.

*3. Class-based addressing is inefficient. We quantify that inefficiency in this problem. Assume that there are N networks and that the fraction of those networks that have k nodes is approximately $p(1 - p)^{k-1}$ for $k = 1, 2, \ldots$. In this expression, p is some number with $0 < p < 1$. Derive an approximate expression for the number of addresses used up by class-based addressing for these networks.

4. A company has three LANs with 500 nodes each. Explain how many of the IP addresses the company uses up if it selects:
    a. Three class B networks.
    b. One class B network with subnetting.
    c. CIDR addressing.
    Justify your answer by describing each addressing scheme.

5. In this example of subnetting we assume that the addresses have only 6 bits (instead of 32) and that they are of one class only with 2 bits for the network and 4 bits for the host. Consider Figure 3.23.
    a. Select addresses and masks for b and c.
    b. What are the addresses of the network that contains b and c.
    c. How many nodes can subnet 1 have?

6. The network of Figure 3.24 uses CIDR. Fill in the routing tables of the two routers.

7. Indicate the successive steps of OSPF for the network of Figure 3.25. Mark "dark" the nodes whose children have all been explored and thicken the lines on the temporary shortest paths.

ᶜ8. Consider once again the network of Figure 3.25. Indicate the successive steps of RIP applied to this network.

ᶜ9. For the network of Figure 3.25, indicate the successive steps of Prim's shortest spanning tree algorithm.

10. Consider the network of Figure 3.26. The routers, represented as circles, are multicast routers. The links are bidirectional and have the same length in both directions. The length of a link is written next to the link.

    a. Construct a tree of shortest paths from the source to the four destinations. Add up the lengths of links traveled by packets when one packet is multicast from the source to the destinations.

    b. Exhibit a multicast tree with a smaller sum of traveled lengths than found in the previous step.

FIGURE 3.23

Network for Problem 5.

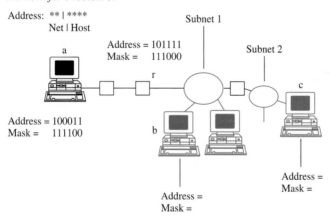

Address: ** | ****
      Net | Host

Subnet 1

Address = 101111
Mask =   111000

Subnet 2

a

r

Address = 100011
Mask =   111100

b

c

Address =
Mask =

Address =
Mask =

Address =
Mask =

FIGURE 3.24

Network for Problem 6.

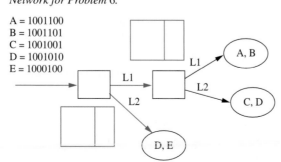

A = 1001100
B = 1001101
C = 1001001
D = 1001010
E = 1000100

L1

L1

L2

L2

L1

L2

A, B

C, D

D, E

FIGURE 3.25

Network for Problems 7 through 9.

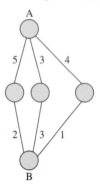

FIGURE 3.26

Network for Problem 10.

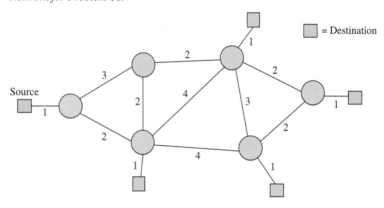

[c]11. We learned in Section 3.8.1 that adaptive algorithms such as the Bellman-Ford shortest path algorithm can oscillate. Consider the network shown in the left part of Figure 3.21. Assume that only 10 percent of the traffic can change direction in each iteration of the algorithm. Indicate the successive steps of the resulting algorithm.

[c]12. Consider the network in Figure 3.20. Propose an algorithm that would react quickly to a link failure and reconstruct valid routing tables.

13. Consider a network with N nodes such that the average number of links out of a node is equal to d. How many packets are transmitted by flooding?

[1]14. What is the role of the three-way handshake in the transport layer? Is it used to:
   a. Detect and correct transmission errors in the initialization process of a new session?
   b. Prevent old packets from creating confusion in the initialization process?

[1]15. Describe two reasons why the three-way handshake may fail to work correctly.

References

Details on the history of Internet can be found at http://www.nsi.wei.com/webinfo/history/timeline.html.

The official references for the Internet are the Requests for Comments (RFCs) published by the Internet Architecture Board. You can find the RFCs at http://ds.internic.net/ds/rfc-index.html. For instance, the following protocols are described in the listed RFCs: PPP: 1661; ARP: 826; IP: 791; ICMP: 792; IGMP: 1112; BGP4: 1771; RIP2: 1723; OSPF2: 1583; UDP: 768; TCP: 793; HTTP: 2068; HTML: 1866. A good starting place (in the summer of 1997) is Postel (1997). This RFC describes the scope of the other RFCs and the current official standards. In most cases, the RFCs are very well written and provide an excellent description of the protocols.

The books by Comer (1991–93) describe the main Internet protocols. An excellent reference for the details of the operations of the protocols is Stevens (1994). Details on Internet routing can be found in Huitema (1995). Peterson (1996) contains a nice presentation of the Internet protocols with an emphasis on what the authors call design philosophy.

Kerschenbaum (1993) is highly recommended for a description of the routing algorithms and of network design algorithms. An excellent mathematical treatment of routing can also be found in Bertsekas (1987), which is a particularly elegant text on data networks.

The original descriptions of the main routing algorithms are Dijkstra (1959), Bellman (1958), and Prim (1957).

Bibliography

Bellman, R., "On a routing problem," *Quart. Appl. Math.,* vol. 16, pp. 87–90, 1958.

Bertsekas, D., and Gallager, R., *Data Networks,* Prentice-Hall, 1987.

Comer, D., and Stevens, D. (for vols. 2 and 3): *Internetworking with TCP/IP,* vols. 1–3, Prentice-Hall, 1991–93.

Dijkstra, E. W., "A note on two problems in connection with graphs," *Numerische Mathematik,* vol. 1, pp. 269–271, 1959.

Huitema, C., *Routing in the Internet,* Prentice-Hall, 1995.

Kershenbaum, Aaron, *Telecommunications Network Design Algorithms,* McGraw-Hill, 1993.

Peterson, Larry L., and Davie, Bruce S., *Computer Networks: A Systems Approach,* Morgan Kaufmann, 1996.

Prim, R. C., "Shortest connection networks and some generalizations," *Bell System Technical Journal,* vol. 36, pp. 1389–1401, 1957.

RFC 2200, IAB, Postel, J., editor, Internet Official Protocol Standards, June 1997.

Stevens, W. Richard, *TCP/IP Illustrated,* vols. 1 and 2, Addison-Wesley, 1994.

Local Area Networks

In this chapter we discuss some popular *local area networks* (LANs). LANs provide inexpensive and fast interconnections of computers that belong to the same organization and are either in the same building or within a few miles of one another. The distance limitation comes from the protocols that govern the sharing of transmission media in a LAN. These protocols would not work well over longer distances because of the propagation delays. LANs also enable users to share printers, modems, file servers, and other resources. Attaching a computer to a LAN requires some hardware and software. In mid-1997, the cost ranges from a few hundred dollars to a thousand dollars, depending on the network. The wiring may account for a substantial portion of the cost of the network.

The most widely used LANs are *Ethernet, token ring,* and *FDDI (fiber distributed data interface)*. These LANs interconnect up to a few hundred nodes. We limit our discussion to these LANs and some wireless LANs. As we explained in Chapter 2, one can interconnect LANs either with switches or with routers. ATM is another interconnection technology, as we discuss below.

The vast majority of LANs are Ethernets, mostly shared 10-Mbps networks (see Chapter 2). Ethernets are inexpensive and easy to set up. Today, many PCs and most workstations come equipped with an Ethernet interface. Many Ethernet installations are being upgraded to switched 10 Mbps or to 100 Mbps (called Fast Ethernet). The specifications of Gigabit Ethernet are scheduled to be published in early 1998.

One selling point of token ring networks is that they deliver packets within a bounded delay, which may be important in some applications. These networks transmit at 4 or 16 Mbps.

When it was introduced in the early 1990s, FDDI was the fastest LAN (100 Mbps). FDDI has other valuable characteristics, in addition to speed. The distance between FDDI nodes can be up to a few kilometers, which is much larger than for Ethernet and token ring and makes FDDI suitable as an interconnection backbone between separate buildings. Moreover, FDDI can be set up so that it keeps on working even after a link or a node fails. Finally, the FDDI network can guarantee that stations get to transmit within a fixed small time, which is useful for real-time applications.

In mid-1997, *asynchronous transfer mode* (ATM) was used—in the LAN environment—mostly to build high-speed backbones that interconnect Ethernets or token rings. An ATM backbone can extend over a wide geographical area and is suitable for campus networks. LAN emulation makes ATM a convenient LAN interconnection technology that also provides a migration path from existing LANs to "native ATM to the desk." However, it is not clear that applications will demand the specific features of ATM for many users. ATM is also used as a link and switching technology for the Internet. We discuss ATM separately in Chapter 5 even though it has a role in local area networks.

There are two types of LANs: shared-medium and switched. We explained that distinction in the case of Ethernet in Chapter 2. Recall that in a shared-medium LAN the nodes all transmit on the same set of wires, whereas in a switched LAN the nodes are attached by dedicated links to a switch.

We start the chapter with a brief look at the protocol architecture of LANs and with a summary of their main characteristics. The subsequent sections explore each LAN in more detail.

4.1   Architecture and Characteristics

This section summarizes the main characteristics of LANs. In particular, we compare the different LANs. We first examine the general protocol architecture of LANs.

4.1.1   Architecture

Figure 4.1 shows the architecture of LANs defined by the working group 802 of IEEE. The physical layer specifies the electrical and mechanical characteristics of the interconnections. For instance, the physical layer specifies the type of twisted pairs to be used, the electrical connectors, and the shape of signals that encode bits. The MAC (*media access control*) and LLC (*logical link control*) sublayers together constitute the data link layer. This layer delivers packets between computers attached to the same LAN.

In a shared LAN, the MAC sublayer regulates the access to the shared physical link. The MAC sublayer converts a shared physical link into virtual point-to-point links between pairs of computers. When the higher layer gives a packet to the MAC with a destination address, the MAC sublayer delivers it to the destination as if there was a dedicated link between the two computers.

The virtual links that the MAC provides are not reliable. The LLC sublayer supervises the transmissions between the nodes. For instance, the LLC sublayer can implement a

FIGURE 4.1

Architecture of LAN protocols.

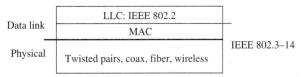

reliable transmission service by retransmitting a packet until its destination acknowledges it. In some applications, the LLC is not expected to provide reliable transmissions because a higher layer arranges for retransmissions (e.g., TCP). In such cases, the LLC is asked to deliver the packets and simply discard erroneous ones.

The IEEE 802.2 standards specify the LLC sublayer, and the IEEE 802.3–14 working groups are defining standards for the physical layer and the MAC sublayer of LANs. The LANs specified by the IEEE 802 standards are compatible above the LLC sublayer even though they differ in the physical layer and the MAC sublayer.

The IEEE 802.1 standard specifies general procedures that all these LANs follow, such as transparent routing. Transparent routing means that, when sending a packet to another computer on the same set of interconnected Ethernets, the sending computer does not need to know that the destination is not on the same Ethernet. (See Chapter 2.)

4.1.2   *Characteristics of LANs*

The main characteristics of a LAN include its throughput, delays, wiring type and distances, security, and reliability. We discuss these characteristics below. We then compare the characteristics of major LANs.

Throughput

The throughput is the average total transmission rate when the LAN is heavily loaded by many nodes.

In a shared LAN, the throughput is a fraction of the rate of the transmitters. This fraction is called the *efficiency* of the MAC protocol. For instance, a network that uses a MAC protocol with an efficiency of 65 percent and 10-Mbps transmitters has a throughput of 6.5 Mbps. These values are typical for a shared 10-Mbps Ethernet. The throughput of a token ring network and that of FDDI are close to their transmission rate.

In a switched LAN, the throughput is some multiple of the rate of the transmitters. The multiple depends on the average number of simultaneous transmissions going through the switch. This average number depends on the (source, destination) pairs of the transmissions. Thus, if the transmission rate of the switched LAN is R bps, then the throughput is nR, where n is some number that depends on the traffic patterns of the network. Unfortunately, the value of n is hard to predict since it is close to 1 when all the computers access the same server most of the time and it is close to half the number of computers when they always communicate in disjoint pairs (an unlikely situation). Thus it may happen that congestion in a switch in fact reduces the throughput of the network instead of increasing it, although this situation is not typical.

The relevance of the throughput is that it determines the average time an application has to wait before it receives a large file from another computer. For instance, consider a shared 10-Mbps Ethernet with a throughput of 6.5 Mbps. Assume that 7 computers are active, typically. The transfer of a 10-Mbyte file between two computers should take about

$$\frac{8 \times 10^7 \text{ bits}}{(6.5 \times 10^6 \text{ bps})/7} \approx 86 \text{ s}$$

This calculation assumes that the computers are fast and are not the bottleneck in the transfer of the file.

Latency

The latency is the time that a packet takes from its arrival at a network interface until it reaches the destination node. We explain in Section 4.8 that this latency is almost always less than a fraction of a second and is often much smaller.

Such latency is usually acceptable for most applications. However, in conversation applications (audio or video), the maximum acceptable delay is about 100 ms and is typically exceeded on a busy token ring or Ethernet. The FDDI MAC protocol has a special feature that guarantees that the latency of "synchronous traffic" does not exceed such a value.

Many distributed processing applications require that delays between computers be very small. New ATM switches offer such small delays but most other LANs are only marginally acceptable for such applications.

Wiring Type and Distances

Some LANs use twisted wire pairs, others use optical fibers. Older LANs and installations in special environments use coaxial cables. (We discuss wireless LANs in Section 4.5.) The maximum length of the connections depends on the LAN and on the technology. Typical values are 100 m for twisted pairs and a few kilometers for multimode fibers.

Security

An optical fiber is more difficult to tap into than a twisted pair, which improves the "physical" security. However, eavesdroppers rarely need to tap directly into a link to gain access to information. A more common attack is to use one of the computers attached on the LAN and configure it to read all the packets that travel on the network (this is called *snooping*). If sensitive pieces of information such as user passwords are transmitted in the clear (nonencrypted), then the eavesdropper obtains entry points into computers. As a rule, a switched LAN is more secure than a shared LAN because a computer on such a network sees only the packets intended for it. (We examine security in Chapter 8.)

Reliability

We depend increasingly on networks. Accordingly, reliability becomes more essential. As we mentioned in the opening remarks of this chapter, FDDI is designed to survive a link or node failure. An Ethernet keeps on working when some of its nodes or links fail. A token ring network can be wired with bypass electronics to have a similar reliability.

Comparison

In Table 4.1 we summarize the main characteristics of LANs. In the Table, Sh. 10 means a shared 10-Mbps Ethernet and Sw. 10 is a switched 10-Mbps Ethernet. Similarly, Sh. 100 and Sw. 100 refer to the 100-Mbps versions of these Ethernets. TR-16 denotes the 16-Mbps token ring. In the throughput expressions, $n10$ or $n100$ mean some multiple of 10 Mbps or 100 Mbps. As we explained above when we discussed throughput, n can range from about 1 to half the number of nodes. The latency values are rough estimates, reflecting the wide variability of delays in LANs. In particular, although the delay of a packet in an Ethernet can be very large, it is rarely so. The delay of packets in an FDDI can be controlled if the synchronous transmission feature is implemented (see the FDDI section).

Snooping indicates that a computer on a shared Ethernet or a ring network can be configured to watch all the packets transmitted on the network, which constitutes a potential

TABLE 4.1 Main Characteristics of LANs
See Text for Comments

Characteristics	LANs					
	Sh. 10	Sw. 10	Sh. 100	Sw. 100	TR-16	FDDI
Throughput, Mbps	8	$n10$	60	$n100$	16	100
Latency, ms	0.1–40	0.1–10	0.01–4	0.01–0.4	0.1–500	0.01–50
Wiring/distance, m	UTP/100 m or fiber/a few kilometers					
Security	Snooping	Good	Snooping	Good	Snooping	Snooping

security risk. One version of shared Ethernet, called secure Ethernet, eliminates that risk by masking the data portion of a packet sent on a port other than the destination port.

4.2 Ethernet and IEEE 802.3

Networks based on the IEEE 802.3 standards compose more than 80 percent of LANs. These networks use a media access control protocol called *carrier sense multiple access with collision detection* (CSMA/CD). We limit the discussion to the most commonly used version of IEEE 802.3 networks: *10BASE-T, 100BASE-T4, 100BASE-TX,* and *100BASE-F*. We briefly comment on Gigabit Ethernet. In mid-1997, about 85 percent of the Ethernets are 10BASE-T networks. These networks use twisted pair interconnections, which is convenient because many networks can use spare telephone wires already installed.

You may come across 10BASE5 Ethernets that use a 1-cm-diameter coaxial cable and 10BASE2 Ethernets that use a 0.5-cm-diameter coaxial cable. The operating principles of these networks are very similar to those of 10BASE-T networks. We decided not to discuss these networks here because such discussion would add little to your understanding of Ethernet. Also, note that optical links can be used to attach computers to an Ethernet hub or switch.

We start by explaining the shared 10BASE-T networks. We first discuss their general layout. We then examine the physical layer and the MAC protocol. Then we explain how the switched 10BASE-T network operates and we discuss 100BASE-T and Gigabit Ethernet.

4.2.1 Layout

Figure 4.2 shows a typical layout of a 10BASE-T network. Each computer contains some *network interface* electronics which implements the MAC sublayer and physical layer. This interface consists of a few integrated circuits; it is standard in most computers and is an inexpensive attachment when optional. The computer network interface is attached to a central hub by two twisted wire pairs: one for receiving signals and the other for transmitting signals. We call these wire pairs the incoming and the outgoing link, respectively. Double-

FIGURE 4.2

Shared 10BASE-T network.

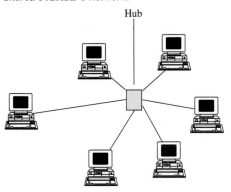

Hub

star arrangements are possible by using intermediate hubs. Also, one can mix different transmission media (twisted pairs, coaxial cable, optical fibers) by using hubs with the appropriate interfaces.

4.2.2   *Physical Layer*

To transmit a packet, the computer first copies it into a buffer of the network interface. The interface waits until its incoming link is idle and then transmits the packet bits on its outgoing link.

To transmit bits on the twisted pair, the interface uses Manchester encoding. That is, the interface transmits a bit 0 by setting the voltage across the wires to -0.85 volts (V) for 0.05 microsecond (μs) and then to $+0.85$ V for 0.05 μs. Similarly, the interface transmits a bit 1 by setting the voltage across the wires to $+0.85$ V for 0.05 μs and then to -0.85 V for 0.05 μs. Each bit transmission takes 0.1 μs, which corresponds to a transmission rate equal to 10 Mbps. The voltage just after the midpoint of a bit transmission interval is positive when the bit is 0 and it is negative when the bit is 1.

The Manchester encoding introduces transitions in each bit transmission interval. The receiver uses these transitions to determine when it should measure the voltage to detect the bits. Specifically, the receiver uses a *phase-locked loop* (PLL). A PLL is a clock that adjusts its speed so that it ticks exactly at the midpoint of the bit transmission intervals. The receiver then measures the voltage just after the clock ticks and decides that it receives a 1 whenever that voltage is negative and a 0 whenever the voltage is positive. To start the PLL, the physical layer appends the 64-bit string $101010\cdots101011$ called the *preamble/start-of-frame delimiter* at the start of the packet. We explain the operations of a PLL in Appendix C.

When it receives a packet, the network interface stores the bits in its buffer and informs the computer when it has received a full packet.

The hub performs the same operations as a computer network interface to receive and retransmit bits. When the hub receives bits on a single incoming pair, it retransmits these bits on all the other outgoing pairs. When it receives bits on more than one incoming pair, the hub sends a "collision detected" signal on all its outgoing pairs. When the network interface of a computer gets that signal on its incoming link while transmitting, it knows that a collision is occurring. We explain below how the MAC protocol reacts to the news of a collision.

4.2.3   MAC

The packets are sent in the form of *frames,* i.e., strings of bits, that have the structure shown in Figure 4.3. The frame is divided into *fields,* which are groups of bits with a precise location inside the frame and a specific signification. The *preamble/start-of-frame delimiter* (PRE/SFD) synchronizes the receiver, as explained above. In addition to the *source address* (SA) and *destination address* (DA), the frame contains an indication of its *length* (LEN), the *cyclic redundancy code* (CRC) bits, and a *padding* (PAD) field that guarantees that valid frames contain at least 512 bits. The addresses (called MAC addresses) are 48-bit strings unique to each computer Ethernet interface worldwide. The first 3 bytes of the address identify the manufacturer of the interface and the remaining 3 identify the specific interface. Some MAC addresses are reserved for broadcast (the all 1's address) and for multicast (group addresses).

Note from Figure 4.3 that the LLC sublayer adds some control information to the data (DSAP, SSAP, and CONT) before it gives the packet to the MAC sublayer. The MAC sublayer adds its own information (DA, SA, LEN). Finally, the physical layer adds the preamble (PRE) and the start-of-frame delimiter (SFD). The process of adding control information to a packet, thereby putting it into a specific framing structure is called *encapsulation.* Thus, a protocol layer encapsulates a packet before giving it to the layer below. Conversely, before giving a packet to the higher layer, a protocol removes its own control information, a process that is called *decapsulation.*

One should note that the Ethernet frame defined by the DEC, Intel, and Xerox (DIX) frame differs from the IEEE 802.3 frame by replacing the length field LEN by a *type* field. The type field in the DIX frame specifies the protocol above the MAC sublayer for which the packet is intended. Indeed, the DIX standard allows for protocols other than the IEEE 802.2 LLC to use the Ethernet frames.

The network interface of the transmitting computer calculates the CRC bits from the other bits in the packet. The receiver performs the same calculation and knows that a transmission error must have occurred if it gets a different result for the CRC bits. We explain CRC in Chapter 6.

FIGURE 4.3

IEEE 802.3 frames.

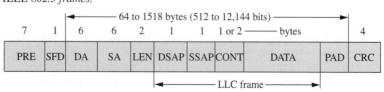

The Ethernet MAC protocol (CSMA/CD) specifies that a node with a packet to transmit must proceed as follows:

1. Wait until the channel is idle.
2. When the channel is idle, transmit and listen while transmitting.
3. In case of a collision, stop the packet transmission, transmit a jam signal, then wait for a random delay and GO TO (1).

The jam signal is a sequence of 32 randomly chosen bits. Its objective is to guarantee that all the nodes become aware of the collision. The protocol gives up the transmission attempt after 16 successive collisions. The size of the random delay after a collision is selected according to the following rule, called the *binary exponential backoff algorithm:* If a packet has collided n successive times, where $n \leq 15$, then the node chooses a random number K with equal probabilities from the set $\{0, 1, 2, 3, \ldots, 2^m - 1\}$ where $m := \min\{10, n\}$; the node then waits for $K \times 512$ bit times. (At 10 Mbps, 1 bit time $= 10^{-7}$ second.)

Thus, after the first collision, a node chooses either the value 0 or 512 bit times for the random delay, with equal probabilities. After two successive collisions, the delay is equally likely to be 0, 512, 1024, or 1536 bit times. After three successive collisions, it is equally likely to be 0, 512, 1024, 1536, 2048, ..., or 3584 bit times. This method for selecting the random delay reduces quickly the chances of repeated collisions by spreading out the range of waiting times.

The value of 512 bit times is chosen because it is larger than the time it takes an electrical signal to travel between any two nodes on the same Ethernet, even if the signal must go across repeaters. Accordingly, if two nodes wait for different multiples of 512 bit times, then the node that chooses the smallest delay transmits successfully. Indeed, the signal from that node reaches the other node before it starts transmitting. According to the protocol, that second node waits before transmitting. The nodes that are not transmitting detect a collision when they observe a packet shorter than the minimum length of a valid frame (512 bits).

In Section 4.8 we analyze the efficiency of the IEEE 802.3 MAC protocol. We explain that this efficiency is typically between 40 and 90 percent for a 10BASE-T network. The main cause of inefficiency of CSMA/CD is that the nodes may be transmitting for some time before they notice that the packets collide. That wasted time increases with the propagation time. Also, once a node transmits successfully, it uses the transmission medium for a time proportional to the length of the packet divided by the transmission rate. These two effects explain that the efficiency of CSMA/CD increases with the average packet length and decreases with the propagation time of packets and with the transmission rate.

4.2.4 *Switched 10BASE-T*

In a switched 10-BASE-T network, the hub is replaced by a switch. When the switch gets an Ethernet packet, it reads its destination address and determines the outgoing link of the packet. If that link is idle, the switch forwards the packet. Otherwise, the switch stores the packet in a buffer until the link is free.

The computer network interface is the same for shared 10BASE-T and switched 10BASE-T. Consequently, the network manager can upgrade the shared Ethernet to a switched Ethernet without the computer users even being aware of the modification until they notice an improvement in performance. The improvement in performance is due to the simultaneous transmissions through the switch.

Figure 4.4 shows a simplified sketch of an Ethernet switch. This switch has an interface for each incoming link. Each interface has a small buffer to store incoming packets and is attached to the same fast bus. When a packet arrives, the link interface obtains its outgoing link number from the CPU and sends the packet to the corresponding link interface over the bus. The outgoing link interface stores the packet in a buffer until it can transmit it.

In some switches, called *store-and-forward switches,* the incoming interface does not forward the packet before it has completely received it and has checked its CRC. In other designs, called *cut-through switches,* the incoming interface starts forwarding the packet as soon as the destination address has arrived.

Some 10BASE-T switches have a 100-Mbps port to attach a server or a backbone. A number of 10BASE-T switches convert the packets into ATM cells before switching them. Some of these switches can be interconnected to an ATM backbone.

4.2.5 100BASE-T

Increasing the transmission rate from 10 Mbps to 100 Mbps reduces the transmission times of packets and improves the performance of Ethernet substantially. This increase in performance is achieved by changing only the physical layer of 10BASE-T.

The 100BASE-T networks can be shared (with a hub) or switched. There are two versions of 100BASE-T: 100BASE-T4 and 100BASE-TX.

100BASE-T4 uses four telephone-grade twisted pairs (category 3 unshielded twisted pairs). Category 5 (data-grade) is recommended in noisy environments. Three pairs transmit data and one pair transmits collision signals. The three data pairs transmit ternary-valued signals (with three different possible values) at 25 MHz. These 27 possible triplets of values on the three pairs encode 4 bits, 25 million times per second, thus providing the 100-Mbps connection.

FIGURE 4.4

Simplified Ethernet switch. (1) A packet arrives. (2) The interface sends the packet's MAC destination address to the CPU. (3) The CPU returns the packet's outgoing link. (4) The interface copies the packet to the interface of its outgoing link.

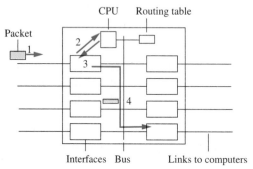

100BASE-TX uses two pairs of data-grade twisted pairs (category 5 unshielded twisted pairs). 100BASE-TX uses the 4B5B encoding that maps groups of 4 bits into symbols of 5 bits chosen to have enough transitions for keeping the receiver synchronized. The bits are transmitted at 100 Mbps on each pair by sending symbol bits at 125 Mbps.

Optical fibers can also be used to attach computers to a 100-Mbps switch (not a hub) with connections that can be up to 2000-m long. This implementation is specified by the 100BASE-FX standard.

4.2.6   Gigabit Ethernet

The IEEE 802.3z working group is currently (in 1997) defining the Gigabit Ethernet standard. Gigabit Ethernet will use the same frame format as the other Ethernets. This network is a promising candidate for Ethernet backbones.

A number of physical layers are being proposed. Optical fibers are the easiest technology to build 100-m-long links at that transmission rate. Engineers are trying to design a twisted pair system that will probably require many pairs and incur serious length limitations. A shared Gigabit Ethernet would be very inefficient. Consequently, it is likely that Gigabit Ethernet will be switched. You should consult the working group's Web page for recent developments.

4.3   Token Ring Networks

Token ring networks are another widely used family of LANs. The token ring networks were developed by IBM in the early 1980s. We limit the discussion to the IEEE 802.5 standard. We start with a look at the physical layout of the network, then we explain the physical layer and the MAC protocol.

4.3.1   Layout

Figure 4.5 is a schematic diagram of a token ring network. The transmission is unidirectional. The actual implementation of a token ring network looks like a star with the center in a wiring closet. The IEEE 802.5 networks transmit at 4 Mbps over twisted pairs and at 16 Mbps over twisted pairs or optical fibers.

4.3.2   Physical Layer

A node that is not transmitting repeats the packets that it receives. To do this, the node recovers the bits from the signal it receives on the input cable and it retransmits these bits on the output cable. A quartz oscillator in the transmitter controls the rate of the transmission. This transmission rate always differs slightly from the reception rate. Indeed, a quartz oscillator in the upstream node controls the reception rate and no two quartz oscillators have exactly the same rate. To accommodate the difference in rates, each node uses a buffer called an *elasticity buffer.* The elasticity buffer stores the bits that accumulate when the input rate is higher than the transmission rate.

FIGURE 4.5

A logical view of a token ring network.

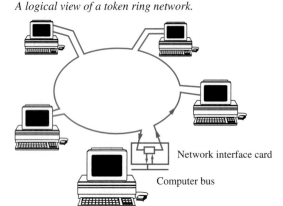

Network interface card

Computer bus

To handle a transmission rate higher than the input rate, the node starts retransmitting incoming bits only after its elasticity buffer has accumulated a specified number B of bits. The value of B is large enough to prevent the buffer's becoming empty during the packet retransmission.

4.3.3   MAC

The MAC protocol of the token ring is as follows: A specific bit pattern, called the *token,* circulates on the ring. When a node wants to transmit, it waits until the token comes by. It then replaces the token with another pattern (SFD) which indicates the *start of frame,* and it appends its packet. The node converts the token into an SFD by monitoring the signal it receives from the ring and by modifying the token while it is stored in the interface buffer (see Figure 4.5) before retransmitting it. The token and the SFD are bit patterns that differ by only the value of the last bit. It is enough for the interface buffer to delay the ring signal by 1 bit to be able to modify the token into an SFD. Once the packet has been transmitted, the node transmits the token, which then becomes available to another node.

Actually, every station can hold the token for up to 10 ms and transmit a number of packets back to back before releasing the token. In addition, the IEEE 802.5 standard provides for multiple packet *priorities.* This feature of the IEEE 802.5 is rarely implemented in actual networks. It can be used to speed up the delivery of urgent packets, such as control packets, by giving them priority over less pressing packets, such as those carrying electronic mail messages. The frame structure is shown in Figure 4.6.

Tokens are identified by a special access control (AC) field that specifies the priority and reservation levels. To transmit a packet at some priority level, a node needs to wait for a token with a lower level and to place a reservation. If a reservation is already indicated on the token, a node can make another reservation at a higher level. After transmitting at a given level, the node lowers the token priority level to its previous value. The frame control (FC) field is used to monitor the ring. One of the nodes is the ring monitor. That node

FIGURE 4.6

IEEE802.5 frames for token and packet.

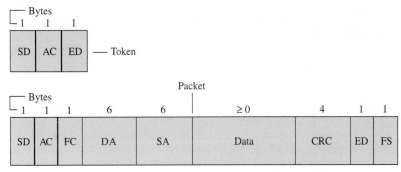

verifies that there is a token and that there is no cycling frame. The possible values of the field FC include *monitor present,* which is issued periodically by the monitor, *reinitialize* to restart the ring, and *I want to be monitor* issued by a node when it realizes that the monitor is down. The frame status (FS) field is set by the destination to signal that it is up and whether it accepted the frame or not.

We analyze the MAC protocol in Section 4.10, where we show that its efficiency is very close to 100 percent. The media access time of the protocol—the time that a packet must wait before it can be transmitted—can be as large as the sum of $N - 1$ token holding times, i.e., $(N - 1) \times 10$ ms.

4.3.4   Interconnecting Token Rings

Multiple token rings can be interconnected together with bridges (switches). When such an arrangement is used, the routing procedure is not transparent as in interconnected Ethernets. Instead, the source first arranges to discover a path through switches to the destination. The source then appends that path description to the packet. That is, the source explicitly indicates the path that the packet will follow, a procedure called *source routing.*

To discover a path to destination B, computer A broadcasts a request "B, please reply." The switches forward this request and append their address to it. When B gets this request, it replies to it, along the reverse route recorded in the request that reaches B. Computer A then reads the route that the reply from B followed and uses that route when sending packets to B. Computer A may get a number of replies from B coming along different routes but uses only the first reply to come back.

4.4   FDDI

The specifications of FDDI are the subject of an ANSI (American National Standard Institute) standard that specifies the layers shown in Figure 4.7. This network was proposed in 1986. FDDI was used mostly for interconnecting fast workstations and LANs. In 1997,

100BASE-T networks are cheaper than FDDI for interconnecting workstations and PCs. However, FDDI maintains advantages as a LAN backbone because of its reliability and long distances between nodes.

The station management protocols supervise the operations of the FDDI and can "repair" the network when a link or a node fails. We start by exploring the layout of an FDDI network. We then discuss the physical layer and the MAC sublayer, and we show how the station management protocol heals the network.

4.4.1   Layout

The *fiber distributed data interface* (FDDI) network is a dual-ring network (see Figure 4.8) with a transmission rate of 100 Mbps on optical fibers. FDDI connects up to 500 nodes, and the maximum length of the fibers is 200 km. The distance between successive nodes cannot exceed a few km.

4.4.2   Physical Layer

FDDI's physical layer has two sublayers: The *physical medium dependent* (PMD) sublayer and the *physical* (PHY) sublayer.

PMD

The PMD sublayer has three versions, shown in Table 4.2. (We study optical links in Chapter 7.) The original PMD specifies the fiber, the optical transmitters and receivers, the connectors, and the optical bypass switches. The transmission is over graded-index fibers with a core diameter of 62.5 μm and a cladding diameter of 125 μm. The attenuation of the fibers is at most 2.5 dB/km. Other fibers may be used. The optical transmitters are LEDs with a wavelength of 1.3 μm. The receivers are PIN diodes. The bit error rate of a link

FIGURE 4.7

Protocols defined by the ANSI standards for FDDI.

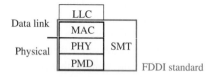

FIGURE 4.8

Layout of FFDI network and main characteristics. Note the two counterrotating rings.

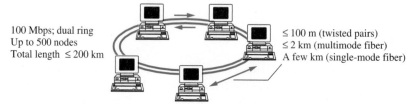

must not exceed 4×10^{-11}. The SMF-PMD is for longer connections; it uses single-mode fibers. FDDI connections over twisted pairs can be up to 100 m long. The pairs can be either type 1 shielded or category 5 unshielded.

PHY

The PHY sublayer specifies the encoding and modulation methods and also the isolation of a malfunctioning station that we will explain below. Over optical fibers, FDDI uses the *4B/5B* encoding, in which a group of 4 bits is encoded into a group of 5 bits called a *symbol*. The symbols of 5 bits are selected to contain at most two successive 0's. FDDI uses 8 of the 16 symbols not used to encode data as *control words*. These control words are used as delimiters and as signaling words. The groups of 5 bits are transmitted using the *nonreturn to zero with inversion* (NRZI) modulation. This modulation method represents the bits by a signal that has two values. The signal makes a transition in an epoch that represents a 1, and it makes no transition in an epoch that represents a 0. Thus, the 4B/5B + NRZI signal makes one transition at least every three bit epochs. A phase-locked loop uses that feature of the signal to synchronize the 125-Mbps clock of the receiver with a 16-bit preamble. Each node uses a 10-bit elasticity buffer. Note that the transition rate of the 4B/5B + NRZI signal is at most 125 MHz, whereas the Manchester encoding would make transitions at 200 MHz.

4.4.3 MAC

We will see shortly that FDDI uses a token-passing MAC protocol. Figure 4.9 shows the structure of a token and an information frame for FDDI. These frames are similar to those of IEEE 802.5.

The traffic is classified into *synchronous* and *asynchronous* traffic. Synchronous traffic, like audio or video information, must be transmitted within a small time. Asynchronous traffic, such as data traffic, usually tolerates large variable delays.

TABLE 4.2   Main Characteristics of LANs
See Text for Comments

PMD	Max. length, m	Medium	Transmission
Original	2000	62.5-μm/125-μm graded-index fiber	1.3 μm
SMF	> 2000	8-μm/125-μm single-mode fiber	1.3 μm
TP	100	Type 1 shielded TP or cat 5 UTP	4B5B

FIGURE 4.9

IEEE 802.5 frames for token and packet.

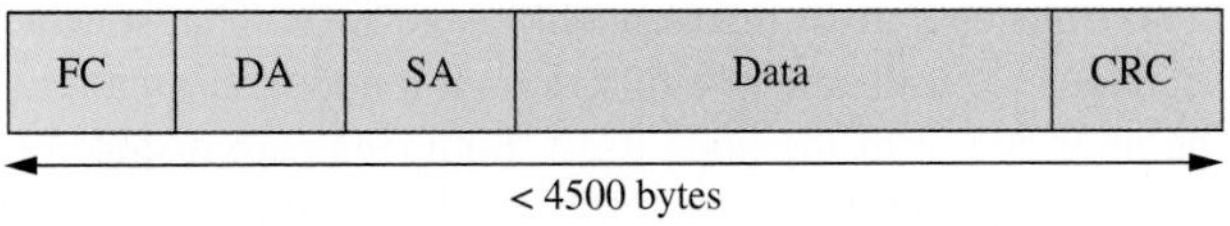

FC	DA	SA	Data	CRC

< 4500 bytes

The nodes start by negotiating the value of a parameter called the *target token rotation time* (TTRT). Each node i is allocated some time $S(i) \geq 0$ for its synchronous traffic. By assumption, $\sum S(i) \leq$ TTRT where the sum is over all the nodes.

The MAC protocol is a token-passing mechanism that makes sure that the time between two successive visits to any node by the token never exceeds $2 \times$ TTRT. Moreover, each time that it gets the token, node i can transmit synchronous traffic for up to $S(i)$ seconds. For instance, the protocol can be set up so that 30 stations with video traffic get to transmit for 1.5 ms after waiting at most 100 ms for the token. In this example, if a station accumulates video at the rate of 1.5 Mbps for up to 100 ms, it needs a buffer that can store up to 150 kbits and it can empty that buffer when it gets to transmit during 1.5 ms at 100 Mbps. This application corresponds to TTRT $= 50$ ms and $S(i) = 1.5$ ms for $i = 1, 2, \ldots, 30$.

Each node uses two timers to implement this FDDI MAC protocol:

- The *token rotation timer* with value TRT. This timer counts up.
- The *token holding time* timer with value THT. This timer counts down.

When node i gets the token, it executes the following steps:

1. It sets THT $=$ TTRT $-$ TRT.
2. It sets TRT $= 0$.
3. Node i transmits synchronous traffic for up to $S(i)$ seconds.
4. If THT is still positive, node i can transmit asynchronous traffic until THT $= 0$.
5. Node i releases the token.

Note that if a node transmits for a long time before releasing the token, then the value of its TRT timer is large when it gets the token the next time around. Consequently, the value THT $=$ TTRT $-$ TRT that the node calculates that next time is small and prevents the node from transmitting many packets. This observation explains why the FDDI MAC protocol prevents a node from hogging the network. We show in Section 4.10 that if all the nodes follow this protocol, then TRT $\leq$ 2TTRT at each node and all the time.

A node is responsible for removing the frames it transmits from the ring. This frame removal is similar to what happens in the IEEE 802.5 MAC protocol. A transmitting node cannot remove everything that is on the ring. Instead, the node must check the source address field *SA* of the frame to see whether it transmitted it. When the node observes one of its frames, it replaces the symbols after the SA field with idle symbols, i.e., the symbols of the preamble. The fragments SD-FC-DA that have not been stripped are removed by a node that starts transmitting after receiving a token.

4.4.4   Station Management

The *station management protocol* (SMT) detects faults and repairs the network. A malfunctioning node is isolated as follows: When all the components of the FDDI network are operational, the two rings operate as independent token rings. That is, when a node wishes to transmit, it can choose to transmit on either ring. If a node malfunctions, either because of a failure of the interface or of the node itself, both rings are disabled (see the left part of Figure 4.10). The nodes on the ring detect that failure and reconfigure the network as a

FIGURE 4.10

Circled station fails at left. At right, the SMT has reconfigured the dual ring into a single ring that isolates the failed station.

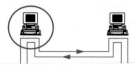
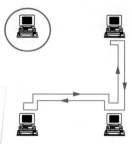

)art of the figure. This self-healing feature is important in
ty and short downtimes.

connect computers equipped with the appropriate wireless
reless LANs eliminate the need for wiring and enable nodes
, the nodes communicate with a hub. Hubs can be organized
tworks are in fact cellular packet networks. In others, the
each other. We discuss two emerging standards: Hiperlan,
European Telecommunications Standards Institute), and

structure shown in Figure 4.11. To transmit data, the packet
hit first assembles its data into packets with a standardized
e large bit error rate of many radio channels, the packets
ed bits. The logical link control then introduces the error
mber that it requires to supervise the transmissions. The
cess to the radio channel. In a wireless LAN, the access
division multiple access not unlike CSMA/CD. We explain
examples later in this section.

Some metropolitan area wireless data services use circuit switching. In such a system, the portable wireless terminal contacts a fixed radio transceiver and sets up a connection with an Internet router, as you would to place a telephone call from a cellular telephone. The connection is maintained for the duration of the data exchange and is then released. Other metropolitan wireless system use a randomized multiple access protocol where the portable wireless terminal sends its packet without first setting up a connection; the LLC or the transport layer uses a retransmission protocol to make sure the packets are received correctly.

The transmitter itself implements the physical layer, which we briefly discuss below.

FIGURE 4.11

Wireless transmitter.

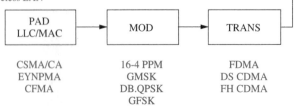

Wireless LAN

CSMA/CA   16-4 PPM   FDMA
EYNPMA   GMSK   DS CDMA
CFMA   DB.QPSK   FH CDMA
            GFSK

┌── Abbreviations ──┐

CSMA/CA = carrier-sense multiple access with collision avoidance
EYNPMA = elimination yield, nonpreemptive priority multiple access
CFMA = collision free multiple access
PPM = pulse-position modulation
GSMK – see Chap. 7
FDMA = frequency division multiple access
CDMA = code division multiple access
FH = frequency hopping
DS = direct sequence

└──┘

4.5.2   Physical Layer

We study the wireless transmission of bits in Chapter 7. Here are the main points you should be aware of when reading the rest of this section.

The wireless transmitter converts bits into radio waves. The energy of these radio waves extends over some range of frequencies. Radio frequencies are a precious commodity because, like land, they don't make any new ones anymore. Accordingly, various governmental agencies (the Federal Communications Commission in the United States) regulate the utilization of radio frequencies. The challenge to the radio engineer is to design a transmission system that utilizes the radio frequencies efficiently and yet achieves a suitable transmission rate, a sufficient transmission distance, and a low bit error rate.

The physical layer of radio wireless LANs is usually time-division multiple access (TDMA) or code-division multiple access (CDMA). These transmission systems differ in how the transmitter converts bits into radio waves.

4.5.3   Hiperlan

We do not elaborate further on the physical layer but we focus instead on the MAC. The MAC of Hiperlan is called *elimination yield, nonpreemptive priority multiple access* (NPMA). The objective of NPMA is to provide nonpreemptive access to high-priority traffic and a fair access to traffic of the same priority. NPMA operates in cycles. Each cycle has three phases, illustrated in Figure 4.12: priority resolution, contention resolution, and transmission. The priority resolution phase guarantees that only the highest-priority stations survive in that cycle. The contention resolution phase selects a subset of the surviving stations. Finally, the surviving stations transmit during the transmission phase.

FIGURE 4.12

Hiperlan's NPMA protocol in operation.

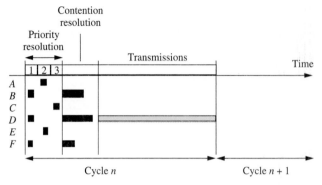

The figure shows stations $A, B, \ldots, F$ competing for the channel. Stations B, D, F have a packet with priority 1 to transmit while stations A and E have a packet with priority 2 and station C has a packet with priority 3.

The priority resolution phase divides time into a number of slots equal to the number of priority levels. The stations transmit during the corresponding slot. By listening to the transmissions, the stations can determine if they have the highest priority. In the figure, at the end of the priority resolution, stations A, C, F know that they don't have packets with the highest priority and they drop out of the cycle.

During the contention resolution phase, the stations transmit bursts with random lengths. A station survives if it senses the channel idle right after its burst ended. In the figure, station D happens to select the largest burst duration and it survives to transmit its packet in the transmission phase.

During the transmission phase, two types of packets can be transmitted: data and acknowledgment. Data packets have a low-bit-rate header that specifies the destination address. The receivers listen to this header and turn on their high-bit-rate receiver only if the packet is destined for them. Acknowledgment packets are transmitted at low bit rate.

The efficiency of Hiperlan depends on the offered load and on the packet lengths. At high load, the efficiency is about 80 percent for 2000-byte packets and 8 percent for 53-byte packets.

4.5.4   IEEE 802.11

The 802.11 MAC protocol offers two services: contention-free and contention. The contention-free service is designed to set up bounded delay transfers. The MAC operates as follows. Time is divided into frames. A frame consists of two parts: the contention-free phase and the contention phase. During the contention-free phase, a control station polls the other stations which then transmit their packets one at a time. During the contention period, the stations transmit using a carrier sense multiple access with a specific collision avoidance (CSMA/CA) protocol.

FIGURE 4.13

IEEE 802.11 MAC protocol in operation.

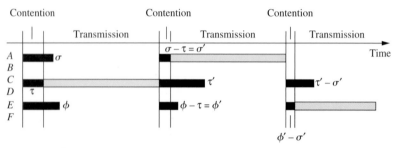

The CSMA/CA protocol, illustrated in Figure 4.13, is similar to CSMA/CD but tries to favor stations that happened to wait longer. In the figure, stations A, C, E collided and select the backoff delays σ, τ, ϕ, respectively. Station C has the shortest time and it transmits. Stations A and E note their residual backoff times $\sigma' = \sigma - \tau$ and $\phi' = \phi - \tau$ and use them for the next backoff period. That procedure is repeated for subsequent collisions. As can be expected, the efficiency of the scheme depends on the propagation delay and on the size of packets.

4.6 Logical Link Control

The MAC layer of LANs provides an unreliable packet transmission service over a shared communication medium. The *logical link control* (LLC) layer converts this unreliable transmission service into packet links between *access points*. The basic functions of LLC are error control, frame multiplexing, and flow control (if required). Frame multiplexing means that distinct communication streams out of or into a computer can coexist and are distinguished by LLC.

The LLC frame structure and the access points are shown in Figure 4.14. The LLC frame control field CONT specifies that the message is either DATA, an XID (exchange identification), or a TEST. No reply to a data packet is expected. An XID is used to find all the members of a given group—which are then expected to reply with an XID—or to broadcast the presence of a node. A TEST packet is used to test a connection to another node. The LLC frame also specifies a service class in the form of a priority that can be used by IEEE 802.4–802.6 networks.

The IEEE 802.2 standard for LLC provides for three types of services: *acknowledged connectionless, connectionless,* and *connection-oriented.* A connection-oriented service delivers a stream of packets in the correct order and without error. An acknowledged connectionless service is used only for point-to-point connections, and it assumes that a frame cannot be sent before the previous one is acknowledged. The connectionless service is a datagram transmission: there is no guarantee the packet will be delivered or that it will arrive without errors.

FIGURE 4.14

IEEE 802.2 LLC frame and service access points.

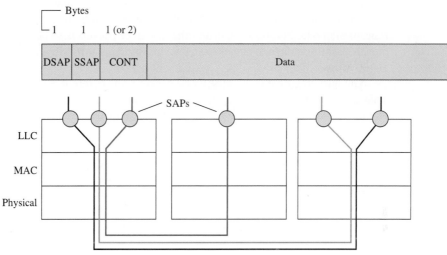

The connection-oriented service is called *HDLC-ABM,* i.e, high-level data link control, asynchronous balanced mode. This is the data link layer protocol of X.25 (see Chapter 6). The normal sequence of operations in a connection-oriented transmission is a *connection setup,* then *data transmission,* and then a *connection release.* HDLC-ABM uses 7-bit sequence numbers.

Some MAC protocols have a built-in *flow control mechanism.* This is the case for the token-passing schemes. Some, such as the CSMA protocols, do not. In the latter case, it may be necessary to control the transmission rate of a node.

4.7 Complement 1: Latency of LANs

As we discussed in Section 4.1, latency is the time that a packet takes from its arrival at a network interface until it reaches the destination node. The latency has four components: the *medium access time,* the *transmission time,* the *queueing time,* and the *propagation time.*

The medium access time is the time that a packet ready to be transmitted has to wait until the node starts transmitting it. In a switched LAN, the medium access time is the time until the line to the switch is available. In some switched LANs, the link between the computer and the switch is half-duplex. That is, the line can be used only in one direction at a time (e.g., 100BASE-T4). Accordingly, in such a LAN, a computer has to wait until the switch stops transmitting to it before it can send a packet toward the switch. In a shared-medium LAN, the media access time depends on the MAC protocol and on the number of other computers that compete for the same link. For instance, if the computers take turns transmitting (as they do in a token ring or an FDDI), this medium access time may be as large as the sum of the transmission times of all the other computers. As another example, in a shared Ethernet, this medium access time is the time it takes until the computer succeeds in transmitting after some random number of collisions and transmissions of other computers.

The transmission time is equal to the number of bits in the packet divided by the transmission rate. For instance, in an Ethernet, since packets have between 500 and 12,000 bits, the transmission time ranges from 0.05 ms to 1.2 ms at 10 Mbps and is 10 times smaller at 100 Mbps.

The queueing time is the time that a packet waits before it is ready to be transmitted. That queueing time is the time to transmit other packets that arrived at the network interface earlier. The average queueing time is typically equal to a few transmission times, depending on the load of the network.

The propagation time is the time signals take to propagate along the transmission medium. This propagation time is equal to the length of the signal path divided by the signal propagation speed. The propagation speed is about 3×10^8 m/s in a twisted pair and 2×10^8 m/s in an optical fiber. These propagation speeds correspond to about 3.3 μs/km in a twisted pair and 5 μs/km in a fiber. Thus, in the time that the signal takes to cover 200 m, the transmitter can send about R bits if the transmission rate is R Mbps.

4.8 Complement 2: Analysis of CSMA/CD Protocol

We first analyze the efficiency $\eta_{\text{CSMA.CD}}$ of a CSMA/CD protocol. We then adapt the analysis to the IEEE 802.3 MAC protocol.

4.8.1 CSMA/CD Protocol

In our CSMA/CD protocol, a node with a packet to transmit must proceed as follows:

1. Wait until the channel is idle.
2. When the channel is idle, transmit and listen while transmitting.
3. In case of a collision, stop the packet transmission, and then wait for a random delay and GO TO (1).

The random delay is a random integer multiple of twice the maximum propagation time ρ of a signal across the shared transmission channel. For instance, if the maximum distance between two nodes is L m, then $\rho = L/\sigma$ where σ is the propagation speed of a signal in the medium. Typically, σ is about 60 percent of the speed $c = 3 \times 10^8$ m/s of light in a vacuum. As an example, if $L = 1000$ m, then $\rho \approx 5$ μs. This random time is chosen in that way so that if two nodes A and B select different integer multiples of 2ρ after a collision, then the node with the smallest multiple gets to transmit its packet without colliding with the other.

That is, we claim that when the later node "goes back to (1)" after its waiting time, it senses the signal from the other node and must then wait until the end of that transmission before transmitting. To verify the claim, assume that A and B start transmitting packets at times T_A and $T_B > T_A$, respectively, and that these packets collide. Note that $T_B - T_A < \rho$, otherwise B would have heard the transmission of A before time T_B and would not have started transmitting before A was done. Say that A detects the collision at time S_A and B at time S_B. Then, $S_B = T_A + \alpha$ and $S_A = T_B + \alpha$, where α is the propagation time of a signal between A and B (plus the small detection time of the interface electronics). Consequently, $0 < S_B - S_A < \rho$. Now assume that A chooses some random time $m\rho$ and that B chooses $n \times 2\rho$ with $0 \leq m \neq n \in \{0, 1, 2, \ldots\}$. Note that if $0 \leq m < n$, then $(S_B + 2n\rho) - (S_A + 2m\rho) > 2(n-m)\rho > \rho$, so that A will complete its transmission before B transmits. Similarly, if $n < m$, then $(S_A + 2m\rho) - (S_B + 2n\rho) > 2(m-n)\rho - \rho > \rho$, so that B completes its transmission before A starts transmitting.

4.8.2   *Efficiency of CSMA/CD*

We define the efficiency as the fraction of time that the nodes using the protocol can transmit new packets under a heavy load imposed by all the nodes. We show that

$$\eta_{\text{CSMA.CD}} \approx \frac{1}{1 + 5a} \qquad \text{with} \qquad a := \frac{\rho}{\tau} \tag{4.1}$$

where τ designates the transmission time of a packet. (Recall that ρ is the maximum propagation time across the shared transmission channel.)

Note that if a single node is active on the network, then that node transmits without ever colliding with other transmissions. In that case, the fraction of time that the node transmits new packets can be close to 100 percent. The efficiency (4.1) is for a congested network where many nodes compete for transmitting on the common channel. It is under such difficult operating conditions that we need to verify how well a protocol performs.

4.8.3   *Analysis*

In this section we derive the efficiency (4.1). The analysis consists of the following steps. We consider that the nodes attempt to transmit at discrete times, in time slots with a duration of 2ρ each. This discretization simplifies the analysis and does not depart significantly from the actual operations of the protocol. The slot duration of 2ρ is used to guarantee that, if nodes select to transmit at the beginning of two different slots, then they cannot collide. We then perform two calculations. First we assume that the system is idle and we determine the probability α that the next time slot is the start of a successful transmission. That is, α is the probability that during that time slot there is no collision and that one node starts transmitting. We find that $\alpha \approx 0.4$. Second, we use the value of α to determine the average number A of time slots that are wasted before a successful transmission. We show that $A \approx 1.5$. To use this result, we argue that the average fraction of the time that the nodes transmit successfully is τ out of $\tau + A \times 2\rho$, which yields an estimate close to Equation (4.1). Finally, we justify (4.1) from that estimate.

First, we calculate α. Say that N ($N \geq 2$) nodes compete for a time-slotted channel by transmitting packets with probability $p \in (0, 1)$, independently of one another, in any given time slot. Denote by $\alpha(p)$ the probability that exactly one node transmits in a given time slot. The claim is that

$$\alpha(p) = Np(1 - p)^{N-1} \tag{4.2}$$

To see (4.2), consider a given time slot. The probability that, during this time slot, a specific node among the N nodes transmits and that the other $N - 1$ nodes do not transmit is equal to

$$p \times (1 - p) \times \cdots \times (1 - p) = p(1 - p)^{N-1}. \tag{4.3}$$

Indeed, the probability that a collection of independent events all occur is the product of the probabilities of the individual events and each node transmits with probability p and does not transmit with probability $1 - p$. (Appendix A explains these notions of probability theory.)

Consequently, the probability that any one of the N nodes transmits and that the other $N - 1$ nodes do not transmit is N times the probability (4.4), which is $\alpha(p)$ given in (4.2). Indeed, there are N mutually exclusive ways for exactly one node to transmit, depending on which of the N nodes transmits; also, the probability that one of a collection of mutually exclusive events occurs is the sum of the probabilities of the individual events.

If the nodes knew the number N of competing nodes, then they could determine the value of p that maximizes $\alpha(p)$ and the efficiency of the protocol. The value of p that maximizes $\alpha(p)$ is $p = 1/N$. Indeed, this value is obtained by setting to zero the derivative of (4.2) with respect to p. One finds

$$\frac{d}{dp}\alpha(p) = N(1 - p)^{N-1} - N(N - 1)p(1 - p)^{N-2},$$

so that the value of p that makes this derivative equal to zero is $p = 1/N$, as claimed.

The corresponding maximum value of $\alpha(p)$ is

$$\alpha\left(\frac{1}{N}\right) = \left(1 - \frac{1}{N}\right)^{N-1} \approx 40\%. \tag{4.4}$$

To see the approximation, you can verify that $\alpha(1/N) \to 1/e \approx 36\%$ as $N \to \infty$. For instance,

$$\alpha(1/4) = 42\%, \qquad \alpha(1/10) = 39\%, \qquad \alpha(1/20) = 38\%.$$

In the CSMA/CD protocol, a node does not know the number N of nodes that have packets to transmit and are competing for the channel. Consequently, the nodes cannot use the value $p = 1/N$. For the purpose of the analysis, we assume that the CSMA/CD protocol selects the random backoff times almost as well as they would if they knew the number of competing nodes. (The exponential backoff mechanism of IEEE 802.3 is effective; attempts to improve it substantially are not convincing.) Thus, we conclude that the probability that a time slot is the start of a successful transmission is $\alpha := 40\%$.

Second, we calculate the average number A of time slots that CSMA/CD wastes before it succeeds in transmitting a packet. With probability $\alpha = 0.4$, the first time slot is successful, so that no time slot is wasted. With probability $1 - \alpha$, the first slot is wasted. In the latter case, after the first wasted slot, we are essentially back to the initial situation, so that CDMA/CD will waste an average number A of slots in addition to the first one before the first successful transmission. Hence,

$$A = \alpha \times 0 + (1 - \alpha)(1 + A).$$

Solving for A, we find $A = \alpha^{-1} - 1$. With $\alpha = 0.4$, this gives $A = 1.5$. (The above calculation is an application of the regenerative method that we explain in Appendix A.)

Using this value $A = 1.5$, we conclude that every successful transmission (with duration τ) is accompanied by a wasted amount of time with an average value equal to 1.5 time slots, or $1.5 \times 2 \times \rho = 3 \times \rho$. Consequently, the efficiency $\eta_{CSMA.CD}$, i.e., the fraction of time when the CSMA/CD protocol transmits successfully, is given by

$$\eta_{CSMA.CD} \approx \frac{\tau}{\tau + 3 \times \rho} = \frac{1}{1 + 3a}.$$

In actuality, the CSMA/CD protocol is not really optimal, and it wastes a larger amount of time than $3 \times \rho$ per packet transmission. Simulations show that the amount of time wasted is closer to $5 \times \rho$. Hence, the efficiency of the CSMA/CD protocol is approximately given by (4.1). We should not be surprised that a simplified analysis gives a result that is not quite exact. The opposite would be miraculous. Nevertheless, the analysis has the merit of explaining how the efficiency is reduced when the parameter a of the network increases.

4.8.4   *Examples*

To develop a concrete feel for the efficiency of the CSMA/CD protocol, let us calculate $\eta_{CSMA.CD}$ for nodes attached to a 2.5-km-long coaxial cable, with a 10-Mbps transmission rate and 620-bit packets. We find that ρ, the one-way propagation time of a signal from one end of the cable to the other, is given by

$$\rho = \frac{2500 \text{ m}}{2.3 \times 10^8 \text{ m/s}} \approx 1.09 \times 10^{-5} \text{ s.}$$

(We used the transmission speed 2.3×10^8 m/s in a coaxial cable.)

The transmission time τ of a packet is

$$\tau = \frac{620 \text{ bits}}{10 \times 10^6 \text{ bps}} = 6.2 \times 10^{-5} \text{ s.}$$

From these two values we conclude that $a = \rho/\tau \approx 0.176$ and, therefore,

$$\eta = \frac{1}{1 + 5a} = \frac{1}{1 + 0.176} \approx 53\%.$$

Consequently, the effective transmission rate of this 10-Mbps network is only 53 percent of 10 Mbps, i.e., 6.3 Mbps, when the network is heavily loaded by many nodes. If the network transmits TCP/IP frames, about 30 bytes, i.e., 240 bits, of the 620 frame bits are not user data. Thus, only a fraction $(620 - 240)/620 \approx 61\%$ of the frame bits is user data bits. Therefore, the maximum rate at which the network can transmit user data is $61\% \times 6.3$ Mbps $= 3.2$ Mbps. Note that this efficiency result is very sensitive to the length of packets, as you can verify easily.

As another example, let us consider a network with twisted pairs that are up to 200 m in length and with a transmission rate of 1 Gbps. These assumptions would correspond to a hypothetical implementation of shared Gigabit Ethernet. Assuming that the average packet length is again 620 bits, we adapt the above calculations and we find

$$\rho = 0.2 \text{ km} \times 3.3 \text{ } \mu\text{s/km} \approx 0.66 \times 10^{-6} \text{ s,}$$

and

$$\tau = \frac{620 \text{ bits}}{10^9 \text{ bps}} = 0.62 \times 10^{-6} \text{ s}.$$

Combining these values, we find $a = \rho/\tau \approx 1$, so that

$$\eta = \frac{1}{1 + 5a} = \frac{1}{1 + 5} \approx 16\%,$$

which shows that a shared Gigabit Ethernet would not be an efficient backbone technology.

4.8.5 Average Medium Access Time

Consider a number of nodes that are using the CSMA/CD protocol to share a transmission channel. How long does one node have to wait, on the average, until it gets to transmit?

If that node is the only active node, then it never has to wait. However, if many nodes are active and are competing for the channel, the medium access time can be large. We assume that N nodes are active and we use the same discrete time model of the protocol as in the calculation of the efficiency. We also assume that the nodes have the same level of activity. That is, we assume that the situation is symmetric, with N active nodes.

We decompose the activity of the network into successive cycles. One cycle starts when a successful transmission ends and consists of an initial amount of wasted time followed by a successful transmission. Each of the N nodes uses a fraction $1/N$ of the cycles to transmit a packet successfully. Accordingly, one node gets to transmit every N cycles, on average. That is, the average time between two successive transmissions by one node is equal to

$$\beta := N(5\rho + \tau).$$

To derive this equality, we assume that the average duration of the wasted time before a successful transmission is about 5ρ, as we argued when we derived the efficiency of CSMA/CD.

We can use β as an estimate of the medium access time. You might point out that half of that value is the typical medium access time for a packet that becomes ready to be transmitted because the packet may become ready at any time during an interval of duration β and gets transmitted at the end of that interval.

As a numerical example, consider a CSMA/CD network with 200-m twisted pairs and with 1000-byte packets and a 100-Mbps transmission rate. We find $\rho = 0.66 \ \mu s$ and $\tau = 1 \ \mu s$. Assume that 20 nodes are active. We then find $\beta = N(5\rho + \tau) = 33 \ \mu s$, which may be bad news for some demanding distributed processing applications.

4.8.6 Efficiency of IEEE 802.3

In IEEE 802.3, the backoff time is a random multiple of 512-bit times, so that the interface board does not have to be programmed by indicating the length of the network it is installed in. Since the IEEE 802.3 backoff delays are longer than necessary, we can anticipate that the efficiency is worse than the value we calculated for CSMA/CD networks.

It is straightforward to adapt the calculation of $\eta_{CSMA/CD}$ to this case. The time slots now have a duration equal to 512 bit times instead of 2ρ. All we need to do to adapt the calculation is then to replace ρ by 216 bit times. Let us denote the duration of 216 bit times by γ. With this notation, the efficiency $\eta_{802.3}$ of the IEEE 802.3 network can be estimated from (4.1) as

$$\eta_{802.3} \approx \frac{1}{1 + 5\gamma/\tau}. \tag{4.5}$$

Note that $\gamma = 216/R$ where R is the transmission rate in bps. Similarly, $\tau = P/R$ where P is the average number of bits in a packet. Hence $\gamma/\tau = 216/P$ and we conclude from (4.7) that

$$\eta_{802.3} \approx \frac{1}{1 + 1000/P}. \tag{4.6}$$

For example, if $P = 620$ we find that the efficiency is only 37 percent.

4.9   Complement 3: Analysis of Token Ring MAC Protocol

In this section we calculate the efficiency of the token ring protocol. We limit the discussion to the protocol that the 16-Mbps token ring uses.

4.9.1   Token Ring MAC Protocol

We start by reviewing the token ring MAC protocol. Recall that a token is circulating around the ring except when a node is transmitting packets. The protocol uses a parameter θ, the maximum token-holding time, that specifies the maximum time that a node can keep the token. For instance, in the IEEE 802.5 protocol $\theta = 10$ ms. To transmit a packet, a node goes through the following steps:

- Wait for the token and grab it.
- Transmit packets for at most θ.
- Release the token.

That is, a node releases the token as soon as it has finished transmitting data. Another version of the protocol is implemented in 4-Mbps token rings. In this version, the node that transmits a packet waits for the last bit of that packet to come back before it releases the token. We do not study the efficiency of this protocol here.

Recall that each node that listens to the ring delays the signal by a few bit times to be able to transform the token into a start-of-frame delimiter when it wants to transmit. We neglect this delay in our analysis because it is much smaller than the average packet transmission time.

4.9.2   Efficiency of Token Ring MAC Protocol

The efficiency η_{TR} of the token ring protocol is defined as the fraction of time that the nodes transmit packets when all the nodes always have packets to transmit. These "heavy load" operating conditions are when the protocol should be efficient. Consequently, it makes sense to define the efficiency under these conditions. We show that

$$\eta_{TR} \approx \frac{1}{1 + \rho/(N\theta)} \tag{4.7}$$

where N is the number of nodes, ρ is the propagation time of a signal around the ring, and θ is the maximum token holding time. For instance, assume that $N = 50$, $\theta = 10$ ms, and $\rho = 8$ μs (which corresponds to a total ring length of 2400 m). We find

$$\eta_{TR} \approx \frac{1}{1 + 8 \times 10^{-6}/(50 \times 10^{-2})} \approx 100\%,$$

which is quite typical.

The point of this example is to show that the efficiency is typically close to 100 percent. Consequently, the interest of the analysis is that it helps us refine our understanding of the operations of the protocol, not that the actual result might be surprising.

4.9.3   Analysis

The analysis is straightforward compared to that of CSMA/CD. Consider the timing diagram in Figure 4.15. At time 0, node 1 starts transmitting packets and the transmission lasts θ. Indeed, we assume that all the nodes always have packets to transmit. Consequently, each node always transmits for the maximum token holding time θ whenever it has the token.

We neglect the transmission time of a token, which is legitimate because the token is much smaller than an average packet. Accordingly, at time θ, node 1 finishes transmitting the last bit of the token. That bit arrives at node 2 at time $\theta + \rho(1, 2)$, where $\rho(1, 2)$ designates the propagation time of a signal from node 1 to node 2. Node 2 starts transmitting at time $\theta + \rho(1, 2)$, as soon as it has received the token and transformed it into a start-of-frame delimiter. Node 2 then proceeds as node 1 just did, and so do nodes $3, \ldots, N$.

Consequently, as Figure 4.15 shows, the N nodes transmit one packet each in time S where

$$S = N\theta + \rho(1, 2) + \cdots + \rho(N, 1) \approx N\theta + \rho.$$

The transmissions of packets occupy $N\theta$ out of this time S. Accordingly, the fraction of time η_{TR} when the nodes transmit packets is equal to

$$\eta_{TR} = \frac{N\theta}{S} = \frac{1}{1 + \rho/(N\theta)},$$

as we stated in (4.7).

FIGURE 4.15

Timing diagram of token ring protocol.

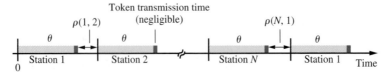

4.9.4   Maximum Medium Access Time

We use the diagram of Figure 4.15 to calculate the *maximum medium access time* (MMAT_{TR}) for the nodes using the token ring protocol. MMAT_{TR} is defined as the maximum time that a node has to wait before it can transmit. The value of MMAT_{TR} is an important design element for networks that are used in control equipment. The maximum medium access time for nodes on an Ethernet network is infinite since a station can be unlucky and keep on colliding and randomly generating backoff times that are longer than those of the other stations. However, MMAT_{TR} is finite.

A node faces the maximum access time when all the other nodes always have packets to transmit. That situation is shown in Figure 4.15. Assume that packet arrives at node 1 and must be transmitted as soon as possible. In the worst case, the packet arrives at node 1 just after that node has started to transmit another packet and when node 1 cannot transmit another packet without exceeding the admissible token holding time θ. The packet must then wait for the N nodes to transmit and for the token to travel around the ring. If the $N - 1$ nodes other than node 1 transmit for the maximum acceptable duration (θ), then the figure shows that the packet must wait for a time MMAT_{TR} given by

$$\text{MMAT}_{TR} = \rho + \tau + (N - 1)\theta \tag{4.8}$$

where τ is the maximum transmission time of a packet.

For instance, in a network with $N = 50$ nodes, with $\theta = 10$ ms, $\tau = 0.75$ ms, and $\rho = 8$ μs, we find that

$$\text{MMAT}_{TR} = 8 \times 10^{-6} + 0.75 \times 10^{-3} + 49 \times 10^{-2} \approx 0.49 \text{ s.}$$

This maximum medium access time is too large for most real-time applications.

4.10   Complement 4: Analysis of FDDI MAC Protocol

The FDDI MAC protocol—which we refer to as the FDDI protocol, for short—is designed to control the medium access time of stations with synchronous traffic. In this section we prove that no station ever has to wait for more than 2 times the target token rotation time before it sees the token. We conclude the section by discussing the efficiency of the protocol.

4.10.1   FDDI Protocol

We start with a brief review of the FDDI protocol. Each station has two timers: a count-down timer with value THT and a count-up timer with value TRT. The stations agree on a value of the target token rotation time TTRT. Let N be the number of stations on the FDDI. For $i = 1, 2, \ldots, N$, station i is allocated some time $S(i) \geq 0$ for transmitting synchronous traffic. By assumption,

$$S(1) + S(2) + \cdots + S(N) \leq \text{TTRT.} \tag{4.9}$$

Fix any $i \in \{1, 2, \ldots, N\}$. When station i gets the token, its timers have the values TRT and THT. Then station i executes the following steps:

1. It sets THT = TTRT − TRT and THT starts decreasing at unit rate.
2. It sets TRT = 0 and TRT starts increasing at unit rate.
3. It transmits synchronous traffic for at most $S(i)$ units of time.
4. It transmits asynchronous traffic for at most as long as THT is positive.
5. It releases the token.

4.10.2   MMAT of FDDI Protocol

We assume that 2TTRT $< \rho$, where ρ is the propagation time of a signal around the ring, including the delays that the nodes introduce in the signal path. We neglect the transmission time of a token. The claim is that the maximum medium access time of the FDDI protocol, $\text{MMAT}_{\text{FDDI}}$ is at most 2TTRT. That is,

$$\text{MMAT}_{\text{FDDI}} \leq 2\text{TTRT}. \tag{4.10}$$

We explained in Section 4.4 how this property makes it possible to use FDDI for real-time applications.

4.10.3   Analysis

We show that

$$\text{TRT} \leq 2\text{TTRT} \tag{4.11}$$

at every station and at all times. The result (4.10) follows from (4.11) since the medium access time is less than the maximum value of TRT. The proof of (4.11) is by induction on the successive arrivals of the token at the stations. Specifically, for $n \geq N + 1$ we consider the nth time T_n that the token arrives at a station. Let us clarify this critical definition by numbering the stations $1, 2, \ldots, N$ in their order of visit by the token around the ring. Say that the token arrives at station 1 at time T_n. The token then arrives at station 2 at time T_{n+1} and at station 3 at time T_{n+2}, and so on.

Let $A(n)$ be the sum of the transmission times of asynchronous traffic by the N stations when they got the token immediately before time T_n, including transmission times equal to 0. Similarly, let $B(n)$ be the corresponding sum of the N transmission times of synchronous traffic.

The proof consists in showing that the following statement is true for all $n \geq N + 1$:

$$\text{If } A(n) + \rho \leq \text{TTRT, then } A(n + 1) + \rho \leq \text{TTRT.} \tag{4.12}$$

We assume that the network is first initialized by having the token rotate freely around the ring with no station transmitting. In that case, the first N asynchronous transmission times are equal to 0, so that $A(N + 1) = 0$. If (4.12) is true for all $n \geq N + 1$, then we can conclude that

$$A(n) + \rho \leq \text{TTRT, for all } n \geq N + 1. \tag{4.13}$$

A look at Figure 4.16 shows that the value of TRT at the station that gets the token at time T_n is equal to $A(n) + B(n) + \rho$.

FIGURE 4.16

Timing diagram of FDDI protocol.

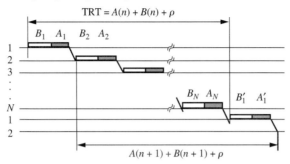

Since $B(n)$ is a sum of transmission times of synchronous traffic by the N stations, we know from the definition of the protocol that

$$B(n) \le S(1) + \cdots + S(N) \le \text{TTRT}. \tag{4.14}$$

Adding up (4.13) and (4.14) we find that $\text{TRT} = A(n) + B(n) + \rho \le 2\text{TTRT}$, which implies (4.11).

To complete the proof we have to show (4.12). Unfortunately, we need some more notation. We denote by $A_1, A_2, \ldots, A_N$ the N transmission times of asynchronous traffic just prior to T_n, including zero times, in consecutive order. We also denote by A_1' the next transmission time of asynchronous traffic after T_n, by the station that gets the token at time T_n. Note that A_1 is the previous asynchronous transmission time by the same station. We use the notation $B_1, B_2, \ldots, B_N, B_1'$ for the corresponding synchronous transmission times.

To draw Figure 4.16 we assume that station 1 gets the token at time T_n. (There is no loss of generality in making that assumption since we can always renumber the stations.) From the figure we see that

$$\text{TRT} = A(n) + B(n) + \rho, \; A(n) = A_1 + \cdots + A_N, \tag{4.15}$$

$$B(n) = B_1 + \cdots + B_N, \; A(n+1) = A_2 + \cdots + A_N + A_1'. \tag{4.16}$$

Fix some $n \ge N + 1$. We assume

$$A(n) + \rho \le \text{TTRT} \tag{4.17}$$

and we want to show that

$$A(n+1) + \rho \le \text{TTRT}. \tag{4.18}$$

We consider two cases: (1) $\text{TRT} + B_1' \ge \text{TTRT}$ and (2) $\text{TRT} + B_1' < \text{TTRT}$. In the first case, station 1 cannot transmit asynchronous traffic at time $T_n + B_1'$ since at that time $\text{THT} \le 0$. Hence, $A_1' = 0$. Consequently,

$$A(n+1) + \rho = A_2 + \cdots + A_N + \rho \le A_1 + \cdots + A_N + \rho = A(n) + \rho \le \text{TTRT}$$

where the last inequality follows from the assumption (4.17). Hence, (4.18) holds.

In the second case, station 1 can transmit asynchronous traffic after T_n. However, it must transmit when its timer THT hits 0. That is, station 1 must ensure that

$$A_1' + B_1' \leq TTRT - TRT.$$

Combining this inequality with the expression for TRT found in (4.15), we conclude that

$$A_1 + \cdots + A_N + B_1 + \cdots + B_N + \rho + A_1' + B_1' \leq TTRT.$$

This inequality implies that

$$A(n+1) + \rho = A_2 + \cdots + A_N + A_1' + \rho \leq TTRT,$$

which is (4.18).

4.10.4 Efficiency of FDDI Protocol

The efficiency of the FDDI MAC protocol can be evaluated by using the same procedure as the one we used for the token ring protocol. The token takes approximately TTRT to rotate once around the ring. As in Figure 4.14, some of that time is the transmission time of frames and the rest of the time is the propagation of the token around the ring, the transmission times of tokens, or the delays inserted by the nodes to be able to modify the token into a start of frame. The inserted delays add up to $N \times D$ if each of the N nodes inserts a delay of D seconds. The token propagation time around the ring is denoted by π. We then find that the efficiency of the FDDI protocol η_{FDDI} is given by

$$\eta_{FDDI} = \frac{TTRT - N \times (D + \sigma) - \rho}{TTRT} \tag{4.19}$$

where σ is the token transmission time. As a numerical illustration, consider an 80-km ring at 100 Mbps, with 300 nodes, a 16-bit delay per node, and a token of 100 bits. We will assume that the fiber has a refractive index of 1.46. For this network,

$$\rho = \frac{80 \text{ km}}{c/1.46} = \frac{1.46 \times 80 \text{ km}}{3 \times 10^5 \text{ km/s}} = 3.9 \times 10^{-4} \text{ s.}$$

Substituting in (4.19), we find

$$\eta_{FDDI} = \frac{TTRT - 300 \times (16 + 100) \times 10^{-8} - 3.9 \times 10^{-4}}{TTRT} = 1 - \frac{7.38 \times 10^{-4}}{TTRT}.$$

With $TTRT = 10$ ms, one finds $\eta_{FDDI} \approx 92.6\%$. This value of TTRT is adequate for voice and video transmissions. We can see from this example that FDDI is able to support a large number of stations with small delays.

4.11 Complement 5: ALOHA

ALOHA is a packet-switched radio communication network that was invented by Norm Abrahmson at the University of Hawaii in the early 1970s. ALOHA is the father of multiple-access protocols. Ethernet is its direct descendant. We discuss ALOHA here not only for historical reasons but mostly because this protocol is regaining some life in wireless packet radio networks. We first describe the ALOHA network. We then study the variations of its protocol.

4.11.1   Description

Figure 4.17 depicts the main components of that network as it was originally implemented. A central node, called the base station, listens to packets transmitted by the other nodes at the radio frequency $f_0 = 407$ MHz and retransmits these packets at the radio frequency $f_1 = 413$ MHz. The nodes in the ALOHA network transmitted the packets at the rate of 9600 bps.

The media access control protocol that ALOHA uses is called the *ALOHA protocol.* This protocol can be used with radio transmitters, as in the original ALOHA network, or with a coaxial cable or a twisted wire pair. The ALOHA protocol and some variants are used in satellite networks.

The nodes transmit packets on a common channel. When two transmissions occur simultaneously, they garble each other. In such an instance, packets are described as *colliding.* In the original ALOHA network, the central node acknowledges the correct packets it receives. When a node does not get an acknowledgment within a specific timeout, it assumes that its packet collided and retransmits it after a random delay.

Note two essential differences between CSMA/CD and the ALOHA protocol: ALOHA does not use carrier sensing and does not stop transmitting a packet when it detects a collision. Carrier sensing is useless since the nodes are far apart and one node may complete its transmission before another notices that it was transmitting. For the same reason, collision detection is too late.

4.11.2   ALOHA Protocols

There are two versions of the ALOHA protocol: *slotted* and *pure.* In the slotted ALOHA protocol, the time axis is divided into time slots with durations equal to the time required to transmit a packet on the channel. Nodes must start their transmissions at the beginning of a time slot. In the pure ALOHA protocol, the nodes can start transmitting at any time.

FIGURE 4.17

ALOHA network.

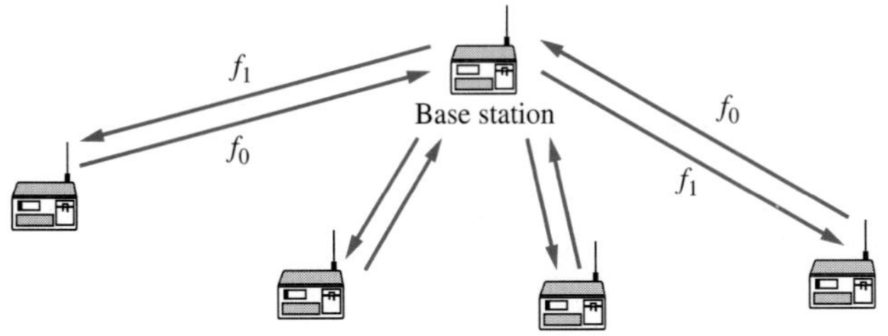

4.11.3   *Efficiency of ALOHA protocols*

How well do the ALOHA protocols perform? The analysis we use to answer this question will determine the *efficiency* of the protocols, i.e., the maximum fraction of time during which packets are transmitted successfully when the ALOHA protocols are used. We show that the efficiency $\eta_{\text{S.ALOHA}}$ of the slotted ALOHA protocol is given by

$$\eta_{\text{S.ALOHA}} = 36\%. \tag{4.20}$$

We also show that the efficiency $\eta_{\text{P.ALOHA}}$ of the pure ALOHA protocol is given by

$$\eta_{\text{P.ALOHA}} = 18\%. \tag{4.21}$$

We can compare the efficiency of the ALOHA protocols with that of the CSMA/CD protocol. Using (4.1), we find that the CSMA/CD protocol is more efficient than the pure ALOHA protocol whenever a is smaller than 89 percent and more efficient than the slotted ALOHA protocol when a is smaller than 34 percent.

4.11.4   *Analysis*

We start with slotted ALOHA and then study pure ALOHA.

Slotted ALOHA

Figure 4.18 represents the flow of packets transmitted by a large number of nodes that use the slotted ALOHA protocol. When a node gets a *new* packet to transmit, it starts transmitting it at the beginning of the next time slot. If the packet is the only one transmitted during that time slot, the transmission is successful. Otherwise, the packet suffers a collision, and the node schedules a retransmission after a random delay. We view the packets that collide as joining a pool of *old* packets.

The protocol wastes some fraction of the time slots because of collisions and some other fraction when all the nodes are postponing their transmissions because of the random delays in scheduling retransmissions. During the remaining fraction of the time slots, the nodes transmit one packet at a time. The transmissions in these time slots are said to be successful.

To carry out the analysis of the efficiency of the slotted ALOHA protocol, we define G as the *total* rate of transmission attempts, in packets per time slot, and S as the rate of successful transmissions, also in packets per time slot (see Figure 4.18). Then,

$$S = G \times p \tag{4.22}$$

where p is the likelihood that the transmission of an arbitrary packet is successful. We

FIGURE 4.18

First transmissions and retransmissions compete for the channel using the ALOHA protocol.

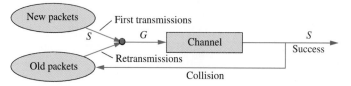

can expect p to decrease when G becomes larger since collisions are more likely when the nodes attempt to transmit more packets. To derive an expression for p as a function of G, we assume that the probability $P(n, T)$ that n packets, either new or old, become ready to be transmitted in a time period with arbitrary duration T (measured in time slots) is given by

$$P(n, T) = \frac{(GT)^n}{n!} e^{-GT} \tag{4.23}$$

for $n \geq 0$. This assumption is justified when there is a large number of nodes that attempt to transmit, each with a small probability, which is the case if the network is congested and all the nodes obey the protocol.

By using this assumption, we find that

$$S = P(1, 1) = Ge^{-G}. \tag{4.24}$$

The maximum value of the right-hand side of (4.24) is $e^{-1} \approx 36\%$. The maximum fraction of time during which the slotted ALOHA protocol can transmit packets successfully is the efficiency $\eta_{S.ALOHA}$ of the slotted ALOHA protocol. Therefore, $\eta_{S.ALOHA} \leq 36\%$. It can be shown that if the nodes randomize their retransmissions suitably, then the efficiency of the slotted ALOHA protocol approaches this upper bound of 36%. Hence,

$$\eta_{S.ALOHA} = 36\%. \tag{4.25}$$

The efficiency and throughput values that we have computed require some interpretation. When a single node tries to transmit on the channel, all its transmissions are successful and the efficiency is 100 percent. This is not the situation captured by the model that we analyzed. Only when many nodes generate the aggregate traffic can one assume that (4.24) holds. Thus, the efficiency $\eta_{S.ALOHA} = 36\%$ corresponds to the case of many nodes having comparable transmission rates and sharing a common channel by using the slotted ALOHA protocol.

Pure ALOHA

The analysis of the efficiency of the pure ALOHA protocol is very similar to that of the slotted ALOHA protocol. The only difference is that the nodes can start transmitting at arbitrary times instead of only at the beginning of time slots. A given packet P that is transmitted at time t will be successful if no other packet's transmission begins in the time interval $(t-1, t+1)$, as illustrated in Figure 4.19. Thus, the probability that the transmission

FIGURE 4.19

Timing of pure ALOHA protocol.

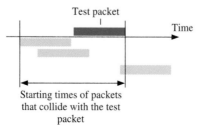

Starting times of packets
that collide with the test
packet

of packet P is successful, when the pure ALOHA protocol is used, is the probability p that no packet's transmission begins during $(t-1, t)$ or $(t, t+1)$, an epoch with total duration 2. According to (4.23), this probability is equal to

$$p = P(0, 2) = e^{-2G}.$$

Therefore, when the nodes use the pure ALOHA protocol,

$$S = Ge^{-2G}.$$

In particular, $S \le 1/(2e) = 18\%$. As for the slotted ALOHA protocol, this upper bound of 18 percent can be achieved if the nodes randomize their retransmissions suitably. Thus, the efficiency of the pure ALOHA protocol is

$$\eta_{\text{P.ALOHA}} = 18\%. \tag{4.26}$$

4.11.5   Reservations

The slotted ALOHA protocol is more difficult to implement than the pure ALOHA protocol because it requires that the nodes have access to a common time reference—a common clock—to identify the beginning of the time slots. The efficiency of the two ALOHA protocols is rather poor. It is desirable to improve the efficiency of these protocols if they share an expensive transmission channel, such as a satellite channel. A simple method for improving the efficiency of the ALOHA protocols is to use *reservations*. Two protocols based on ALOHA that use reservations are SPADE and the R.ALOHA protocol.

SPADE is the reservation protocol used by the Intelsat telecommunications satellites. In SPADE, there are 397 bidirectional 64-kbps channels for voice transmissions and one 128-kbps channel for reservations. The reservation channel is divided into frames of 50 small slots, where 50 is an upper bound on the number of nodes allowed to use the satellite. Slot k is allocated to node k, which can use it to reserve one (chosen at random) of the 397 bidirectional channels. Each communication channel is then used for voice transmission by the nodes in the order in which it was reserved. After the completion of its voice transmission, the node indicates in its slot on the reservation channel that it no longer needs the communication channel. Note that, under heavy usage, the voice channels are used with perfect efficiency since they are reserved by the time they become free again. As a result, the 397 bidirectional channels are used with 100 percent efficiency and the 128-kbps reservation channel is wasted in that it cannot be used for voice transmission. Since the reservation channel has the same transmission rate as one of the 397 bidirectional channels, we conclude that the efficiency of SPADE is 397/398.

The R.ALOHA (reservation.ALOHA) protocol is for nodes that use the same channel for the reservations and for the transmissions of packets. The protocol begins with a *reservation phase*. During the reservation phase, the nodes use the slotted ALOHA protocol to attempt to access the channel. A node which succeeds in accessing the channel broadcasts a reservation by transmitting a packet that contains the node identification number. At the end of the reservation phase, the nodes which made reservations transmit in the order of the reservations. At the end of the transmission phase, a new reservation phase starts, which, in turn, is followed by a new transmission phase, and so on. The R.ALOHA protocol uses the reservation phases with an efficiency of about 36 percent and the transmission phases with perfect efficiency. Thus, a reservation takes an average of $1/0.36$ reservation slots, since each time slot has the probability 0.36 of being used successfully. Thus, if a reservation slot has duration TRES and if a packet transmission lasts τ, the efficiency $\eta_{R.ALOHA}$ of the R.ALOHA protocol is

$$\eta_{R.ALOHA} = \frac{\tau}{\dfrac{TRES}{0.36} + \tau} \approx \frac{1}{1 + 2.8 \times \dfrac{TRES}{\tau}}. \tag{4.27}$$

As an illustration, assume that the duration of a reservation slot is 5 percent of that of a packet transmission. We find that the efficiency of the R.ALOHA protocol is $\eta_{R.ALOHA} = 1/(1 + 2.8 \times 0.05) \approx 88\%$.

The throughput of the pure, slotted, and R.ALOHA protocols is the product of the efficiency of the protocol multiplied by the transmission rate. The transmission rate of radio transmitters is limited by the bandwidth of the radio channel available to the transmitters. The frequencies used for radio transmissions are strictly controlled by governmental agencies [the *Federal Communications Commission* (FCC) in the United States] in order to avoid interference and to maintain an orderly utilization of that valuable resource. The restrictions on the available frequencies limit the transmission rate, and, therefore, the number of nodes that can use radio transmitters simultaneously. When a large number of users must be accommodated, as in the case of the *mobile telephones* used in automobiles in urban areas, a solution is to divide the geographical area into cells. The users must transmit on frequencies determined by their locations and reserved when the calls are placed. The figure assumes that the power of the transmitters is small enough for cells that use the same frequencies not to interfere. Each cell has a base node, and the mobile users communicate with the base node. Every mobile transmitter identifies the cell it belongs to by comparing the powers of the signals it receives from the different base nodes. The mobile transmitter receives the greatest power from the base node of the cell it belongs to.

Analog cellular telephones use frequency modulation, with a bandwidth of 30 kHz per channel, and cells with a radius from 10 km to 40 km. Digital cellular telephones support a larger number of users because they use audio compression to utilize the bandwidth more efficiently, and they have better connections. The allocation of frequencies to cells can be applied to small cells that cover only a few rooms in a building or a fraction of a city block. Such a division of a region into *microcells* permits even higher transmission rates over the same total range of frequencies. Such a system can be used for portable computer terminals with connections that have bit rates high enough for applications that require small response times and that use graphics or video.

Summary

- The data link layer of a LAN is decomposed into the MAC sublayer, which regulates the access to a common channel, and the LLC sublayer, which implements packet transmission services.
- The ALOHA protocols, slotted and pure, are multiple-access protocols in which nodes transmit and, after a random delay, retransmit the packets that collide. The efficiencies of these protocols are 36 percent and 18 percent, respectively.

- The IEEE 802.3 standards cover the physical layer and the MAC sublayer for Ethernet. Shared Ethernet uses the CSMA/CD protocol. The efficiency of that protocol is approximately $1/(1 + 5a)$, where a is the ratio of the one-way propagation time of a signal on the medium to the transmission time of a packet. Ethernet is suitable for fast delivery of packets when the load is not too heavy and when the nodes do not require bounded delays.
- Nodes in a token ring network use a token-passing protocol. The efficiency of that protocol is close to 100 percent. The IEEE 802.5 implementation can handle multiple priorities.
- FDDI is the ANSI standard for 100-Mbps networks. The nodes are attached in a dual-ring configuration. The encoding is 4B/5B + NRZI. The medium access time is bounded by $2 \times$ TTRT, where TTRT is a parameter that the nodes agree upon. The efficiency of FDDI is given by (4.19).
- Wireless LANs are used to avoid the cost of wiring and in applications where mobility is desirable. We discussed two emerging standards for wireless LANs.
- The IEEE 802.2 standards for LLC provide for three types of packet transmission services: acknowledged connectionless for point-to-point connections, unacknowledged connectionless, and the HDLC-ABM connection-oriented services.

Problems

Problems with a * are somewhat more challenging. Those with a ^{c} are based on material in a complement. Problems with a ^{1} are borrowed from the first edition.

1. Consider a 10Base-T Ethernet and assume that its throughput is 6 Mbps. Fifty computers are attached to the network. What is the maximum average transmission rate per computer?

2. The maximum distance between two computers attached to a 10Base-T hub is 200 m. Assume that the propagation speed of electrical signals in the twisted pairs is at least 1.75×10^8 m/s. Assume also that the hub takes at most 0.1 μs to detect a collision. What is the maximum time between when a node starts transmitting a packet and when it learns that the packet has collided?

3. When we examined the format of an Ethernet frame we noted that a number of bits are not data bits. Determine the average fraction of data bits in a packet as a function the average packet length. Even if no time was wasted because of collisions, the data rate of the network would be limited by this fraction of the transmission rate.

^{1}4. The 200 computers in a research laboratory are attached to a 10Base-T Ethernet with an efficiency of 65 percent. The packets have 800 data bits. On the average, how many packets can each computer send every second? (*Hint:* See Problem 3.)

*5. Consider a switched Ethernet network with 50 computers attached to a switch with 10-Mbps links and one file server attached to the switch with a 100-Mbps link. Analyze the throughput of the network as a function of the fractions of packets going between computers and between computers and the server.

[1,c]6. Consider a large number of nodes that use the slotted ALOHA protocol with the following retransmission algorithm. A new packet arrives and is transmitted in a time slot with probability ϵ. Every old packet is transmitted with probability p. All these transmissions are independent of one another. Show that if a random sequence of events leads the number of old packets to reach a large value N, then the average number of old packets after the next time slot is larger than N.

[1,c]7. We know that the time that a packet may have to wait until it is transmitted by a CSMA/CD network is unbounded. Should this deter us from using CSMA/CD in time-critical applications? We explore this question in this problem. There are 50 nodes on a CSMA/CD network. Assume that two transmissions collide. What is the probability that these two nodes would collide during their five subsequent attempts? How large can the delay be in that case? Can you repeat these calculations to include the possibility that another node would transmit in a time slot with duration $2 \times \pi$ with probability 0.01 percent?

*8. In this problem we explore the average rate seen by a user on a shared network. Imagine that N active users share a link with transmission rate R bps. The behavior of a user is random and we model it as follows. An active user requests a file, then waits for the file. After receiving the file, the user waits for a random time, then repeats the procedure. Each file has an exponential length, with mean $A \times R$. Each waiting time is also exponentially distributed but with mean B. All these random variables are independent. How would you simulate this system to determine the average time it takes a user to get a file? Can you estimate the answer without performing the simulation?

*,c9. We explore a variation of the ALOHA protocol called *inhibit sense multiple access* (ISMA) that is used by the cellular packet radio network CDPD. A simplified model of this protocol is that it behaves like slotted ALOHA except that the base station uses a control channel to inhibit transmissions. The rule of the protocol is that if a station is transmitting during one time slot, it indicates whether it wants to transmit in the next slot. If that is the case, then the base station sends an inhibit signal and the other stations do not try to transmit in the next slot. Otherwise, the other stations compete for the next slot. Derive the efficiency of this protocol as a function of the average number of back-to-back packets that the stations send.

References

The official references for the Ethernet and token ring are the IEEE standards. See the IEEE 802 Web page ftp://stdsbbs.ieee.org/pub/802-main for a list of those references. This web site also gives you access to the latest drafts and presentations.

Hegering (1993) describes wiring strategies for Ethernet. Details of 100-Mbps Ethernet technology are in Johnson (1996). The original paper on Ethernet, by its inventors, is Metcalfe (1976). Göhring (1992) is a useful reference on token ring networks. The development of Gigabit Ethernet is promoted by an open forum called the Gigabit Ethernet Alliance. The Web page is http://www.gigabit-ethernet.org/.

Bibliography

Göhring, H-G, and Kauffels, F-J, *Token Ring—Principles, Perspectives and Strategies,* Addison-Wesley, 1992.

Hegering, H-G, and Läpple, A., *Ethernet: Building a Communications Infrastructure,* Addison-Wesley, 1993.

Johnson, H. W., *Fast Ethernet—Dawn of a New Network,* Prentice-Hall, 1996.

Metcalfe, R. M., and Boggs, D. R., "Ethernet: distributed packet switching for local computer networks," *Comm. ACM,* 19, 395–404, 1976.

CHAPTER 5 Asynchronous Transfer Mode

We introduced asynchronous transfer mode (ATM) in Chapter 2. In this chapter we explore the ATM architecture, protocols, and technology in more detail. Section 5.1 explains the architecture of ATM protocols and reviews the basic operating principles of ATM. In Section 5.2 we examine routing in ATM networks. Section 5.3 explores the end-to-end transmission services that ATM networks provide. Internetworking with ATM—either through LAN emulation or as a transport mechanism for IP—is explained in Section 5.4. The chapter concludes with a brief complementary section on the analysis of delays in switches.

5.1 Architecture

The ATM is being designed by a consortium of a few hundred companies regrouped under the ATM Forum. The ATM Forum publishes *recommendations* that define the ATM protocols.

5.1.1 Protocol Layers

We start with a look at the architecture of the ATM protocols shown in Figure 5.1 The physical layer transports bits between devices attached to the same link. The ATM layer implements *end-to-end communication services.* These services belong to a wide range of *service classes* that differ greatly by their *quality of service* (QoS). The service classes go from "best effort" to "low delay and low loss." The best effort service offers no guarantee on the delay nor the loss rate. This service is similar to the service that TCP offers today. The "low delay and low loss" quality is suitable for video conferences or other demanding applications. The AAL, for ATM adaptation layer, adapts the information stream that the higher layers produce to the ATM communication service. This adaptation involves packaging the information into cells. The AAL may also add control information the higher layers require to maintain the timing of a bit stream or to control errors. The higher layers perform the additional tasks that applications need.

Note that the AAL and higher layers are divided into a *control plane* and a *user plane.* The control plane sets up the connection and the user plane delivers the user information. Our figure is not complete: we did not draw the path that the control information follows. This path may be different from that of the user information. This separation of the control and user information is similar to the separation of the call control messages and the voice bits in the telephone network. For simplicity, we did not draw the operations and maintenance (OAM) protocols that supervise the operations of the network.

5.1.2 Three Application Examples

Figure 5.2 sketches the protocol stacks of three applications implemented on an ATM network. We did not draw the control plane, for simplicity.

The HTTP application *a* (see Chapter 3) is used by a World Wide Web browser to get a Web page or other document when the user clicks on a link in another Web page. In this example, HTTP runs over an ATM network that is set up to transport the IP packets. AAL-5 is a specific implementation of the AAL.

Application *b* is a video conference that uses the compression algorithm MPEG-2 (see Chapter 9) and forward error correction algorithms. In contrast to example *a,* the video

FIGURE 5.1

Simplified architecture of ATM protocols.

Control
User
Control
User
Higher
AAL
ATM
Physical
End system
Switch
End system

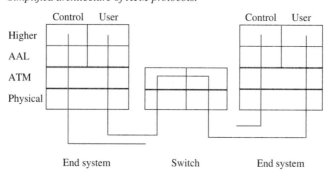

FIGURE 5.2

Examples of applications in an ATM network: (a) HTTP; (b) video conferences; (c) LAN emulation.

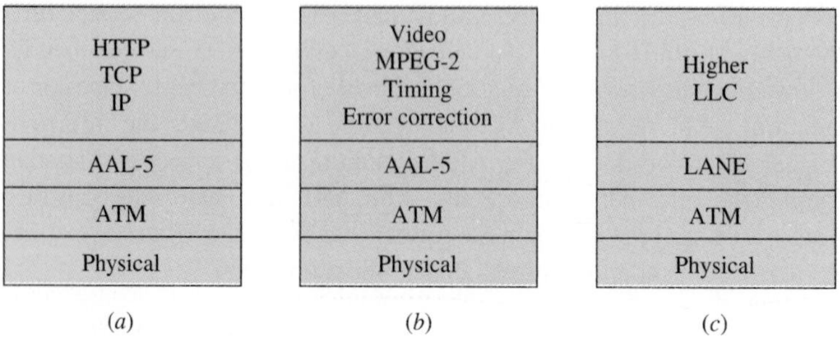

HTTP
TCP
IP
AAL-5
ATM
Physical
(a)
Video
MPEG-2
Timing
Error correction
AAL-5
ATM
Physical
(b)
Higher
LLC
LANE
ATM
Physical
(c)

application uses ATM directly without going through IP. With this model, the application can take full advantage of the quality of service that ATM offers.

Applications *a* and *b* might run in a single computer, with the web and video connections complementing each other in a distance learning or telemedicine application, for instance.

Application *c* is the emulation of a LAN such as a Fast Ethernet with an ATM network. The computer that runs the protocols shown in the figure can use its standard LAN software as if it were on that LAN. A special protocol, called the LAN emulation (LANE) protocol, performs the tasks the ATM network needs to mimic the LAN.

5.1.3 *Design Philosophy of ATM*

The design of ATM was initiated by AT&T, and that telephone heritage is reflected in the virtual circuit approach. As in the telephone network, every ATM connection must be set up and controlled by the network.

An important advantage of virtual circuit routing is that it enables the network to reserve the resources that each connection needs and to block a connection request when sufficient resources are not available. Although the network reserves resources for connections, these reservations are not for the peak transmission rates, as would be the case in a circuit switched network. We elaborate on this important distinction next.

A link loses cells when the buffer of its transmitter overflows. The buffer overflows when its input rate exceeds the transmitter rate for long enough. When the network reserves a transmission rate, say 2 Mbps, for a connection on a given link, it bases that reservation on the average bit rate of the connection. For example, consider a 100-Mbps link. It is probably safe for the network to accept connections through that link if the sum of their average rates is less than 80 Mbps, say. Indeed, even though the connection rates fluctuate, it is unlikely that their sum exceeds 100 Mbps long enough for the link to lose cells.

In contrast, when it accepts a connection, a circuit switched network reserves the peak rate of the connection on every link from the source to the destination. Accordingly, a circuit switched network reserves the sum of the peak rates for a set of connections through a link. Since the sum of the peak rates is typically much larger than the sum of the average rates, we say that virtual circuit switching achieves a *statistical multiplexing gain.*

From the above discussion, we see that virtual circuit switching enjoys the main benefit of packet switching: statistical multiplexing. Moreover, virtual circuit switching has also the main advantage of circuit switching: tight control of the resource allocation and therefore, presumably, of the quality of service.

One might then conclude that ATM is a good synthesis of decades of experience with circuit switching in the telephone network and with packet switching in data networks. ATM should be well suited to provide economically a wide range of services from email to high-quality video conferences or distributed virtual-reality applications.

However, four aspects of ATM may slow down or limit its deployment:

- *Compatibility:* There is a large installed base of TCP/IP-based applications and other applications that cannot exploit the benefits of ATM without nontrivial modifications.
- *Complexity:* Since the ATM network must be aware of the connections, the "state" of a switch is complex and so are the connection control protocols.

- *Reliability:* The network elements must remember a vast amount of information. Consequently, a failure has a large impact on many connections.
- *Need:* IP engineers argue that the sophistication of ATM is not required. They claim that it is preferable to build a fast (lean and mean) IP network instead of building a meticulous but heavy and necessarily slower (in raw speed) ATM network.

In spite of these questions about the prospects for end-to-end ATM systems, one should note that LANs, IP, and ATM are complementary. Many IP routers and LAN switches have an ATM core and many IP links and LAN backbones use ATM connections.

5.1.4 Operating Principles

We explain the ATM layers further in this chapter. For now, let us review the main operating principles of ATM that we sketched in Chapter 2. ATM is connection-oriented and transports data in 53-byte cells along *virtual circuits.* That is, if Alice wants to send information to Bob, she asks the network to set up a connection between her computer and Bob's with a specific quality of service for a given application. The network then selects a path along a set of links and switches between the two computers. The links and switches along that path must have enough spare capacity and buffer space to be able to carry Alice's cells with the quality of service she requests. Alice's computer then fragments the information into ATM cells and injects them into the network at a suitable pace. The network carries the cells along the path to Bob's computer so as to meet the specified quality of service. Bob's computer puts the information back in the original format. After the information transfer, the application may release the path, thus freeing the resources (link capacity and buffer space) for other connections.

As you see, many tasks are involved in an information transfer. Let us highlight these tasks to help us situate them later.

1. Source requests connection with specific quality of service for given application.
2. Network locates path with sufficient spare resources.
3. Source places information into ATM cells.
4. Source injects cells into network at a suitable pace.
5. Network carries cells along virtual circuit to meet quality of service.
6. Destination computer puts cells back into original information format.
7. The application releases the connection.

Tasks 1, 2, and 7 constitute the call setup and release. In task 1, the application must know the QoS it needs. For task 2, the network decides whether it can admit a call or must block it because it does not have sufficient spare resources. If it admits the call, the network selects the path by using a routing algorithm. Task 3 is the job of the AAL in the source. The ATM layer in the source paces the connections (task 4). This pacing prevents excessive congestion from building up in the network. To perform task 5, the network switches may allocate different priorities to cells. The AAL in the destination computer performs task 6. We discuss a number of aspects of these tasks in the chapter, but first we take a closer look at ATM cells. In particular we discuss the header that carries the control information.

5.1.5   ATM Cell Format

Figure 5.3 shows the format of ATM cells across the user-network interface. The 5-byte header has the following fields: GFC (generic flow control) is used by the network to tell the source to adjust its pace of transmission; VPI and VCI are the virtual path and virtual circuit identifiers (see below); the payload type (PT) indicates whether the cell is a user cell or a control cell such as a cell that the network uses to monitor the connection; CLP is the cell loss priority (see below). Finally, HEC is the header checksum computed on the previous 4 bytes of the header.

The connection is identified by the pair VPI, VCI. The routing is based only on the VPI. That is, the virtual path identified by the VPI is a collection of virtual circuits that are routed together. When a cell does not conform to the traffic description that the user gave when the connection was set up, the network can mark that cell as being eligible for discard in case of congestion. That discard eligibility is indicated by the CLP. The understanding is that the network does its best to carry such cells but may drop them.

Cells between network switches do not have the GFC field; instead, their VPI field also uses those 4 bits. Consequently, the network can set up a large number of internal virtual circuits between network devices. The network uses these virtual circuits to exchange routing information and other control information.

Why 53 Bytes?

ATM cells have a 5-byte header and a 48-byte payload. To waste a smaller fraction of the link capacity in transmitting headers, it would seem desirable to use a larger payload. However, a larger payload takes longer to fill.

For instance, consider an "ATM phone" application (Figure 5.4) where two telephones are connected with an ATM link. Assume that the link C between the telephone sets has an infinite transmission rate and a negligible propagation delay. Then the 64-kbps bit stream that one telephone set produces is delayed exactly by 6 ms before it reaches the other telephone set. Indeed, it takes 6 ms to accumulate 48 bytes of voice payload, since $(48 \times 8 \text{ bits})/(64 \text{ kbps}) = 6 \text{ ms}$.

At the outset, the ATM designers agreed that 6 ms was the maximum tolerable packetization delay for voice signals. Accordingly, they decided that the ATM cells should have a 48-byte payload.

5.1.6   AAL

The ATM adaptation layer converts the information stream into ATM cells (see Figure 5.5). The AAL is divided into two sublayers: the convergence sublayer (CS) and the segmentation

FIGURE 5.3

Format of ATM cells across the user-network interface. See text for details.

4	8	16	3	1	8	384 ← Bits
GFC	VPI	VCI	PT	CLP	HEC	Data
ATM header: 5 bytes					ATM payload: 48 bytes	

FIGURE 5.4

Packetization delays the digital voice signal.

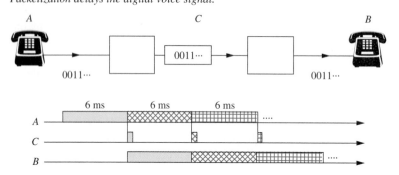

FIGURE 5.5

The ATM adaptation layer (AAL) converts the information stream into 48-byte cells. The AAL consists of the convergence sublayer and the segmentation/reassembly sublayer.

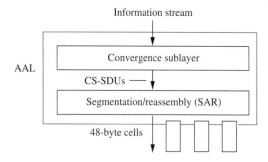

and reassembly sublayer (SAR). The CS prepares the information stream by decomposing it into packets called CS-SDUs (service data units). The CS-SDUs have a format and control information that are adapted to the application. The SAR packages the CS-SDUs into ATM cells, possibly adding additional control information that the destination may need.

There are five versions of the AAL that are designed for five classes of traffic: AAL-1 for constant-bit-rate real time, AAL-2 for variable-bit-rate real time, AAL-3 for connection-oriented packet streams, AAL-4 for datagrams, and AAL-5 for IP packets. In 1997, developments of ATM indicate that AAL-5 will be used also for most datagram applications and for variable-bit-rate applications such as high-definition television (HDTV). Connection-oriented applications may also be implemented over AAL-5. We describe AAL-1, which will be used for telephone transmissions and for constant-bit-rate video, and AAL-5. Please refer to the literature for the other versions.

AAL-1 is designed to transport constant-bit-rate traffic generated by telephone calls or constant-bit-rate video. Figure 5.6 shows that the convergence sublayer of AAL-1 groups the bit stream into 47-byte packets. The SAR sublayer adds a 1-byte header that consists of a convergence sublayer indication (CSI), a sequence number (SN), and a sequence number protection (SNP). The use of CSI is optional. The destination uses the SN to detect missing

FIGURE 5.6

AAL-5 and AAL-1.

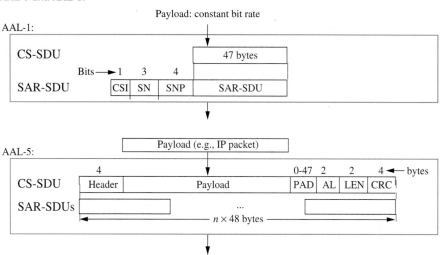

cells. The SNP can correct single bit errors and detect multiple bit errors in the sequence number.

AAL-5 carries IP packets or other payloads with little overhead. The CS packages the information into CS-SDUs with a length equal to a multiple of 48 bytes. The CS-SDU contains a header whose use is under study. The 2-byte alignment field (AL) is a filler. The padding (PAD) and length (LEN) are required to accommodate variable-length payloads. The CRC is calculated over the complete CS-SDU. The SAR then puts the CS-SDU into back-to-back 48-byte cells with no additional control information. Lost cells are detected by the higher layer (e.g., TCP or UDP).

5.1.7 Network Operations and Maintenance

Consider a simple network that consists of a single ATM switch attached to a number of computers. The network manager may set up permanent virtual paths between the computers through the switch. The manager allocates a specific VPI for each pair of computers. For instance, say that the virtual path between computer *A* and computer *B* uses VPI = 17. To communicate with *B*, computer *A* uses VPI = 17 and some VCI. The virtual path is set up permanently with resources reserved for it in the switch. For instance, VPI 17 may be allocated a transmission rate of 10 Mbps and 2 Mbytes of buffer space inside the switch to absorb traffic fluctuations.

In a large network, it is wasteful to allocate resources permanently to all possible virtual paths that might conceivably be set up. In such a network, it is preferable to set up virtual paths on demand. A computer communicates with the network to request a connection, using a VPI/VCI reserved for call setup requests. Internally, the network switches use reserved VPI/VCIs to exchange the control information they need to set up and monitor connections. We explain some of the procedures that the network uses to set up connections in Section 5.2. Here we explain the basic ideas behind network maintenance.

The objective of network maintenance is to monitor the connections and to take corrective actions as needed. The maintenance protocols supervise the connections by using OAM (operations and maintenance) cells. Some OAM cells indicate an alarm, others signal that the destination fails to receive the user cells, others are "loopback" cells that an intermediate switch must loop back.

We examine a concrete scenario to illustrate the OAM protocols. In Figure 5.7, a virtual path is set up between the computers A and B with VPI = 17. A few virtual circuit connections belong to this virtual path; these connections do not use the reserved VCIs 3 and 4. An OAM connection with VPI = 17 and VCI = 4 monitors the virtual path VCI = 17 from end to end. Another OAM connection (VPI = 17, VCI = 3) supervises the connection between the end user *A* and the first network switch *V*.

At some time, *B* notices that the user cells from *A* do not arrive as they should. The OAM protocol implemented in *B* sends a "far end receive failure" (FERF) cell to *A* along the (17, 4) OAM connection. When it gets that cell, the OAM protocol in *A* starts a procedure to locate the faulty network element. The OAM sends a loopback cell along the OAM connection (17, 4), asking switch *W* to loop it back. We assume that the faulty element is between *V* and *W* and prevents the loopback cell from coming back to *A*. The OAM protocol inside *A* then sends a loopback cell along (17, 4) asking *V* to loop it back. The cell comes back to *A*. The OAM protocol then knows that the failure is somewhere between *V* and *W,* and the OAM protocol sends an alarm describing the problem to the network operator.

We sketched the maintenance procedure for a virtual path. The ATM Forum has defined similar procedures, with the corresponding format of OAM cells, for maintaining a virtual circuit, a transmission path, a communication link, and transmission equipment between electronic regenerators.

In addition to defining network maintenance protocols, the ATM Forum has defined network management protocols. These protocols classify the attributes of the different

FIGURE 5.7

OAM cells that monitor a VPC.

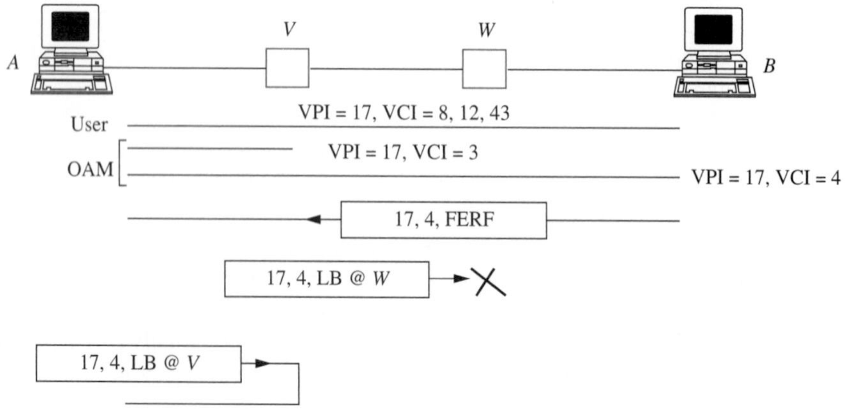

network elements. For instance, the attributes of a virtual circuit connection are its status (whether it is up or down), the current quality of service (cell loss rate, delay statistics), and descriptors of the traffic that it carries. The attributes of an ATM link include the maximum numbers of virtual circuits and virtual paths it can carry and the current numbers it carries. The network management protocols also specify how to read these attributes and how to modify those that can be controlled (such as turning off a faulty transmitter for repair).

Using the management protocols, a network management software can construct a management information base (MIB) which maintains an up-to-date picture of the status of the network. The network operator uses this MIB to take corrective actions and to plan network modifications and upgrades.

5.2 Routing in ATM

We explain how the switches allocate VPIs and the routing tables they maintain. We then outline some of the main ideas in the routing algorithms that the ATM Forum is considering for ATM. Finally, we discuss two commonly used designs of ATM switches.

5.2.1 Routing Tables

Each ATM switch maintains a routing table that indicates on which link it should send the cells. A cell belongs to a virtual path that is identified by the VPI of the cell. Thus, the outgoing link is determined by the VPI of the cell.

However, there is a slight twist to this procedure. The VPIs of different virtual paths need only be unique on each link to differentiate them. Consider the situation in Figure 5.8. In that network, three switches interconnect the five computers $A, B, \ldots, E$. Three virtual paths—represented by thick lines—are set up successively: from A to C, from B to E, and from C to D, in that order. These virtual paths each carry a number of virtual circuit connections.

FIGURE 5.8

Three virtual paths and their VPIs along the links of an ATM network. Note that the VPIs are unique per link but not globally. Each switch maintains a routing table that specifies the outgoing link and the new VPI of each cell.

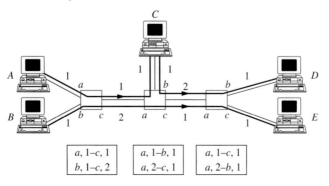

In the figure, the numbers on the thick lines are the VPI of each virtual path along each link of the network. When the virtual path from A to C is first set up, the network allocates the VPI number 1 to that path along each link. When the virtual path from B to E is set up next, the VPI number 1 is not used between B and the first switch, so that the network allocates that number to the path. Although the virtual path from A to C already uses the number 1 on other links, choosing 1 for this new path does not create any risk of confusion. On the next link, the path from B to E cannot use the number 1, which is already used on that link. The network allocates the smallest free number on that link to the path, i.e., the number 2. The procedure continues for the other links and the other virtual path from C to D.

Note that the VPI of the path from B to E is 1 before the first switch, then 2 on the next link, then 1 again for the next two links. Each switch maintains a routing table that it updates whenever a new connection is set up or a connection terminates. The table has one entry per connection. The entry has the following format: incoming link, incoming VPI–outgoing link, outgoing VPI (see figure).

5.2.2   Network Node Interface

The ATM Forum is considering a protocol called PNNI, for private network-node interface, for routing in private ATM networks. This protocol, or something like it, will probably be recommended for public ATM networks as well. The two main objectives of PNNI are that the routing should be based on the quality of service that connections request and that the algorithms and protocols should be scalable. PNNI has two parts: addressing and routing. We address these topics separately.

Routing

Consider the left part in Figure 5.9. The circles represent subnetworks and the lines are ATM links or some more complex connections such as virtual paths. Node A wants to set up a connection to D.

In a nutshell, the steps of the routing algorithm are as follows. Each link and each connection through a subnetwork is described by a vector of attributes $x(i)$. These attributes are distributed to all the nodes by flooding. Node A uses the attributes $\{x(1), \ldots, x(9)\}$ to calculate a preferred path to D. We assume that this preferred path is (A, B, C, D).

FIGURE 5.9

Left: a network of subnetworks. Right: crankback during call setup.

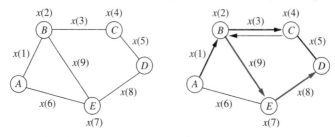

Node *A* then sends a connection request toward *D* along that preferred path using source routing. As it gets that request, *B* checks whether it can accept it. That is, *B* verifies it can carry the new call with traffic descriptors and QoS attributes specified in the request message. Assume that *B* accepts the request; it forwards it to *C*. Assume that *C* does not accept the request. Node *C* notifies *B* of its rejection.

Node *B* then calculates a preferred path to *D*, based on network status information, as *A* did earlier, but with more-current information. Say that *B* prefers the path (*B, E, D*). Node *B* then sends the request toward *D* along that path and the procedure continues. In the figure, we assume that this second step succeeds and that the final path is (*A, B, E, D*). This mechanism is called *crankback*.

How does *A* calculate a preferred path? The current ATM Forum recommendation specifies that the attributes $x(i)$ should describe for connection *i* the delay (maximum and maximum variation), the cell loss ratio, the "spare capacity," and some measure of desirability. Node *A* estimates the effect of adding the connection that it is trying to route on the attributes of the links. Node *A* then uses these modified attributes to execute an algorithm similar to Dijkstra's shortest path algorithm that determines the preferred path. You will note that since the paths have multiple attributes, the algorithm must weigh the different attributes to select a preferred path.

Addressing

Addressing is inspired from the hierarchical addressing of Internet, except that the number of levels in the hierarchy can be much larger (up to 105!) than just subnets and autonomous systems. Consider Figure 5.10. The network administrators assign ATM addresses to the nodes. The address of a node indicates its membership in the hierarchy. For instance, in Figure 5.10, the node with address 1.1.2 belongs to the subgroup 1 of group 1. We call this subgroup "group 1.1." Similarly, node 2.2.3 belongs to the group 2.2. (We assumed a two-level hierarchy to keep the figure simple.)

The network is self-organizing. Each node talks to its neighbors to identify the members of the group of the same level, starting from the lowest level. Then the group members elect a leader which represents the group for the next level and the procedure continues until the top level is reached.

In the example of the figure, the nodes 1.1.1, 1.1.2, and 1.1.3 discover that they all belong to group 1.1. Group 1.1 then elects a leader, say node 1.1.1, that we call "node 1.1." By flooding their link metrics inside group 1.1, the members of that group learn path

FIGURE 5.10

Network nodes with hierarchy of groups.

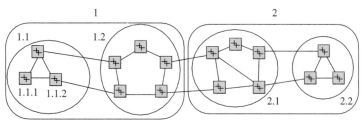

metrics to each other and node 1.1 can calculate path metrics to other groups of the same hierarchical levels. Thus node 1.1 finds out it can reach node 1.2 and it can estimate the metric of the shortest path to that node. The nodes 2.1 and 2.2 perform the same steps. The procedure then continues at the next level up. That is, nodes 1.1 and 1.2 learn that they both belong to group 1 and they elect a group leader, say node 1.1. We call this leader "node 1." (Thus, node 1 is in fact node 1.1.1 in our example.) Node 1 discovers it can reach node 2 and it can calculate the shortest path.

In this process, node 1.1 has a table with its group members and attributes of the paths to them. Node 1.1 gives that information to node 1. Node 1 gives the aggregate information it gets from nodes 1.1 and 1.2 to node 2. The information is then fed down to the group members. Eventually, node 1.1.2 gets the attributes of paths to 2.2.3. Node 1.1.2 can then use the procedure we explained above to select a path to 2.2.3 and send a request. The request for a connection from 1.1.2 to 2.2.3 may then be accepted, possibly after crankback.

5.2.3   Switch Designs

A number of designs have been studied for ATM switches. In this section we explain two designs that are used for fast switches with a limited number of ports (a few dozens). Other designs are better suited for switches with a very large number of ports (thousands). Please refer to the literature if you want a discussion of those designs. The two most commonly used ATM switch designs are shown in Figure 5.11: input buffer and output buffer.

Input Buffer Switch

In an input buffer switch, cells enter a buffer as they arrive at the switch. Cells wait in line in the buffers until the switching fabric can send them out on their output link. The advantage of this design is that the buffer memory runs only at the line rate. For instance, if the links operate at 2.4 Gbps, then the input buffers need to be able to store bits at that rate, not faster.

In an actual implementation, some fast electronic device converts the serial bit stream that arrives on a link into N parallel bit streams. This serial to parallel conversion divides the speed of the electronics of the switch by a factor N. A parallel to serial conversion takes place at the output of the switch.

FIGURE 5.11

Input buffer and output buffer switch.

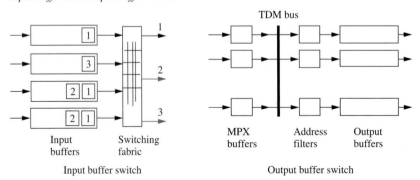

The disadvantage of the input buffer design is that the throughput of the switch is reduced by "head-of-line blocking." This blocking effect is illustrated in Figure 5.11. Three cells at the head of their input buffer are destined to output link 1. The switching fabric can send only one of these three cells at a time to output link 1. Assume that the switching fabric transmits the cell from input buffer 1 to output link 1 and that at the same time it sends the cell from the second input buffer to the third output link. In that case, cells in the third and fourth output buffers that want to go to the second output link are blocked. The switching fabric could send one of these two cells if they were not blocked by the cells at the head of their waiting line.

To reduce head-of-line blocking in an input buffer switch, one may replace the single input buffer of each input line by a few parallel input buffers. For instance, assume that, in the input buffer switch in Figure 5.11, we put three buffers at each input line, one for each output line. As a cell arrives, we send it to the input buffer that corresponds to its output link. With this arrangement, we can eliminate head-of-line blocking altogether. The switching fabric must now choose which cells it should send to the output links. At any given time, the switching fabric can choose among a potentially large number of such possibilities. The switch must make this choice in one cell transmission time. Researchers have designed fast scheduling algorithms that maximize the throughput of the switch and achieve a very low delay.

Output Buffer Switch

The output buffer switch in Figure 5.11 transmits all the arriving cells on an internal bus. At each output link, an output buffer copies the cells destined for that output. Such a design does not suffer from head-of-line blocking. However, since a number of cells may arrive at the same time for the same output link, the speed of the output buffers must be a multiple of the line rate.

Input and Output Buffer Switch

Some designs combine input and output buffers by implementing an input buffer switch with an internal speedup. Specifically, these switches can transfer S cells (typically $S = 2, 3$, or 4) during one cell transmission time. Accordingly, up to S cells may arrive at an output link while the link can transmit only one cell. Consequently, a buffer is needed at the output. These switches reduce the effect of head-of-line blocking while limiting the speed of the electronics they require to S times the line rate instead of M times for an $M \times N$ output buffer switch.

Priorities and Multicast Switching

The input buffer and output buffer switches can be modified to handle multiple priorities. For instance, each output buffer of an output buffer switch can be replicated for each priority level, and the transmitter then serves the cells in the corresponding priority order. The switches can also be designed to multicast cells, i.e., to copy a cell to different output links. In addition, the switches can be modified to copy one cell to a number of output links.

We explain in Section 5.5 how to estimate the delay through an ATM switch. The upshot of the analysis is that the delay rarely exceeds a few cell transmission times, provided that the sources pace their transmissions suitably.

5.3   End-to-End Services

The main design objective of ATM is to provide a wide range of qualities of service. The ATM Forum specifies five service classes. These classes differ in the attributes of the quality of service and in the descriptors of the traffic that they carry. That is, a given service class carries traffic that must conform to some traffic descriptors. The service class then delivers the traffic with a quality of service that meets specific attributes. We explain the attributes of the quality of service and the traffic descriptors. We then describe the classes.

5.3.1   Quality of Service Attributes

The quality of service (QoS) attributes are the following:

- Cell loss ratio (CLR): fraction of cells lost during transmission.
- Cell delay variation (CDV): maximum difference between end-to-end cell delays.
- Maximum cell transfer delay (max CTD): maximum end-to-end cell delay.
- Mean cell transfer delay (mean CTD): average end-to-end cell delay.
- Minimum cell rate (MCR): minimum rate at which the network delivers cells.

Table 5.1 gives a few plausible attributes that representative applications might request.

5.3.2   Traffic Descriptors

The traffic descriptors are defined by an algorithm: the generalized cell rate algorithm (GCRA). The GCRA—also called *leaky bucket*—controls the cell transmission times. Figure 5.12 explains the leaky bucket. The figure shows two buffers with different transmission rates and buffer capacities. The cells that arrive are immediately duplicated and enter the two buffers. We say that the cell traffic that arrives conforms to the parameters (PCR, SCR, CDVT, BT) if the buffers never overflow. We describe these parameters below. The parameters are defined jointly for two different traffic types, which we discuss next.

TABLE 5.1   Representative QoS Attributes for Different Applications (Not Standardized)

Applications	Attributes				
	CLR	CDV	Max CTD	Mean CTD	MCR
Video conference	10^{-5}	50 ms	120 ms	100 ms	NA
Telephone	10^{-5}	2 ms	41 ms	40 ms	NA
Email	10^{-5}		Not specified		NA
File transfer	10^{-8}		Not specified		20 cells/s

NA means not applicable

FIGURE 5.12

Traffic policer. The traffic conforms to the traffic descriptors if the buffers never overflow.

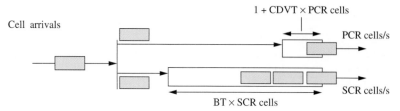

Constant Bit Rate

Constant-bit-rate (CBR) traffic must not overflow the top buffer in Figure 5.12. PCR is the peak cell rate and CDVT is the cell delay variation tolerance of the traffic. For instance, if CDVT $= 0$, then the minimum time between two cell arrivals is 1/PCR. Indeed, when the first cell arrives, the buffer that can contain exactly one cell is full. That cell must be completely out of the buffer before the next cell arrives for the buffer not to overflow and the cell takes 1/PCR to be transmitted. If CDVT $\times$ PCR $= 0.1$, then the minimum time between two cell arrivals is 0.9/PCR.

Essentially, PCR is the maximum arrival rate of cells, and small fluctuations are possible when CDVT > 0. These fluctuations may be caused by the multiplexing or framing of the ATM cells by the physical layer. This is why the traffic descriptor allows for such fluctuations.

Variable Bit Rate

Variable-bit-rate (VBR) traffic must not overflow either of the two buffers in Figure 5.12. The parameter SCR, the sustained cell rate, is smaller than PCR. SCR is an upper bound on the long-term arrival rate of cells. The parameter BT, burst tolerance, allows for the cells to arrive faster than at rate SCR for some time. The larger BT, the larger this time.

Note that the source can use the traffic policer shown in Figure 5.12 to make sure that it does not send cells that do not conform to the descriptors. Similarly, the network can use the policer to verify that the source conforms to the descriptors. The ATM Forum specifies that the switch can set the CLP of nonconformant cells (that make one of the buffers overflow) and discard such cells as needed.

We are now ready to define the service classes.

5.3.3   Service Classes

Here is a summary of the service classes listed in Table 5.2:

- CBR (constant-bit-rate service) transports CBR traffic with specified loss rate and delays.
- VBR-RT (variable-bit-rate–real-time service) transports VBR traffic with specified loss rate and delays.

TABLE 5.2   **Service Classes**

	Definition	
Service class	Traffic descriptors	Specified QoS attributes
CBR	CBR	CLR, CDV, max CTD, mean CTD
VBR-RT	VBR	CLR, CDV, max CTD, mean CTD
VBR-NRT	VBR	CLR
ABR	NA	MCR
UBR	NA	None

NA means not applicable

- VBR-NRT (variable-bit-rate–non-real-time service) transports VBR traffic with specified loss rate.
- ABR (available-bit-rate service) delivers cells at a minimum rate; the understanding is that the network gets the cells from the source as fast as it can when it has spare capacity.
- UBR (unspecified-bit-rate service) is a best-effort service that attempts to deliver the cells without making any QoS commitment.

To provide CBR and VBR services, the network reserves resources for the connection when the service is set up. In ABR service, the network regulates the flow of cells by sending information to the sources indicating the rate at which they can send cells.

5.4 Internetworking with ATM

ATM is designed to complement Internet and LAN technologies. This complementarity is well illustrated by the use of ATM as a network to transport IP packets or LAN packets. We discuss these two important applications next.

5.4.1 IP over ATM

Figure 5.13 shows computers interconnected with Ethernets and with an ATM network. We want the computers to be able to run the TCP/IP-based applications as if the packets were delivered with TCP/IP. In such a network, the ATM technology is used to take advantage of the flexibility in allocating quality of service to connections. Thus, a few of the computers may be using ATM in its "native mode." For instance, a few computers may be set up for high-speed and low-latency data delivery for distributed computing applications. The other computers are set up to use their familiar TCP/IP applications. It is also possible to have some applications in one computer use IP while others use "native ATM." We do not discuss here the pros and cons of this approach compared to an IP-only network. Let us simply remark that today IP is not capable of guaranteeing the quality of service that some applications might conceivably require and that ATM can provide that guarantee.

FIGURE 5.13

IP over ATM.

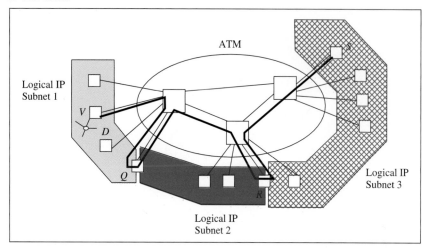

For simplicity, we look only at the TCP/IP computers. That is, we assume that all the computers in Figure 5.13 are running IP over the ATM network. We explain the simplest scheme that can be used for transporting packets over ATM. In this scheme, the network is divided logically into autonomous systems that are thought of as being interconnected by border gateways. That is, the logical arrangement is identical to that of an IP network. The delivery of packets takes place as in an IP network, as we show next.

Consider that computer S wants to send an IP packet to computer D. These computers use the IP addresses S and D. Computer S sends the packet to its router R. Computer R consults its routing table and finds that the next hop is to router Q and that this router is at ATM address q. Router R sets up an ATM virtual circuit to router Q and fragments the IP packet into ATM cells that it sends over the virtual circuit. When Q gets the cells, it reassembles the IP packet, looks at its destination address, finds that the next hop is to router V, sends the packet over to V by proceeding as R did. When it gets the packet, V sends it as an Ethernet packet to the destination D. In practice, the virtual circuits between the routers are set up permanently and the routers exchange information to maintain their routing tables.

The strategy we just described is called *classical IP over ATM*. As you have noticed, a more efficient transfer can take place by having R find out the ATM address of V and set up a virtual circuit directly from R to V. Such a mechanism—known as the *shortcut model*—is further away from the classical IP routing model but can be implemented.

5.4.2   LAN Emulation over ATM

Figure 5.14 shows a number of computers attached to Ethernets and to an ATM network. We want to make it possible for all the computers to be able to assume that they are on a common Ethernet. That is, we want to make the ATM network transparent. In other words, we want to use the ATM network to emulate an Ethernet.

FIGURE 5.14

Ethernet emulation over ATM.

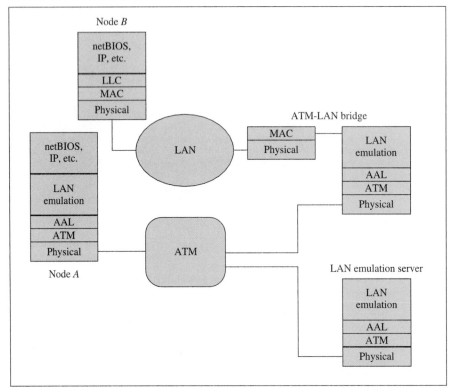

We describe a procedure that performs the emulation. The computers attached to the ATM network run a software called LAN emulation (LANE). In addition, a computer Q attached to the LAN acts as the LANE server. Consider a packet that computer S wants to send to computer D. Computer S assumes that D is on the same Ethernet and prepares a packet $[s, d|data]$ where s and d are the Ethernet addresses of S and D. This packet is intercepted by the LANE in S. The LANE looks in a table to find the ATM address to which it should send the packet. If the table has no entry for d, the LANE asks the server Q for the ATM address that corresponds to d. The server returns the address r. The LANE in S then sets up an ATM virtual circuit to r and sends the Ethernet packet $[s, d|data]$ as a sequence of *ATM* cells. The router R reassembles the packet and sends it on the Ethernet E. Permanent virtual circuits can be set up between ATM nodes, to reduce the processing and delay.

This LAN emulation strategy enables network managers to interconnect existing Ethernets with an ATM backbone. In addition, this strategy makes it possible to have these Ethernet coexist with native ATM connections. The need for a LANE server is viewed by network managers as a worrisome point of failure of the network. The arrangement can be made more reliable by duplicating this server role.

5.5 Complement: Delay in Simple Switch

We calculate the average delay through an ATM switch. Our objective is to show that the delay of cells across a well-managed ATM network is very small because of the small size of the cells. By well-managed, we mean that the network must limit the connections it accepts so that the load on the links is not excessive. The analysis that we present does not require complex methods and can be understood by readers who have read Appendix A.

Our model of the switch is an M/D/1 queue. That is, during the nth cell transmission time, a random number $A(n)$ of cells arrive at the buffer. The random variables $A(n)$ are independent and Poisson-distributed with mean λ. Thus, $E\{A(n)\} = \rho$, and one can show that $E\{A(n)^2\} = \rho + \rho^2$. Let X_n be the number of cells in the buffer at the beginning of the nth slot time (one slot time is equal to one cell transmission time). Then,

$$X_{n+1} = (X_n - 1)^+ + A_n = X_n + A_n - 1(X_n > 0), \ n \ge 0, \tag{5.1}$$

where we use the notation that for any number z, $z^+ = \max\{z, 0\}$ and $1(\cdot)$ is the indicator function, so $1(z > 0) = 1$ if $z > 0$, and 0 otherwise. The term $(X_n - 1)^+$ accounts for the fact that if $X_n > 0$, then one cell will be transmitted, leaving a list of size $(X_n - 1)$. Assume that we have reached steady state, so that $E\{X(n + 1)\} = E\{X(n)\}$ and $E\{X(n + 1)^2\} = E\{X(n)^2\}$. Taking expectations on both sides of (5.1) we get

$$E(X) = E(X) + E(A) - P(X > 0),$$

so

$$P(X > 0) = E(A) =: \rho.$$

Next we square both sides of (5.1) and take expectations to get

$$E(A^2) + \rho + 2\rho E(X) - 2E(X) - 2\rho^2 = 0.$$

Since $E(A^2) = \rho^2 + \rho$, we find

$$E(X) = \frac{2\rho - \rho^2}{2(1 - \rho)}. \tag{5.2}$$

Summary

- Asynchronous transfer mode was designed by AT&T as a method for transporting integrated services. In the early 1990s, a number of network designers proposed using ATM as a local area network technology. An industry consortium, the ATM Forum, coordinates the standardization efforts.
- ATM transports 53-byte cells over virtual circuits. The user specifies the characteristics of the traffic and the quality of service that it requests from the network. The network then searches for a path with sufficient spare resources to meet that request.
- With its small cells, ATM can switch and transport information with very small delays and delay variations. Using virtual circuits, the ATM network can reserve the bandwidth that connections need and can monitor the traffic along a connection and price it accordingly.

- IP over ATM and LAN Emulation provide ways of integrating legacy networks in an ATM network that can provide native ATM connectivity for demanding applications.
- ATM is an excellent link and switching technology. The penetration of native ATM will depend on the capability of standard frame-based technologies (e.g., Ethernet and TCP/IP) to provide the quality of service that new applications may require.

Problems

A problem marked with ^{c} is based on material in the complement. The * indicates that the problem is more complicated than the others.

1. How long is the cell packetizing delay for 1.5-Mbps video?
2. How many 64-kbps audio signals can be carried over ATM by a 1.5-Mbps link? Assume that AAL-1 is used and that the 1.5-Mbps reserves 5 percent of the bits for framing.
3. Imagine an ATM network where a virtual circuit goes over 5 links, on average. Estimate the number of virtual circuits that can be set up at any one time. (Let K be the maximum number of virtual circuits on a link.)
4. Consider the network shown in Figure 5.8. Assume that N virtual circuits are set up between A and C, C and D, and B and E. What is the maximum value of N? (Let K be the maximum number of virtual circuits on a link.) Repeat the problem if the virtual circuit number must be unique end-to-end instead of being only unique on each link.
5. Examine the output buffer switch of Figure 5.11. The internal TDM bus has 424 lines. Determine the speed of the memory needed to prevent all losses during the address filtering as a function of the number of input ports of the switch.
6. Consider the $N \times N$ output buffer switch shown in Figure 5.11. Assume that 75 percent of the traffic from each input port is destined for output port number 1. Assume that the average traffic rate on each input link is equal to λ. What is the maximum value of λ if the output links transmit at 155 Mbps?
7. Consider a VBR traffic with the following parameters: PCR = 10,000; CDVT ≈ 0; SCR = 2000; BT = 40. Explain why an output buffer ATM switch with line rate 100 Mbps and a buffer that can hold 256 kbytes can carry 100 such connections without losing a cell.
8. Twenty 1.5-Mbps video signals are transported by ATM cells, using AAL-1. The signals are multiplexed on one signal at 40 Mbps. The signal consists of frames of 10 ATM cells separated by 95 bytes of framing overhead. Because of the framing structure, the cell stream is not perfectly periodic. Provide the CBR traffic specifications of that stream.
^{c,*}9. We study a 4×4 output buffer switch, as in Figure 5.11. In one cell time, one cell destined for output port 1 arrives at the switch on input link k with probability λ, for $k = 1, \ldots, 4$, where $\lambda < 1/4$. The arrivals on the four input ports are independent. Adapt the analysis of Section 5.5 to calculate the average delay through the buffer of output port 1.

References

The official references for the asynchronous transfer mode networks are the ATM Forum Recommendations. See the Web page of the ATM Forum for information about the standardization activities of this industry consortium (http://www.atmforum.com/). The ATM Forum regularly publishes recommendations. The currently approved technical specifications can be obtained from http://www.atmforum.com/atmforum/specs/approved.html. These specifications include the following:

- ATM User-Network Interface Specification V3.1, 1994. UNI Signaling 4.0, July 1996.
- Traffic Management 4.0, January 1997.
- UNI Signaling 4.0, July 1996.
- P-NNI V1.0, March 1996.
- Physical Interface Specification for 25.6 Mb/s over Twisted Pair, November 1995.
- ATM Physical Medium Dependent Interface Specification for 155 Mb/s over Twisted Pair Cable, September 1994.
- LAN Emulation over ATM 1.0, January 1995
- ILMI 4.0, September 1996.

De Prycker (1993) is a very good presentation of ATM. The tutorials Le Boudec (1992) and Alles (1995) are highly recommended. Varaiya (1996) explores ATM in more detail than the current chapter.

Bibliography

Alles, A., "ATM Internetworking" Cisco Systems, http://www.cisco.com/, May 1995.

De Prycker, M., *Asynchronous Transfer Mode,* 2nd ed., Ellis Horwood, 1993.

Le Boudec, J-Y, "The asynchronous transfer mode: A tutorial," *Computer Networks and ISDN Systems,* 24, 279–309, 1992.

Varaiya, P., and Walrand, J., *High-Performance Communication Networks,* Morgan Kaufmann, 1996.

CHAPTER 6

Data Link Layer and Retransmission Protocols

Moving bits from one node to another is the most basic step in establishing network communication and in offering network users communication services. The transmission of bits is an elementary service performed by the physical layer. In this chapter, we assume that such a bit transmission service is implemented between two computers. We study the physical layer in Chapter 7. More complex services are built on top of this elementary service. The transmission of *packets* between nodes attached to the same transmission link is one such service. This chapter explains how packet transmissions are built from the bit-transmission service provided by the physical layer.

The data link layer performs two main steps to implement a packet transmission: (1) it frames the packets into a specific format at the sending node and extracts the packets at the receiving node and (2) it supervises the resulting packet delivery. The framing enables the receiving node to locate the packets in the bit stream.

Many applications require reliable transmission of information. However, the transmission of bits is never perfectly reliable. Random errors occur even with the most carefully designed transmission link. The physical layer provides an unreliable bit-transmission facility commonly called an *unreliable bit pipe*. We learn in this chapter that the network nodes follow special procedures to transmit packets *reliably* over unreliable bit pipes. The procedures specify how the nodes can detect transmission errors and correct them by retransmissions. The procedures are a distributed script called a *retransmission protocol*. The network nodes implement the retransmission protocol either at the data link layer or at the transport layer.

In Section 6.1, we explain framing and the error control codes. Our presentation of the retransmission protocols starts with a discussion of their main objectives and features in Section 6.2. In the subsequent four sections, we examine four commonly used retransmission protocols: the *stop-and-wait protocol* (SWP), the *alternating bit protocol* (ABP), the *selective repeat protocol* (SRP), and the *GO BACK N* (GBN) protocol. We will see that some protocols transmit packets faster than others. The faster protocols are more complex. The increased complexity is justified when the application demands fast packet transmissions. We will explain that the packet transmission rate of a retransmission protocol depends on some parameters of the protocol and that the network designer must adapt these protocol parameters to the transmission link. Each section starts with a description of the operations of the corresponding protocol. The section is then followed by a discussion of the correctness and by an analysis of the efficiency of the protocol.

In Section 6.7, we see how the retransmission protocols are implemented in popular networks. The network designer who implements a data link protocol must choose the protocol and its parameters as well as the structure of the packets. Section 6.7 indicates the choices that designer made in popular networks.

We conclude the chapter with complementary sections on error control codes, protocol verification methods, the analysis of the efficiency of protocols in the presence of errors, and the congestion avoidance mechanisms of Internet.

6.1 Framing

Imagine a receiver that gets a string of bits from its physical layer. This string contains a group of bits called a *packet*. The receiver must extract the packet from the string, identify that the packet is intended for itself if the link is attached to a number of receivers, and determine if the packet was corrupted by transmission errors.

6.1.1 Encapsulation

To make these operations possible, the data link layer encapsulates (envelops) the packet into a *frame* that contains some control information in addition to the packet.

Format

The frame format depends on the protocol but generally consists of three parts: a header, the packet, and a trailer. The header identifies the start of the packet and the address in the case of a shared link. The trailer contains the error control bits. If the packet has a variable length, then either the header specifies the packet length or the trailer contains a mark that indicates the end of the packet. We comment on a few representative frame formats below.

HDLC

The *high-level data link control* (HDLC) is a data link protocol used by many networks. HDLC transports information units called *frames* that have the format shown in Figure 6.1. These frames start and end with a start/end flag that is the bit pattern 01111110. HDLC evolved from the *synchronous data link control* (SDLC) developed by IBM. We describe the roles of the three types of frames below.

To prevent the flag from occurring inside the packet (and thereby confusing the receiver), a USART (universal synchronous and asynchronous receiver and transmitter) in the sender and a USART in the receiver modify the bit stream as follows. Within the body of the frame, the sending USART inserts a 0 bit any time five consecutive 1 bits are transmitted (regardless of whether the next "real" bit to be transmitted is 0 or 1). The receiving USART discards every arriving 0 bit that follows five consecutive 1 bits. Figure 6.2 shows the modifications that the USARTS perform at the sender and at the receiver.

FIGURE 6.1

HDLC frame (top) and control fields (bottom).

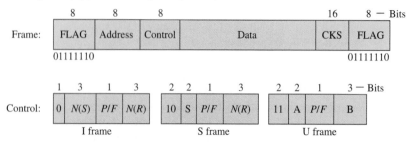

FIGURE 6.2

Bit stuffing and destuffing operations.

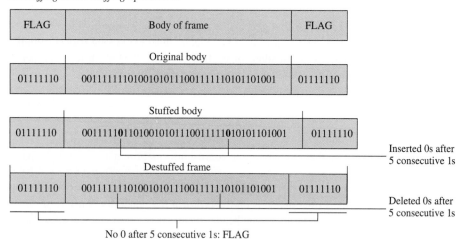

HDLC has three modes of operation: normal response mode, asynchronous response mode, and asynchronous balanced mode. The normal response mode is used when a primary node polls secondary nodes attached to it. In this mode, a secondary node cannot initiate a transfer, but can only respond to a poll request. In the asynchronous response mode, the secondary station can initiate a transfer. Two stations connected in the asynchronous balanced mode have symmetric roles. We limit our discussion to the asynchronous balanced mode, which is the most widely used in communication networks.

The three types of frames (see Figure 6.1) of HDLC are *information transfer* (I), *supervisory* (S), and *unnumbered* (U). The S and U frames carry only control information. The $N(S)$ number in an I frame is the sequence number of the information frame modulo 8. The $N(R)$ number in an I frame or an S frame is the sequence number of the acknowledgment. By convention, the number $N(R)$ signifies that the receiver is expecting the I frame with sequence number $N(R)$. That is, this number $N(R)$ acknowledges the I frames up to $N(R) - 1$ (in the number modulo 8).

U frames are used to connect and disconnect. At the time of the connection setup, the sender can request that the I frames and their acknowledgments be numbered modulo 128 instead of modulo 8.

The S frames can indicate that the receiver is ready to receive (RR), not ready to receive (RNR), or that it rejects (REJ) all frames with sequence number following $N(R)$. The sender must then retransmit these frames.

HDLC uses the P/F bit as follows. An I frame or an S frame RR with $P = 1$ requires a response with an S frame with $F = 1$. Thus, the RR frame with $P = 1$ forces the receiver to send an acknowledgment, which speeds up error detection.

Figure 6.3 shows a typical sequence of frames when the two computers A and B use the asynchronous balanced mode of HDLC. In the top part of the figure, computer A is sending data to computer B. The first frame that A sends has sequence number 0 and requests an immediate acknowledgment. This first frame is an I frame with $P = 1$. When it gets this frame, computer B sends back an acknowledgment with sequence number 1 that acknowledges the frame with sequence number 0. This reply is a supervisory frame which indicates that B is ready to receive more data. Computer A then sends two data frames with sequence numbers 1 and 2. Finally, computer B sends a supervisory frame with $N(R) = 3$ that acknowledges the frames with sequence numbers up to 2.

In the bottom part of Figure 6.3 computers A and B both send data to each other. Computer B sends data in two frames with sequence numbers 0 and 1. Computer A sends data with sequence number 0 and requests an immediate acknowledgment. The procedure continues as in the top part except that the data frames contain acknowledgments.

LAPB

LAPB (*balanced link access procedures*) is a subset of HDLC. The frame format is the same as for HDLC (see Figure 6.1). LAPB is essentially the asynchronous balanced mode of HDLC.

FIGURE 6.3

Sequences of frame exchanges in HDLC connections.

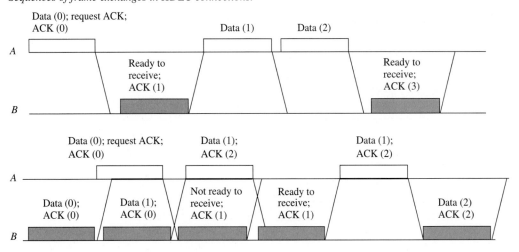

FDDI

We explained in Section 4.4 that FDDI uses the 4B/5B encoding. This encoding converts groups of 4 bits into 5-bit symbols chosen to introduce enough transitions to keep the receiver synchronized.

Some of the 16 possible 5-bit symbols that are not used to encode data are reserved as frame delimiters. Specifically, the stations keep on transmitting IDLE symbols between frames to keep the receivers synchronized. The first symbol different from IDLE indicates the start of the frame. The stations use other reserved symbols to exchange control information.

ATM

We learned in Chapter 5 that ATM cells have 53 bytes: a 5-byte header and a 48-byte information payload. The header contains 4 bytes of routing and flow control information plus a header error control byte. The error control byte HEC, the fifth header byte, is the result of an algebraic operation performed on the previous 4 bytes. Specifically, $HEC = CRC \oplus 01010101$, where CRC is the cyclic redundancy code calculated on the previous 4 bytes with the generator $1'0000'0111$ and where $x \oplus y$ indicates the addition modulo 2 without carry. We explain the CRC calculation later in "Error Detection and Retransmission."

We examine two contrasting examples of how the ATM cells are delineated within a bit stream. We first look at ATM cells transported by a DS-3 line. The DS-3 (digital system—level 3) is designed to transport 672 telephone conversations over a digital link that has a total rate of 44.736 Mbps. When the ATM cells are transported over a DS-3 line, the cells have a fixed position within the DS-3 frames, as shown in Figure 6.4. The receiver locates the successive ATM cells from the boundaries of the DS-3 frames.

Some physical layers transport ATM cells with a movable location within the bit stream. This is the case for the synchronous optical transmission system (SONET) that the telephone network uses for long distance transmissions. (See Chapter 7.) In such a system, to locate the cells within the bit stream, the receiver uses the HEC byte as we explain next. The receiver guesses the start of the cell and checks whether the fifth byte is the checksum of the previous four. Depending on the result of this check, the receiver may start searching for a different start of frame. The algorithm is shown in Figure 6.5. Note that the receiver does not think that it has the wrong cell locations at the first wrong checksum and, conversely, it does

FIGURE 6.4

ATM cells in a DS-3 frame. The bits are read from left to right and top to bottom. The 12 × 4 byte column on the left contains overhead framing bytes.

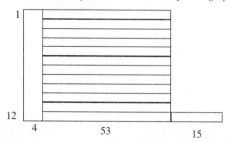

FIGURE 6.5

Algorithm for locating ATM cells from their HEC bytes.

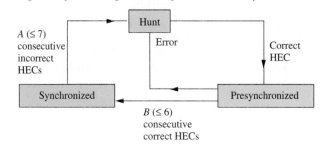

not assume that it has the correct location at the first correct checksum. This mechanism is necessary because of the possible transmission errors and the possibility that a fifth byte can happen to be, by pure chance, the checksum of the previous four.

6.1.2   Error Control

We have learned how bits are transmitted as electromagnetic waves. In addition to signal distortions, random errors can be introduced during the transmission. We explain in this section how such random errors are controlled.

Error Detection or Correction

Some applications require a reliable transfer of user bits across the network. Since transmission errors are unavoidable, such applications necessitate the correction of the errors. Errors are due to missing packets or to packets corrupted by transmission errors.

Networks use two types of error control mechanisms: error correction and error detection with retransmissions. When using error correction, the packets contain enough redundant information—called an *error correction code*—for the receiver to be able to correct the erroneous packets and to reconstruct the missing packets. CD players also use error correction codes to correct errors caused by dust or scratches on the disc.

When error detection is used, each packet contains additional bits—called an *error detection code*—that enable the receiver to detect that transmission errors corrupted the packet. When the receiver gets an incorrect packet, it arranges for the sender to send another copy of the same packet. The sender and receiver follow specific rules, called a *retransmission protocol,* for supervising these retransmissions.

Error correction must be used when it is not possible to retransmit erroneous packets. For instance, in video conferences or in telephone conversations, a retransmission would introduce excessive delays.

Error Correction

A simple error correction coding method is to send every packet three times. Three values are then received for each bit and the receiver uses a majority vote on each bit. Let us evaluate how effective this method is assuming that all the bit errors are independent. The probability that a given bit is incorrectly received two or three times is equal to

$$\epsilon = 3 \times \text{BER}^2(1 - \text{BER}) + \text{BER}^3.$$

Therefore, the probability that at least one of the N bits of the packet is received incorrectly two or three times is equal to

$$\text{PER}^c = 1 - (1 - \epsilon)^N \approx N\epsilon \approx 3N \times \text{BER}^2.$$

For $N = 10^5$ and $\text{BER} = 10^{-7}$, one finds

$$\text{PER}^c \approx 3 \times 10^{-9}.$$

The calculation shows that by repeating every packet three times the packet error rate is reduced from 10^{-2} to 3×10^{-9}. The price paid is that the useful transmission rate has effectively been reduced by a factor of 3. There exist much more efficient error correction coding techniques than simply repeating the packet. Such techniques are based on the following idea. The encoder calculates M bits on the basis of the N packet bits. The resulting $N+M$ bits are called a *codeword*. There are 2^N possible codewords $C(1), C(2), \ldots, C(2^N)$: one for each packet of N bits. Define the *distance* between two codewords as the number of bits which differ in those codewords. The M bits are calculated so that the 2^N codewords are as far apart as possible. Specifically, the M bits are chosen so as to maximize the distance between the two closest codewords. Assume that two codewords differ by at least 5 bits. When the receiver gets the string W of $N + M$ bits, it finds the codeword $C(i)$ which is the closest to W. You can check that if at most 2 bits are incorrectly received, then the codeword $C(i)$ is precisely the codeword which was sent. One says that the code can correct 2 bit errors. More generally, assume that any two codewords differ by at least d bits, as shown in Figure 6.6. This code can correct errors that affect fewer than $d/2$ bits. If errors corrupt more than $d/2$ but fewer than d bits, then the code detects the errors but cannot correct them.

The *Bose-Chaudhuri-Hocquenghem* (BCH) codes are commonly used error correction codes. Let p be a prime number and q an arbitrary power of p. For any $m, t \geq 1$ there exists a BCH code which can correct t bit errors for packets with $N + M = qm - 1$ and

FIGURE 6.6

Code with minimun distance d. This code corrects errors that affect fewer than $d/2$ bits.

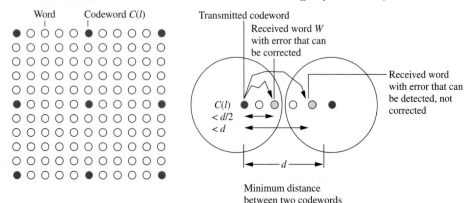

$M < 2mt$. Such a BCH code is designated by BCH$(N + M, N, t)$. Examples of BCH codes are BCH$(1023, 1003, 2)$ and BCH$(1023, 923, 10)$. The BCH codes with $m = 1$ are called the *Reed-Solomon* (RS) codes. RS codes are used in compact discs.

The BCH and turbo codes are block codes: They group the source bits into blocks of M bits that they encode one at a time into blocks of $N + M$ bits. The decoder decodes one block at a time. In a sense, this mode of operation is similar to isolated word recognition when one tries to read a poorly handwritten letter. In contrast, the *convolutional codes* use some memory of previously seen messages to decode the current message. This approach is closer to how we actually recognize handwriting.

We explain the BCH codes and convolutional codes in Section 6.8. We also explain the general structure of turbo codes, a recently developed class of codes and decoding algorithms that exhibit remarkable performance when used over very noisy channels.

Error Detection and Retransmission

Instead of adding enough redundancy to be able to *correct* transmission errors, one can add just enough redundancy to be able to *detect* transmission errors. This method is called *error detection coding*. The advantage of error detection over error correction is that it requires fewer additional bits. For instance, a single parity bit can detect that a bit is not correct in a packet but cannot identify which one. The disadvantage is that the packets which are received incorrectly must be retransmitted. Retransmission may not always be possible, as in a compact disc. Also, retransmission may be too time-consuming, as in video and audio applications and also in data transmissions from deep space probes. Error detection is the error-control method used by all computer networks. Future high-speed networks may use error correction. The packet with its error detection bits is a codeword. If two codewords differ by at least d bits, then the code can detect all errors that corrupt up to $d - 1$ bits. Indeed, a codeword that is transmitted cannot be converted into another codeword by corrupting fewer than d bits. Notice that such a code could correct only errors that corrupt fewer than $d/2$ bits, as we saw in the discussion of error-correcting codes.

The error detection code used by almost all communication networks is the *cyclic redundancy code* (CRC). Such codes detect the occurrence of transmission errors by adding a few bits, called CRC *bits* or *checksum bits,* to each packet. The CRC is a specific method for calculating the bits that are added to a packet. We will explain how the calculations are performed, and we will then discuss the error detection properties of the code.

Let M be a string of m bits. (Typically, m is a few hundreds or thousands.) The string M is the *message* to be transmitted. Another string R of r bits will be added to M. (Typically, $r = 16$ or $r = 32$.) The bits R are the CRC *bits*. One may consider the strings M, R, and $T = MR$ as being binary words, with the leftmost bit being the most significant. For instance, the string 10001011 corresponds to the value $2^7 + 2^3 + 2^1 + 2^0 = 139$. With this interpretation, one sees that the string $T = MR$ corresponds to the value

$$T = M \times 2^r + R.$$

We call this T to remind us that it is *transmitted.*

The procedure for choosing R is as follows:

1. A string G of $r + 1$ bits is first agreed upon. (G is called the *generator*).
2. R is then chosen so that

$$T = A \times G \text{ for some } A.$$

Thus, R is chosen so that T is an exact multiple of the generator G. The idea is that if the received string $\hat{T}$ is found not to be a multiple of G, then the receiver knows that a transmission error must have occurred.

In order to explain the calculations, let us adopt the convention that all the operations on strings of 1s and 0s are done modulo 2 without carry. With this convention, one has $1 + 1 = 0 + 0 = 0$ and $1 + 0 = 0 + 1 = 1$. An addition modulo 2 is the same as an *exclusive or*. Since the operations are performed digit by digit without carry, one has

$$10001011 + 10100010 = 00101001 \quad \text{and} \quad 10001011 + 10001011 = 00000000.$$

Note that these rules are not the same as those of additions in base 2. In base 2, $10+11 = 101$, whereas in operations modulo 2 without carry, one has $10 + 11 = 01$.

Thus, to indicate that the received string $\hat{T}$ differs from the transmitted string T in the two leftmost bits and in the rightmost bit, one can write that $\hat{T} = T + E$ with $E = 1100\cdots001$. The string E is called the *error pattern* since its 1s indicate which bits of T were changed by noise. Now,

$$\hat{T} = T + E = A \times G + E,$$

so that the received message $\hat{T}$ is a multiple of the generator G if and only if the error pattern E is a multiple of G. Indeed, the rules of operations are such that $E = \hat{T} + A \times G$, so that $\hat{T} = B \times G$ if and only if $E = (A + B) \times G$.

The generator G should be chosen so as to minimize the likelihood that the error pattern E is a multiple of G. It turns out that if $G = 1'0001'0000'0010'0001$, then E cannot be a multiple of G if E has fewer than 32,768 bits and if it has one, two, or three 1s. Thus, this choice of G allows us to detect all the single, double, and triple bit errors in messages M of up to 32,752 bits. (In addition, this G detects all the errors that involve an odd number of bits and most errors that affect up to 18 consecutive bits.) The code obtained by using this particular G is called CRC-CCITT.

The calculation of R so that $T = M \times 2^r + R = A \times G$ is based on the observation that this equation is equivalent to

$$M \times 2^r = A \times G + R,$$

so that R is the remainder of the division of $M \times 2^r$ by G (using addition modulo 2 without carry). This remainder can be determined by long division. As a numerical example, Figure 6.16 details the calculation that shows that if $r = 3, G = 1011$, and $M = 1101$, then $R = 001$ so that the transmitted string is $T = 1101001$.

In communication devices, these calculations can be performed by dedicated integrated circuits, as we explain in Section 6.8.

Error Control Summary

- Noise introduces transmission errors. Such errors limit the rate at which bits can be sent reliably over a given channel. The maximum rate of reliable transmissions is called the *channel capacity*.
- *Error detection* is performed by adding enough error-control bits for the receiver to detect the occurrence of transmission errors. Incorrectly received packets can be retransmitted. The CRC is a commonly used error detection code.

- *Error correction* is performed by adding more error control bits so that the receiver can correct the corrupted packet. Error correction is preferable to error detection when the bit error probability is large or when retransmissions are not possible or practical.

6.2 Retransmission Protocols

In Chapter 2 we described how a store-and-forward packet switching network transmits a bit stream or a bit file from one network user to another. The source first divides the bit stream or bit file into *packets.* A packet is a group of bits with a specific format. The format specifies the position of the control and user bits in the packet and the meaning of the control bits. The data link layer performs the *framing* of the packets.

Each link along a path from the sending user to the receiving user transmits the packet in turn. The data link layer uses the physical layer of the link to transmit the bits of the packet. Thus, the data link layer converts the bit pipe that the physical layer implements into a *packet link,* i.e., a facility for transmitting packets. The data link layer implements this packet link between nodes that are attached directly to the same physical layer and therefore to the same physical transmission medium.

6.2.1 Link or End-to-end Control

Communication networks control the errors either link by link or end to end. When using link error control, the data link layer of each link implements a reliable packet delivery. The transmitter of each link calculates error detection bits that the receiver verifies. If the packet it receives is incorrect, the receiver arranges for the transmitter to send a copy.

When using end-to-end control, the links simply discard incorrect packets without arranging for retransmissions. Consequently, some packets may not arrive at the destination. The destination asks the source to retransmit the missing or incorrect packets. In the OSI terminology, the transport layer implements the end-to-end error control.

As a rule, end-to-end control is preferable to link control when the packet error rate of the links is very small. Indeed, in that case, the verification of the packets by each link along the path is an excessive burden; it is more efficient to retransmit from end to end the rare packet that was corrupted or missing at the destination instead of slowing down every packet at every link.

If the packet error rate is significant, then most packets would not make it to the destination without errors and it is preferable to check every packet at every link, like a careful accountant verifying every step of a long calculation.

Wireless links generally have a large error rate. It is therefore sensible to perform error control on such a link. Optical fiber links usually have a very small error rate. Consequently, networks that use such links often use end-to-end error control. If a connection goes over some wireless links and other optical links, the sensible error control strategy is to implement a link error control on each wireless link in addition to an end-to-end control of the connection.

Networks use the same protocols for end-to-end and for link control. Thus, although we discuss these protocols in the data link chapter, keep in mind that some networks use them end to end at the transport layer. We call these protocols *retransmission protocols* to dissociate them from the data link layer and to emphasize the possibility of implementing them end to end.

6.2.2   *Retransmission Protocols: Preview and Summary*

Most networks use one of four retransmission protocols: the *stop-and-wait protocol* (SWP), the *alternating bit protocol* (ABP), the *selective repeat protocol* (SRP), and the *GO BACK N* protocol (GBN). These four protocols are based on the same mechanisms: timers and acknowledgments. To know which packets it should retransmit, the sender starts a count-down timer whenever it transmits a packet. The receiver transmits an acknowledgment for every correct packet it receives. The sender assumes that transmissions corrupted a packet when its acknowledgment is late to arrive—that is, if the acknowledgment does not reach the sender before the timer that corresponds to that packet expires. Some retransmission protocols use variations on these basic mechanisms. We discuss some of these variations.

Correctness and Efficiency

Two characteristics of a retransmission protocol are important: *correctness* and *efficiency*. A retransmission protocol is correct if it delivers exactly one correct copy of each packet. The receiver may get multiple correct copies of a packet but it must be able to recognize the copies so that it forwards only one copy to the next node or to the higher protocol layer. When proving that a retransmission protocol is correct we always assume that *transmissions are eventually successful*. That is, if a sender keeps on sending copies of a packet, then the receiver eventually gets a correct copy. Similarly, if the receiver keeps on sending acknowledgments, then the sender eventually gets a correct acknowledgment.

The efficiency is the maximum average rate at which the protocol delivers correct packets divided by the channel packet transmission rate. For instance, a SONET transmission link at 155 Mbps can transmit approximately 330,000 ATM cells per second, taking into account the fraction of the rate used up by the SONET overhead bytes. A given retransmission protocol might transmit only 200,000 ATM cells per second, even when the source always has packets to transmit. In this example, the efficiency of the protocol is close to 60 percent. One should be aware that the efficiency of a retransmission protocol depends on the channel rate. Thus, a protocol whose efficiency is 85 percent when the channel rate is 149 Mbps might have an efficiency of only 60 percent once we replace the transmitters to have a channel rate of 596 Mbps.

The efficiency depends on various parameters such as packet lengths, channel rates and delays, and processing times. The efficiency also depends on the packet error rate (PER) of the channels. Most channels have a small PER and the efficiency of a retransmission protocol is then approximately the same as if there were never any errors. In practical terms, the PER reduces the efficiency by only a few percent. Thus you do not miss out much by only studying the efficiency of retransmission protocols assuming that PER $= 0$. However, it is essential for the protocol not to fail when an occasional transmission error happens.

Hence, it is important to understand how to verify that a retransmission protocol is correct in the presence of transmission errors.

The four protocols that we examine differ in how the sender and receiver keep track of the acknowledgments and packets they are waiting for. The protocols differ in complexity and in efficiency. The increasing order of complexity and efficiency of the protocols is SWP, ABP, GO BACK N, SRP.

In the subsequent sections we examine the details of these protocols and we explain how to prove their correctness and to evaluate their efficiency. Each section starts with a summary of the operations of the protocol. We then discuss the correctness and the efficiency of the protocol.

Retransmission Protocol Summary
Here is a summary of our discussion:

- The objective of a retransmission protocol is to implement a reliable packet channel over an unreliable one.
- Packets that are not received correctly are retransmitted. The sender is informed of transmission errors by timers and acknowledgments.
- Retransmission protocols are correct if they enable the receiver to get exactly one correct copy of each packet.
- The efficiency of a retransmission protocol is the maximum average rate at which it delivers correct packets divided by the channel packet transmission rate.

In Sections 6.3 through 6.6, we describe four commonly used protocols. In addition to describing how packets are transmitted, received, and acknowledged between two nodes in the protocol, we measure the efficiency of each protocol and compare their advantages and disadvantages.

6.3 Stop-and-Wait Protocol (SWP)

6.3.1 Summary of Operations

The simplest retransmission protocol is the *stop-and-wait protocol* (SWP). Whenever the receiver gets a correct packet, it transmits an acknowledgment back to the sender. The sender sends a copy of the packet if it does not get the acknowledgment within T seconds. The acknowledgments are unnumbered.

We explain that this protocol is correct when used over a channel with a known bound T on the acknowledgment delay. That is, the acknowledgment of a packet never reaches the sender more than T seconds after the sender started sending the packet. The sender knows the value of T. It may be that an acknowledgment does not come back, because transmission errors corrupted either the packet or the acknowledgment or because a node dropped the packet or the acknowledgment.

Figure 6.7 shows a typical sequence of operations of SWP. This protocol is an academic version of the actual protocol that some networks use. For instance, TFTP in Internet uses this protocol but it numbers the packets and their acknowledgments. In practice, an upper

FIGURE 6.7

Typical operations of SWP.

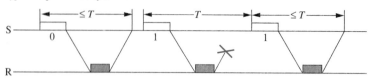

bound on the delay of acknowledgment is dangerous to assume, for obvious reasons. We describe this protocol because its analysis provides a simple illustration of the main issues that are relevant for all the retransmission protocols.

Note that we call sender (S) the node that transmits the packets and receiver (R) the node that waits for the packets and transmits the acknowledgments. The channel between the sender and the receiver is either half-duplex or full-duplex. If the channel is half-duplex, then the packets and acknowledgments cannot propagate at the same time. The situation is then similar to trains sharing a single track between two stations.

Using the stop-and-wait protocol, the sender appends a sequence number to each packet. The sequence numbers can be alternately 0 and 1. The sender transmits a packet and waits for up to T seconds for the acknowledgment. When it gets the acknowledgment, the sender moves on to the next packet. If the acknowledgment does not arrive within T seconds, then the acknowledgment will never arrive (by assumption). In that case, the sender repeats the procedure with a copy of the packet: it transmits that copy and waits for the acknowledgment for up to T seconds. The sender would be confused if an acknowledgment were to arrive after T seconds. Note that the receiver would be confused if the packets were not numbered. For instance, in Figure 6.7, the receiver could not tell whether the third packet is new or is a copy of the previous packet if the packets were not numbered.

6.3.2 Correctness

As a warm-up exercise for the rest of the chapter let us show that SWP is correct provided that transmissions are eventually successful and that the sender knows a bound T on the acknowledgment delay.

We don't need a formal description to understand why the protocol is correct. Eventually, the sender gets an acknowledgment of packet 1 and can then send packet 2. In the meantime, it may happen that the receiver gets a few correct copies of packet 1. When that happens, the receiver knows that these are all copies of the first packet because they all have the same sequence number (0). Thus, the receiver can keep (or forward) exactly one copy of packet 1 and discard the others. The situation then repeats for the subsequent packets. The sender cannot get confused because either it gets an acknowledgment within T seconds or it never gets the acknowledgment. Thus, after T seconds, the system is back to square one, so to speak: the situation is the same as at time 0, except that the receiver may have received a correct version of packet 1. By attributing alternatively the sequence numbers 0 and 1 to the packets, the receiver can distinguish copies of a packet it received correctly previously from a new packet.

As you can imagine, this sort of verbal argument quickly becomes less than convincing as soon as the protocol is more complicated. It is then easy to overlook possible sequences of events. Protocol experts have developed systematic methods for specifying protocols and for verifying that they are correct. We use this particularly simple protocol to explain one of those methods.

The method consists in listing the possible *states* of the system and to explore the possible transitions between those states. This method is called the *finite-state machine* analysis of the protocol. In principle, this method can be used to analyze a protocol with an arbitrary complexity.

For SWP, we can summarize the state by a pair (x, y) of numbers 0 or 1. The state is observed when the sender transmits a new packet or a copy. The value of x is 0 if the sender last transmitted packet n for some odd value of n; $x = 1$ otherwise. The value of y is the sequence number of the packet that the receiver is currently waiting for. When the sender transmits the first packet with sequence number 0, one has $(x, y) = (0, 0)$. If the receiver gets the packet correctly, then y jumps to 1 since the receiver now waits for the next packet with sequence number 1. Thus, the state jumps from $(0, 0)$ to $(0, 1)$ when the receiver gets the first correct copy of the packet with sequence number 0. (See Figure 6.8.) If the acknowledgment of that packet reaches the sender correctly, then eventually the state jumps to $(1, 1)$ when the sender transmits the next packet. If the acknowledgment does not reach the sender, then after T seconds the sender transmits a copy of the packet with the same sequence number. If the first packet does not reach the receiver, then the state is again $(0, 0)$ when the sender transmits another copy of the packet. By exploring the similar possibilities when the sender is sending a packet with sequence number 1, one concludes that the possible transitions between the four states are as shown in Figure 6.8.

The figure shows that SWP is correct. Indeed, for $x = 0$ and $x = 1$, let $S(x)$ denote the event "sender transmits new packet with sequence number x" and $R(x)$ the event "Receiver gets first copy of new packet with sequence number x." The figure shows that the events follow each other in the sequence $S(0), R(0), S(1), R(1), S(0), R(0), \ldots$. The receiver would miss a packet if the sequence $S(0), S(1), R(1)$ could occur. The receiver would get two copies of a packet and mistake them for new packets if the sequence $S(0), R(0), R(0)$ could occur.

FIGURE 6.8

Finite-state machine model of SWP.

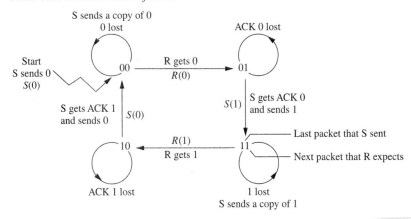

When the protocol is complex, checking that the sequences of transitions are all acceptable as we just did becomes tedious and protocol researchers have designed automatic ways of doing the verification. We explain one such method here. We construct an auxiliary system that monitors the evolution of SWP. We call this system a *monitor*. The monitor has three states: A, B, and E. The state of the monitor changes when SWP makes a transition $S(0)$, $R(0)$, $S(1)$, or $R(1)$, according to the rules shown in Figure 6.9. Initially, the monitor is in state A. The monitor detects an illicit sequence of states of SRP whenever its state moves to E. The meaning of state A is that the sender last transmitted a new packet with sequence number 0. Thus, the protocol is incorrect if, when the monitor is in state A, the receiver gets a new packet with sequence number 1. The situation is symmetric for state B. To check that the monitor cannot reach state E, we combine the descriptions of SWP and of its monitor into a graph with states (x, y, z) where z is the state of the monitor. We show that graph in the center part of Figure 6.9. We see from the figure that there is no state (x, y, E), so that there is no possible error.

Even checking that some states cannot be reached can be a complicated operation that requires an automatic procedure. Here is an algorithm that produces the set R of all the states that can be reached. First R contains only the initial state. We then add to R all the states that can be reached in one step from the states in R. In this basic form the algorithm never stops. We need to keep track of states that we have seen before so that we do not examine them forever. One way to do this checking is to use an auxiliary set V of states that the algorithm has yet to explore. The steps of the algorithm are listed below:

$R = V = \{\theta_0\}$.

Repeat until $V = \emptyset$: Pick $\theta \in V$. Let $L(\theta)$ be the set of states that can reached in one step from θ. $V = V \cup L(\theta) - R$, $R = R \cup L(\theta)$.

For the system in Figure 6.9, the reachability algorithm has the following steps:

$R = V = \{00A\}$.
$\theta = 00A, L(\theta) = \{01A\}, R = \{00A, 01A\}, V = \{01A\}$.
$\theta = 01A, L(\theta) = \{11B\}, R = \{00A, 0, A, 11B\}, V = \{11B\}$.
$\theta = 11B, L(\theta) = \{10B\}, R = \{00A, 01A, 11B, 10B\}, V = \{10B\}$.
$\theta = 10B, L(\theta) = \{00A\}, R = \{00A, 01A, 11B, 10B\}, V = \emptyset$.

These steps are shown in the right part of Figure 6.9.

FIGURE 6.9

Monitor of SWP (left), joint FSM (center), and reachability algorithm (right).

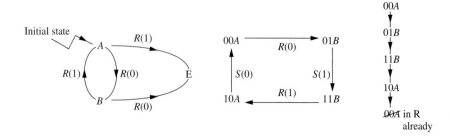

To summarize, we described the evolution of SWP by a graph that shows the possible state transitions. We added a monitor whose state changes depending on the transitions of SWP. Finally, we used a reachability algorithm to check that the monitor cannot reach an error state.

6.3.3 Efficiency

The efficiency depends on the packet error rate and on the actual delay of the acknowledgments. Assume first that there are no transmission errors and that every acknowledgment arrives at the sender exactly S seconds after the start of the transmission of the packet. Then the protocol transmits one packet every S seconds.

The impact of transmission errors is analyzed as follows. Assume that with probability $1 - p$ either the packet does not reach the receiver or the acknowledgment does not make it back to the sender. Assume also that if the acknowledgment gets to the sender (with probability p), then it does so in exactly S seconds. Under these assumptions, the time X that the sender takes between the first transmission of packet n and that of packet $n + 1$ is a random variable. To analyze X, we note that $X = S$ with probability p, when the acknowledgment gets back to the sender. With probability $1 - p$, the acknowledgment does not come back. In that case, after T seconds, the sender restarts its attempt to deliver packet n. These possibilities are illustrated in the left part of Figure 6.10. That is, with probability p, one has $X = T + Y$, where Y is the time that the sender takes from that second transmission of packet n until the transmission of packet $n + 1$. Consequently, we have

$$E(X) = pS + (1 - p)\{T + E(Y)\}.$$

Since the transmission errors after the second transmission have the same likelihood as after the first one, the random time Y had the same statistics as the random variable X. Indeed, both X and Y are the random time until the acknowledgment of a packet is correctly received, measured from the the start of the transmission of the packet. In particular, $E(X) = E(Y)$ and we can rewrite the previous equation as

$$E(X) = pS + (1 - p)\{T + E(X)\}.$$

Solving this equation for $E(X)$, we find

$$E(X) = S + T\frac{1 - p}{p}.$$

FIGURE 6.10

Analysis of the efficiency of SWP.

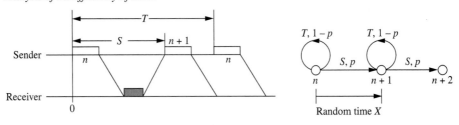

Thus, under the assumptions SWP transmits one packet every $E(X)$ seconds, on average. Consequently, the transmission rate—or *throughput*—is $1/E(X)$.

The efficiency of SWP is equal to its throughput $1/E(X)$ divided by the line rate $1/\tau$, where τ is the transmission time of a packet. That is, the efficiency η_{SWP} is equal to

$$\eta_{\text{SWP}} = \frac{\tau}{S + T(1-p)/p}. \tag{6.1}$$

The minimum value of T for SWP is equal to S because the transmitter waits until it knows that no acknowledgment can arrive before it retransmits a packet.

6.4 Alternating Bit Protocol (ABP)

6.4.1 Summary of Operations

The alternating bit protocol is similar to SWP. The key difference is that the sender does not need a bound T on the arrival time of acknowledgments. The trick to make ABP work is to add a sequence number to the acknowledgments.

The sender numbers the packets successively $0, 1, 0, 1, \ldots$ by adding a one bit sequence number to every packet. When the receiver gets a correct packet with sequence number 0 or 1, it transmits an acknowledgment with the same sequence number 0 or 1 to the sender. The sender chooses some time value T and starts a countdown timer with that initial value T, then transmits a packet with number 0 while keeping a copy. The sender then waits for the acknowledgment with number 0. If such an acknowledgment does not arrive within T seconds, then the sender repeats the procedure with a copy of the same packet. If the acknowledgment with number 0 arrives within T seconds, then the sender repeats the procedure with another packet to which it appends the sequence number 1. Figure 6.11 shows a sequence of operations of ABP.

The sender uses the sequence numbers to distinguish the acknowledgment of the last packet it transmitted from a late acknowledgment of a packet it transmitted previously. The receiver uses the numbering to distinguish a new packet from subsequent copies that the sender transmits because of late acknowledgments.

6.4.2 Correctness

Although it is not a surprising result, it is slightly harder than for SWP to show that ABP is correct. We must assume that the packets arrive at the receiver in the order that the sender transmits them, and similarly for the acknowledgments. Also, we assume that the receiver acknowledges the packets in the order it receives them. We call these assumptions the *FIFO assumptions* (first in–first out). Finally, we assume that the transmissions are eventually successful. Note that these assumptions prevent using ABP across a datagram network that is not FIFO such as Internet. Since the result is intuitive, we give its proof as a complement in Section 6.9.

ABP is not correct when used over a channel that is not FIFO. To verify this fact, consider the following sequence of events:

$$P1,\ P1',\ A1,\ P2,\ A2,\ P3,\ L3,\ A1'.$$

FIGURE 6.11

Alternating bit protocol. Packets and acknowledgments are numbered alternately 0 and 1.

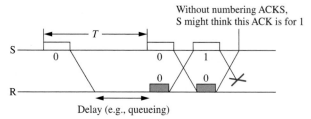

In this sequence, Pn means that the sender transmits packet n, $P1'$ that the sender transmits a copy of packet 1 (after not getting the acknowledgment within T), An that the sender gets the acknowledgment of packet n, and $L3$ that packet 3 is lost. Using the ABP numbering, the sender would confuse $A1'$ with $A3$ and would not retransmit packet 3. In this example, the PTA is not FIFO since its inputs are 1123 and its outputs are 121.

If the retransmission protocol is implemented on a single link, then the PTA is FIFO. The PTA is also FIFO if the retransmission protocol is implemented from end-to-end in a network that uses virtual circuit switching, provided that the receiver acknowledges the packets in the order it receives them. Transmission channels that are not FIFO can arise in datagram networks when the nodes implement the retransmission protocol end to end. Given that such networks are widely used, retransmission protocols that are correct for such channels are important. The ABP protocol is correct when the sender and transmitter number the packets and ACKs sequentially. The sender and receiver can use such a numbering by adding the sequence number (m, n) to the nth packet of a connection m. This procedure is correct provided that the sender and receiver have a fail-safe procedure for agreeing on the connection number m. One such procedure is the *three-way handshake* as we explained when we discussed TCP.

6.4.3 *Efficiency*

Since the sender can select a smaller value of T when using ABP than when using SWP, it does not have to wait as long before retransmitting a packet. Consequently, ABP has a larger efficiency than SWP. For instance, if the channel is full-duplex then the sender can choose T equal to a transmission time. With this choice, the sender transmits back-to-back copies of the first packet with sequence number 0. As soon as it gets an acknowledgment of packet 0, the sender starts sending back to back copies of the next packet with sequence number 1.

To calculate the efficiency of ABP, we assume that when the sender sends a packet, either the acknowledgment comes back exactly after S time units or it does not ever come back. With these assumptions, we obtain the same result as for SWP. That is, we find

$$\eta_{ABP} = \frac{\tau}{S + T \times (1 - p)/p}. \tag{6.2}$$

However, using ABP, the sender does not have to choose T large enough to exceed the maximum delay of an acknowledgment. Accordingly, by choosing a smaller value of T, the ABP protocol is more efficient than SWP. For instance, if the line is full-duplex, then the transmitter can choose $T = \tau$ and send packets back to back.

In an actual network, the delay between the transmission of a packet and the return of its acknowledgment is not always equal to the same constant S. Instead, that delay varies from packet to packet and we can model it as some random variable.

From our discussion of ABP, you can see that we could make better use of the network resources with a protocol that can transmit a packet without having to wait for the preceding one to be acknowledged. The GO BACK N protocol has that feature; we discuss it in the next section.

6.5 GO BACK N (GBN)

6.5.1 *Summary of Operations*

The GO BACK N protocol has a larger efficiency than ABP because the sender sends a number of packets before getting the first acknowledgment. Using GBN, the sender chooses an integer W called the *window size* and a time value $T \geq W \times \tau$ where τ is the transmission time of a packet. The transmitter must send the packets 0, 1, 2, ... and appends the sequence number n to packet n ($n \geq 0$); when the receiver gets a correct packet with sequence number n, it transmits an acknowledgment with sequence number n to the sender.

The sender first transmits packets $0, 1, 2, \ldots, W - 1$ and waits for up to T seconds for each of their acknowledgments. As soon as the sender gets the acknowledgment with sequence number 0, it transmits packet W. When it gets the acknowledgment of packet 1, the sender transmits packet $W + 1$, and so on. If the sender fails to get the acknowledgment of packet number $n + 1$ within T seconds after sending that packet, it repeats the procedure starting with packet $n + 1$. That is, the sender transmits the packets $n + 1, n + 2, \ldots, n + W$ and then waits for their acknowledgments. Figure 6.12 shows a sequence of transmissions of GBN.

Since the sender does not have to wait for the first acknowledgment before sending the second packet, it is clear that the efficiency of GBN is larger than that of ABP.

FIGURE 6.12

GO BACK N. The window size is equal to 4. When the acknowledgment of a packet (here packet 0) fails to arrive within a timeout value, the sender starts retransmitting that packet and all the subsequent packets.

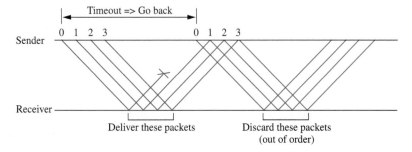

We analyze the correctness of GBN in Section 6.10. We show that if the channels between the sender and the receiver are FIFO, then the sender can number the packets successively $0, 1, \ldots, W, 0, 1, \ldots, W, 0, 1 \ldots$ instead of numbering them sequentially $0, 1, 2, \ldots, W - 1, W, W + 1, W + 2, \ldots$. That is, the sender and receiver can number the packets modulo $W + 1$.

If GBN is used over channels that are not FIFO, then the sender can number the packets modulo $W + 2M + 1$ where M is a bound on the degree of overtaking of the channels. Note that GBN for $W = 1$ is ABP.

Negative acknowledgments (NACKs) can be used to speed up the GBN protocol. A negative acknowledgment (NACK) is transmitted as soon as the receiver realizes that it received an incorrect packet. This occurs when the receiver notices a gap in the numbering of the correct packets it receives. The sender can then retransmit a copy of the corrupted packet immediately without having to wait for the timeout, as would be the case without NACKs.

6.5.2   Efficiency

We expect GBN to be more efficient than ABP. Assume that every acknowledgment of a packet reaches the sender exactly S seconds after the sender started sending the packet. Denote by τ the transmission time of a packet. We first analyze the efficiency of the protocol assuming that there are no transmission errors. The sender transmits W packets in $W \times \tau$ seconds.

If $S \geq W \times \tau$, the sender transmits packets $0, 1, \ldots, W - 1$ then waits for the first acknowledgment which arrives after S seconds. In the absence of errors, the sequence of operations then restarts with packet W at time S. Thus, every S seconds, the sender transmits W packets. The average transmission time per packet is therefore equal to S/W instead of being equal to τ in the ideal case. We conclude that the efficiency of GBN is equal to $W \times \tau/S$.

If $S < W \times \tau$, then the sender keeps on sending packets and the efficiency of the protocol is 100 percent. Consequently, the efficiency η_{GBN} of GBN, in the absence of errors, is equal to

$$\eta_{\text{GBN}} = \min\left\{1, \frac{W \times \tau}{S}\right\}. \tag{6.3}$$

Note that, as long as $T \geq W$, the value of T does not affect the efficiency when there are no errors since the transmissions of new packets are governed by when the acknowledgments reach the sender.

Transmission errors reduce the efficiency, obviously. We analyze the efficiency of GBN in the presence of errors in Section 6.12. As we stated in the first section of this chapter, the packet error rate of most modern networks that use GBN is usually very small so that the efficiency is close to its zero-error value. An exception to this rule is a noisy or fading wireless link.

6.5.3   Concrete Examples: Choosing W

To appreciate the relevance of the formula (6.3), we examine two concrete examples. First consider a wireless link with a distance of 20 km and a transmission rate of 13 kbps. Let us assume that the packets have 1000 bits and acknowledgments 100 bits. With these characteristics, the one-way propagation time is equal to $(20 \text{ km})/(3 \times 10^5 \text{ km/s}) = 6.67 \times 10^{-5}$ s. This propagation time corresponds to $(6.67 \times 10^{-5}) \times (13 \times 10^3) = 0.86$ bit transmission times. Accordingly, $S = 1000 + 100 + 2 \times 0.86 \approx 1102$ bit transmission times. Also, $\tau = 1000$ bit transmission times. Consequently, we find that the efficiency of GO BACK N is equal to

$$\min\left\{1, \frac{W \times 1000}{1102}\right\}.$$

That is, if $W = 1$, then the efficiency is approximately equal to 90 percent. If $W = 2$, then the efficiency is equal to 100 percent. For this system, ABP would be 90 percent efficient and the more complex GO BACK N would increase the efficiency to 100 percent.

As another example, consider an optical link of 50 km with transmission rate of 2.4 Gbps and with packets and acknowledgments of 424 bits. For this link, τ corresponds to $(50 \text{ km}) \times (5 \text{ }\mu\text{s/km}) \times (2.4 \times 10^9 \text{ b/s}) = 6 \times 10^5$ bit transmission times. Accordingly, $S = 2 \times 424 + 2 \times 6 \times 10^5 \approx 1.2 \times 10^6$ bit transmission times. The efficiency of GO BACK N or SRP with window size W for this link is

$$\min\left\{1, \frac{W \times 424}{1.2 \times 10^6}\right\}.$$

This formula shows that the efficiency of 100 percent is achieved for a window size W equal to $1.2^6/424 = 2831$.

The conclusion of these examples is clear. GO BACK N is much more efficient than ABP in situations where the round-trip time S is significantly larger than the transmission time τ. This is more likely to be the case when the transmission rate is large, packets are rather small, and the propagation distance is large. If the errors are controlled from end to end in a store-and-forward network instead of on a link-by-link basis, then GO BACK N is always substantially more efficient than ABP.

6.5.4 *Adapting to Network Delays*

From our analysis of the efficiency we learned that the window size W should be at least large enough for the sender to keep transmitting when there is no transmission error. That is, the window size should be at least equal to S/τ. In end-to-end implementations, S is never known ahead of time. The protocol must adapt the window size to actual operating conditions.

6.5.5 *GO BACK N Summary*

Let us summarize what we have learned about GO BACK N:

- The efficiency of GO BACK N is larger than that of ABP. No buffering is required at the receiver.
- The packets are numbered consecutively $0, 1, 2, \ldots, W, 0, 1, 2, \ldots, W$, and so on (if the channels are FIFO).
- The receiver acknowledges a correct packet with an ACK that has the same number as the packet. The receiver delivers packets in order, and it discards packets received out of order.

- The sender can have up to W unacknowledged packets.
- When the sender does not receive the acknowledgment of a packet before a timeout, it retransmits copies of that packet and of all the subsequent packets.

6.6 Selective Repeat Protocol (SRP)

6.6.1 Summary of Operations

You may have noticed that the GO BACK N protocol may retransmit packets that the receiver gets correctly and whose acknowledgments reach the sender correctly. For instance, if packet 1 is corrupted but the subsequent packets $2, 3, \ldots, W$ and their acknowledgments are correct, the sender nevertheless retransmits all these packets.

These retransmissions are not necessary and slow down the protocol. The last protocol, SRP, eliminates the unnecessary retransmissions. Here is how it works. The protocol uses some window size W and a time value $T \geq W \times \tau$. Packet n is labeled with the sequence number n and so is its acknowledgment that the receiver transmits whenever it gets packet n correctly ($n \geq 0$).

The sender first transmits the packets $0, 1, \ldots, W - 1$. (See Figure 6.13.) The sender keeps track of each acknowledgment it gets within T seconds after it transmitted the packet. At some arbitrary time, denote by L the largest integer so that the sender has received all the acknowledgments with sequence numbers $0, 1, \ldots, L$. At that time, the sender is allowed to send the packets $L + 1, L + 2, \ldots, L + W$. If the sender fails to get the acknowledgment of packet n within T seconds, it retransmits that packet. (But not the subsequent ones as it would using GBN.)

Denote by R the largest integer so that the receiver has received packets $0, 1, \ldots, R$ correctly. At that time, if the receiver gets a packet with sequence number in $\{R + 1, R + 2, \ldots, R + W\}$, then it stores the packet if it is out of order and delivers it if it arrives in order. If the receiver gets a correct packet whose sequence number is not in that set, then it should acknowledge that packet but discard it. Thus, the receiver needs to be able to store up to $W - 1$ packets that arrive out of order.

FIGURE 6.13

Selective repeat protocol. The receiver stores correct packets that it gets out of order and delivers them in order.

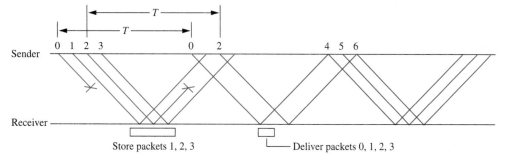

If the channels between the sender and the receiver are FIFO, then the nodes can number the packets and acknowledgments modulo $2W$. That is, the sequence numbers are $0, 1, \ldots, 2W - 1$. Figure 6.14 shows that a smaller set of sequence numbers does not work. In that example $W = 4$ and the sender numbers the packets modulo 7 instead of modulo 8. The receiver cannot tell whether the last packet it receives is a new packet (left) or a copy of a previous packet (right).

We explain in Section 6.9 how the numbering must be modified when the channels are not FIFO.

6.6.2   Efficiency

In the absence of errors, and under the same assumptions, we find that the efficiency of SRP is identical to that of GO BACK N, i.e., is given by (6.3), which we repeat here for convenience:

$$\eta_{\text{GBN}} = \min\left\{1, \frac{W \times \tau}{S}\right\}.$$

Transmission errors reduce the efficiency of the protocol.

6.6.3   Correctness

Consider SRP with explicit numbering. Assume that the channels between the sender and receiver are FIFO. Our induction hypothesis $H(L + W)$ is that if the sender last received the acknowledgment $L+W$, then the sender can transmit packets $\{L+W+1, L+W+2, \ldots, L+2W\}$, the channels can contain at most packets $\{L, L+2, \ldots, L+2W\}$. If $H(L+W)$ is true, then eventually the sender gets acknowledgment $L+W+1$. At that time, the transmitter can transmit packets $\{L+W+2, L+W+2, \ldots, L+2W+1\}$ and there cannot be any packet L in the channels. Indeed, the sender transmitted packet $L + W + 1$ after all copies of packet L. Thus, the channels can now contain at most packets $\{L + 1, L + 3, \ldots, L + 2W + 1\}$. Hence $H(L + W)$ implies $H(L + W + 1)$. Since $H(W)$ is trivially true, it follows that $H(n)$ is true for all n. Since the numbering modulo $2W$ can distinguish all the packets in the channels, this numbering suffices to prevent confusions. (We leave the details to the reader.) Hence, SRP is correct with numbering modulo $2W$ if the channels are FIFO.

FIGURE 6.14

Failure of SRP when the set of sequence numbers is too small.

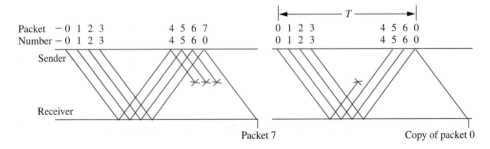

6.6.4   Selective Repeat Protocol Summary

Let us summarize what we have learned about SRP:

- SRP is more efficient than ABP because the sender does not have to wait for a packet to be acknowledged before sending the next one. SRP is more efficient than GO BACK N because the sender retransmits only packets whose acknowledgments were late.
- The packets are numbered modulo $2W$.
- The $(n + W)$th packet cannot be transmitted while the nth packet is still unacknowledged.
- The sender retransmits a copy of a packet that has not been acknowledged before a timeout.
- The receiver acknowledges the reception of a correct packet with an ACK that has the same number as the packet.
- The receiver stores up to W packets it receives out of order, and it delivers the packets in order. More precisely, if the receiver got the packets correctly up to packet R, then it can store the packets $\{R + 1, \ldots, R + W\}$.

6.7   Examples

We conclude this chapter by discussing the data link layers of some widespread communication networks. We learned that the data link layer converts the bit pipe provided by the physical layer into a packet link. We have focused on the *data link protocol,* which regulates the transfer of packets and supervises the retransmissions of copies of corrupted packets. We saw that packets must be numbered and that the receiver transmits numbered acknowledgments to the sender. Implicit in that discussion is the need for specifying the *structure* of the packets. That is, the network designer must specify the number of bits in the packet, the position of the bits that encode the packet number, and how acknowledgments are distinguished from data. We will indicate the structure of the packets in some networks.

6.7.1   Data Link of SNA

SNA is IBM's *Systems Network Architecture.* The first version was released in 1974. SNA, a store-and-forward network, provides communication between IBM computers and peripherals. An SNA network consists of a collection of interconnected *nodes* (terminals, device controllers, communication processors, and computers). Each node contains one or more *network addressable units* (NAUs). The NAUs are grouped into *domains.* Each domain contains one special NAU called a *system services control point* (SSCP) that supervises it.

A user process, i.e., an active program, that wishes to use the network must connect itself to an NAU, which can then be thought of as being a "socket" or plug into the network. Different NAUs have different addresses.

SNA was designed before ISO's OSI model, and as a result, its architecture differs from that of the OSI model. The *data link control* of SNA converts a stream of bits into packets and takes care of error recovery. It produces a reliable sequence of packets from an unreliable bit stream. Thus, the functions of the data link control are similar to those of OSI's data link layer. SNA's data link control protocol is *SDLC* (synchronous data link control). It is GO BACK N with positive and negative acknowledgments. The SDLC packet structure is indicated in Figure 6.1. It is a bit-oriented synchronous packet (see Section 6.1).

The *control field* is used for the sequence number and for acknowledgments and for specifying the type of the packet. The CRC (cyclic redundancy code, see Section 6.1) field contains error detection bits. Neighboring nodes in SNA are usually connected by multiple channels (called a *transmission group*) for reliability and increased bandwidth. The data link control selects one available link for each packet. Packets are resequenced at each node.

6.7.2   Data Link Layer of Public Data Networks (X.25)

By X.25 one denotes the CCITT standards for layers 1, 2, and 3 of *public data networks.* These are networks that are accessible to the general public, in contrast to private networks. *Telenet* and *Tymnet* are examples of U.S. public data networks. The X.25 recommendations, first published in 1974 and periodically revised since, specify the interface required between the computer and the network and the protocols to be used at layers 2 and 3. The recommended physical layer is X.21, which is similar to RS-232-C. The physical layer RS-232-C is also supported by X.25.

The data link layer of X.25 is known under the name *LAPB* (link access protocol B), and it is similar to SNA's data link protocol. The data link layer of X.25 uses the GO BACK N protocol with both positive and negative acknowledgments. The window size is 8 or 128. The window size 128 is used for satellite links and for slow radio channels. X.25 uses synchronous bit-oriented transmissions with the frames that we explained in Section 6.1.

6.7.3   Retransmission Protocol in Internet

Internet uses mostly optical links with a very low bit error probability. Consequently, the Internet does not perform error control on each link. Instead, the Internet TCP/IP protocol suite specifies that the retransmission protocol is implemented by TCP and not at the link layer. (This should not prevent wireless links from implementing a link level error control, obviously.)

The TCP retransmission protocol is essentially SRP with a window that is adapted to the network delays. TCP acknowledges previous packets by indicating the sequence number of the next byte that the receiver expects. We examine the specific retransmission protocol of TCP in more detail in Section 6.11.

6.7.4   Data Link Layer in Frame Relay

As in Internet, frame relay drops packets with CRC errors. It is up to a higher protocol to retransmit.

6.7.5   *XMODEM*

XMODEM is a popular data link protocol for communications via modems between two personal computers or between a personal computer and a mainframe. XMODEM is similar to ABP. It is rather primitive in that it does not permit variable packet lengths and does not make use of the full-duplex link. However, we discuss it because its relative universality and ease of implementation make its use widespread. The packet structure is indicated in Figure 6.15*a*. The SOH (start of header) is the character 00000001. The NUM field contains the packet number (modulo 256). The packets are numbered successively, starting with 1. The field CNUM contains 2's complement of NUM, i.e., the character obtained by flipping every bit in NUM. The DATA take up the next 128 bytes. The last byte, CKS (checksum), is equal to the sum (modulo 2 without carry) of the 128 data bytes.

The receiver transmits a NACK (character 00010101) every 15 seconds when it is not receiving data. If the sender has data to transmit, it waits for the next NACK, then starts transmitting. After receiving each packet, the receiver checks the CKS and transmits an ACK (character 00000110) if it was correct or a NACK if it was not.

If the packet was not correctly received (indicated by a NACK), then the sender retransmits the same data. At the end of the file transmission, the sender transmits EOT (character 00000100) and waits for it to be acknowledged (retransmitting it if it is not).

A number of variations have been developed. One is XMODEM-CRC, which uses a 16-bit CRC instead of CKS. The receiver indicates that the CRC option is to be used by sending the ASCII character *C* instead of NACK to initiate the transmission. Another variation is YMODEM, which permits the transmission of blocks of 1024 bytes of data in addition to blocks of 128 bytes. It also makes it possible to abort a transfer by sending two consecutive 11000100 characters. Batch file transfers are also made possible by YMODEM.

6.7.6   *Kermit*

Kermit is a protocol designed to provide error-free file transfers between mainframe computers, workstations, and personal computers attached by a point-to-point link, usually with modems. It uses variable-length packets with a maximum length of 94 characters. The protocol is similar to ABP, but a version of SRP can be supported.

Kermit converts the 256 possible 8-bit bytes into *printable* 7-bit ASCII characters. The ASCII characters correspond to the decimal values 0 through 127 (using only 7 bits). Of

FIGURE 6.15

XMODEM (a) and Kermit (b) packet structure. The 128-byte packets of XMODEM are numbered modulo 256 by NUM. CNUM is 2's complement of NUM and is used to control errors. Kermit converts data into ASCII characters. In both protocols, packets are acknowledged positively or negatively.

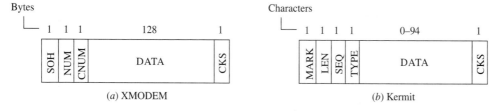

(a) XMODEM   (b) Kermit

these 128 characters, only those from 32 to 126 are printable (uppercase and lowercase letters, numbers, punctuation marks, and other common symbols, such as "space," @, +, ∗, and &). The characters corresponding to the values 0 through 31 and to 127 are called *control characters* and have special meanings, such as line feed and carriage return.

The packet structure in Kermit is illustrated in Figure 6.15*b*. The fields of that packet are as follows:

- MARK is the start of the packet; it is usually the ASCII character "Control A" transmitted as is (not converted).
- LEN indicates the length of the packet in ASCII characters (values 0 to 94 encoded as explained above).
- SEQ is the packet sequence number (modulo 64) starting at 0 (encoded as before).
- TYPE indicates the packet type. Here are the most common types: data, acknowledge (ACK), negative acknowledge (NACK), send initiate, receive initiate, break transmission, file header, end of file, error, generic command (the data field then contains the specifics: L for log out, F for finish but don't log out, D for directory query, U for disk query, E for erase file, T for type, Q for query server status), host command, text display (indicates text to be displayed on screen), initialize (exchange parameters but do not start a file transfer), and file attributes.
- DATA is the data (if any).
- CKS is the checksum.

The Kermit protocol is the following:

Step 1—Initialization
In order to start a transaction, the sender transmits an initiation packet which contains the following parameters:

- *MAXL:* maximum length packet that can be transmitted
- *TIME:* timeout for other device
- *NPAD:* number of padding characters before each packet
- *PADC:* control character required for padding
- *EOL:* character to terminate packets (if any)
- *QTCL:* character to precede control character (usually #).

Some other optional information may also be contained in this packet indicating that CKS is replaced by a CRC or how repeated characters are indicated. A special CAPAS field is used to indicate special capabilities, such as long packets and selective repeat.

The receiver then acknowledges with an ACK packet which contains the same parameters indicating the values that it supports.

Step 2—Data Transfer
Each packet is acknowledged with an ACK (correctly received) or a NACK (incorrectly received). Retransmissions use the original packet number. Received copies are acknowledged but discarded.

When the connection is full-duplex, Kermit can use a selective repeat protocol. This protocol is requested in the initialization phase by marking the corresponding bit in the CAPAS field. The window size is also negotiated during the initialization. The protocol used is then SRP.

6.8   Complement 1: Error Control Codes

In this section we explain the encoding of the CRC that we discussed in Section 6.1. Recall that the CRC is an error detection code that most data communication networks use. For instance, in Ethernet, the CRC computes 32 bits from the bits in the packet and appends those 32 bits at the end of the packet. The receiving node verifies that it gets the same value for the 32 CRC bits. If it does not, then it knows that some transmission error corrupted the packet.

We then discuss the BCH and the Reed-Solomon codes. These codes have good error correction properties and can be encoded and decoded efficiently. They are used in many applications. For instance, the Reed-Solomon code is used in CD-ROMs and audio CDs to correct errors caused by scratches or dust on the disc. In these systems, the code is combined with a permutation that spreads bits so that a burst of errors affects only a few bits in any given packet protected by a Reed-Solomon codes. This scrambling makes the encoding robust against long error bursts. The same coding method can be used to protect packetized real-time transmissions against packet losses. BCH codes are used in satellite and deep-space transmissions and in pagers. We conclude the section by explaining the main ideas of convolutional codes (trellis codes).

6.8.1   *Calculating the CRC*

Roughly speaking, the CRC error detection bits are the remainder of a long division of the packet bits by some given number. The calculation, done in hardware at the data link layer but in software at higher layers, is therefore implemented by a long-division algorithm.

Recall from section 6.1 that to define the CRC one first chooses a generator

$$\mathbf{g} = g_m g_{m-1} \cdots g_2 g_1 g_0$$

which is a binary word with $m + 1$ digits, the most- and least-significant bits g_m and g_0 being 1. The message $\mathbf{u} = u_{k-1} u_{k-2} \cdots u_1 u_0$ is a binary word of up to k digits. The CRC $\mathbf{r}$ is the binary word with m digits such that

$$2^m \mathbf{u} = \mathbf{ag} + \mathbf{r}$$

where the operations are modulo 2 without carry. The transmitted word is the codeword

$$[\mathbf{u}|\mathbf{r}] = 2^m \mathbf{u} + \mathbf{r} = \mathbf{ag}.$$

That is, R is the remainder of the division of $2^m \mathbf{u}$ by $\mathbf{g}$.

Circuit
The circuit in Figure 6.16 calculates the long division of $\mathbf{u} \times 2^3 = (1101) \times 2^3 = 1101000$ by $\mathbf{g} = 1011$. Consider the step in the long division indicated by the shaded box where we

calculate $1110 + 1011$. Write the term 1110 as $u_n u_{n-1} u_{n-2} u_{n-3} + a_2 a_1 a_0 0$. In the next step, write the term 1010 as $u_{n-1} u_{n-2} u_{n-3} u_{n-4} + a_2' a_1' a_0' 0$. With this notation, the step $1110 + 1011 = 101$ can be written as

$$u_{n-1} u_{n-2} u_{n-3} + a_2' a_1' a_0' = u_n u_{n-1} u_{n-2} u_{n-3} + a_2 a_1 a_0 0 + g_3 g_2 g_1 g_0 \times (u_n + a_2).$$

Indeed, during that step, we add $\mathbf{g}$ to $u_n u_{n-1} u_{n-2} u_{n-3} + a_2 a_1 a_0 0$ if and only if the leading bit $u_n + a_2$ is equal to 1. Consequently, we see that

$$a_2' a_1' a_0' = a_1 a_0 0 + g_2 g_1 g_0 \times (u_n + a_2). \tag{6.4}$$

The circuit in the left-hand part of the figure performs the long division by implementing the calculation of (6.4). The bits of $\mathbf{u}$ enter the circuit one at a time, most-significant bit first. Initially, the registers hold the value 0. As bit u_n of $\mathbf{u}$ enters the circuit, the adder computes $a_2 + u_n$, the registers shift by one position to the left. In the rightmost register, a_0 is replaced by $a_0' = a_2 + u_n$. At the same time, a_1 is replaced by $a_1' = a_0 + (a_2 + u_n)$, and so on. These are precisely the equations (6.4). Note that the wiring of the feedback specifies $\mathbf{g}$: there is a feedback line into register a_n if $g_n = 1$. Just after bit u_0 has entered the circuit and the shift has taken place, the registers a_2, a_1, a_0 contain the remainder r_2, r_1, r_0 of the long division.

Figure 6.17 shows the circuit that calculates the CRC for $\mathbf{g} = 1 g_{m-1} g_{m-2} \cdots g_2 g_1 1$.

FIGURE 6.16

CRC circuit (left) and corresponding long division (right).

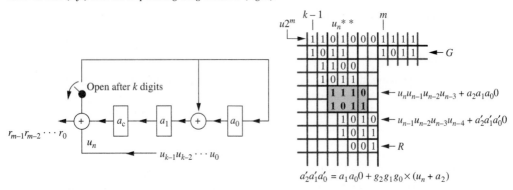

FIGURE 6.17

CRC circuit for a general generator.

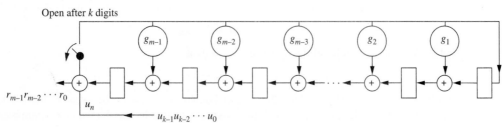

Decoding

The same circuit can check that the received word **v** is a multiple of **g**. If **v** is not a multiple of **g**, it is desirable to find the codeword $\hat{\mathbf{c}}$ closest to **v**. Indeed, if the code has a minimum distance equal to $2t + 1$ and if v differs from the original codeword **c** by at most t errors, then $\hat{\mathbf{c}} = \mathbf{c}$. Some efficient algorithms and circuits have been developed that find $\hat{\mathbf{c}}$. The references of this section describe these error correction procedures.

6.8.2   *Bose-Chaudhuri-Hocquenghem and Reed-Solomon Codes*

How should one choose the generator g? A good choice yields a code with a large minimum distance. There is one class of generators whose distance properties are well understood. These are the BCH codes (Bose, Chaudhuri, and Hocquenghem). The Reed-Solomon are a particular subclass of those codes. We explain how to calculate these codes in this section, which is rather technical. Remember that our ultimate objective is to exhibit codes that have good distance properties. The approach is algebraic and constructs matrices that are shown to be nonsingular because a polynomial of degree n cannot have more than n roots.

Let q be some power of a prime number. That is, $q = p^b$ where p is prime and b is some integer. We consider sourcewords and codewords that are strings of letters from an alphabet **GF**(q) with q different symbols. For any integer $m \geq 1$ and any integer t with $mt < n := q^m - 1$, one can construct a BCH(n, t) code with source words of $k = n - 2mt$ symbols and codewords of n symbols and a minimum distance of $2t + 1$.

The generator of these codewords is $\mathbf{g} = 1g_{m-1} \ldots g_2 g_1 1$, where

$$g(x) := x^m + g_{m-1}x^{m-1} + \cdots + g_1 x + 1$$

is the polynomial of lowest degree with coefficients in **F** which has the roots $\alpha, \alpha^2, \ldots, \alpha^{2t}$. In this definition, α is a primitive element in **GF**(q^m).

To clarify this definition, we need to specify what we mean by the sets **GF**(q) and **GF**(q^m), how the addition and multiplication are defined in these sets, and the meaning of a primitive element α in **GF**(q^m). We explain these notions in the next section.

Galois Fields

The set **GF**(p^b) with its rules of operations is called the Galois field with $q = p^b$ elements. The remarkable property of Galois fields is that the operations satisfy the same properties as for real numbers. These sets were discovered by the French mathematician Evariste Galois not long before he died in a duel at the age of 21 (in 1832).

We consider only the case $p = 2$ so that our set **GF**$(q) = $ **GF**(2^b) has 2^b elements that can be represented as binary words of b bits. The operations on **GF**(2^b) are defined by identifying its elements with polynomials of degree at most $b - 1$. That is, we identify the binary word $\beta = \beta_{b-1} \cdots \beta_1 \beta_0$ with the polynomial $\beta(x) := \beta_{b-1}x^{b-1} + \beta_{b-2}x^{b-2} + \cdots + \beta_2 x^2 + \beta_1 x + \beta_0$. We add and multiply the polynomials as we do with real polynomials and we then perform two simplifications. We first simplify the coefficients modulo 2. We then perform a second simplification by evaluating the polynomial modulo $f(x)$ where $f(x)$ is a given prime polynomial of degree b. A prime polynomial is one that cannot be factored into polynomials of lower degree.

For instance, with $b = 2$ we can choose $f(x) = 1 + x + x^2$ and we then have $\mathbf{GF}(4) = \{0, 1, x, 1 + x\}$. The rules of operations that we just explained imply that

$$(1 + x)(1 + x) = 1 + 2x + x^2, \text{ after the usual multiplication}$$
$$= 1 + x^2, \text{ after evaluating the coefficients modulo 2}$$
$$= x, \text{ after evaluating the polynomial modulo } f(x).$$

We also find for this example that, with $\alpha := x$,

$$\mathbf{GF}(4) = \{0, 1, \alpha, \alpha^2\}.$$

The element $\alpha = x$ is said to be primitive in $\mathbf{GF}(4)$ because its powers generate all the elements of that set. The binary word representation of $\mathbf{GF}(4)$ is obviously $\mathbf{GF}(4) = \{00, 01, 10, 11\}$. With this representation, $0 = 00, 1 = 01, \alpha = x = 10, \alpha^2 = x^2 = 1 + x = 11$.

As another example, let $b = 4$. We choose $f(x) = 1 + x + x^4$. According to the definitions of addition and multiplication modulo $f(x)$, one finds that

$$x^{10} = x^4 \times x^4 \times x^2 = (1 + x)(1 + x)x^2 = (1 + 2x + x^2)x^2$$
$$= (1 + x^2)x^2 = x^2 + x^4 = 1 + x + x^2.$$

We can again choose $\alpha = x$. In the binary representation, this corresponds to $\mathbf{GF}(2^4) = \{0, 1\}^4$ and the above calculation can be written as $(0010)^{10} = 0111$. Note that these operations are not the same as addition and multiplication modulo 2 without carry. For instance, in $\mathbf{GF}(2^4)$ one has

$$(1000) \times (0010) = x^3 \times x = x^4 = 1 + x = 0011.$$

The operations modulo 2 without carry would give $(1000) \times (0010) = 10000$. With $\alpha = x = 0010$ we find that

$$\mathbf{GF}(2^4) = \{0, 1, \alpha, \ldots, \alpha^{14}\}$$
$$= \{0000, 0001, 0010, 0100, 1000, 0011, 0110, 1100,$$
$$1011, 0101, 1010, 0111, 1110, 1111, 1101, 1001\}.$$

Binary BCH(n, t)

Let $n = 2^b - 1$ and t be such that $n - bt \geq 1$. We define the binary BCH codes with n-bit binary codewords and a minimum distance of $2t + 1$. There are at least 2^k such codewords that can encode k-bit sourcewords, where $k = n - bt$.

For instance, we can choose $b = 4$ and $t = 2$ to produce the BCH(15,2) code that can correct up to two errors. The codewords have 15 bits and the sourcewords have 7 bits. To construct this code we must find the lowest-degree polynomial $g(x)$ with coefficients in $\{0, 1\}$ that admits the roots $\alpha, \alpha^2, \alpha^3, \alpha^{2t} = \alpha^4$ where α is primitive in $\mathbf{GF}(2^4)$. To construct this polynomial, we find, for $i = 1$ and $i = 3$, the smallest-degree polynomial $\phi_i(x)$ with coefficients in $\{0, 1\}$ such that $\phi_i(\alpha^i) = 0$. Since $1 + \alpha + \alpha^4 = 0$, we see that $\phi_1(x) = 1 + x + x^4$. For $i = 3$ you can verify that $\phi_3(x) = 1 + x + x^2 + x^3 + x^4$. We then calculate $g(x)$ the least common multiple of $\phi_1(x)$ and $\phi_3(x)$. Since these two polynomials are irreducible,

$$g(x) = \phi_1(x)\phi_3(x) = 1 + x^4 + x^6 + x^7 + x^8.$$

More generally, for an arbitrary value of t, we define $g(x)$ as the least common multiple of $\{\phi_i(x), i = 1, 3, \ldots, 2t - 1\}$. How does this work? That is, why does this $g(x)$ produce a code with distance t? We explain this fact next. First note that $g(\alpha^i) = 0$ for $i = 1, 3, \ldots, 2t - 1$. Also, because the coefficients of $g(x)$ are in $\{0, 1\}$,

$$g(x)^2 = \left(\sum_{i=1}^{m} g_i x\right)^2 = \sum_{i=1}^{m} g_i x^2 = g(x^2).$$

(Indeed, the double products vanish modulo 2 so that only the sum of the squares remains.) Consequently, $g(\alpha^i) = 0$ for $i = 2$ and $i = 4$.

Now, the codewords produced with this generator polynomial are the multiples of $g(x)$ and are therefore the words $c_{n-1}c_{n-2}\cdots c_1 c_0$ such that $c(x) = c_{n-1}x^{n-1} + c_{n-2}x^{n-2} + \cdots + c_1 x + c_0$ admits α^i as roots for $i = 1, 3, \ldots, 2t - 1$. Consequently,

$$M\mathbf{c}^T = \begin{bmatrix} \alpha_0 & \alpha_1 & \cdot & \cdot & \alpha_{n-1} \\ \alpha_0^3 & \alpha_1^3 & \cdot & \cdot & \alpha_{n-1}^3 \\ \alpha_0^5 & \alpha_1^5 & \cdot & \cdot & \alpha_{n-1}^5 \\ \cdot & & & & \cdot \\ \alpha_0^{2t-1} & \alpha_1^{2t-1} & \cdot & \cdot & \alpha_{n-1}^{2t-1} \end{bmatrix} \begin{bmatrix} c_0 \\ c_1 \\ \cdots \\ c_{n-1} \end{bmatrix} = \begin{bmatrix} c(\alpha) \\ c(\alpha^3) \\ \cdots \\ c(\alpha^{2t-1}) \end{bmatrix} = \mathbf{0}.$$

We claim that these equations are linearly independent and impose $2t$ constraints on the n symbols of $\mathbf{c}$, which has therefore $k = n - 2t$ free symbols. To prove the claim, consider any $(2t) \times (2t)$ submatrix M of H obtained by selecting the columns $i_1, i_2, \ldots, i_{2t}$ of H. This submatrix is nonsingular. Indeed, assume that $\mathbf{f} = (f_0, f_1, \ldots, f_{2t-1}) \in \mathbf{F}^{2t}$ is such that $\mathbf{f}M = \mathbf{0}$. Then

$$\mathbf{0} = \mathbf{f}M = \begin{bmatrix} \alpha^{i_1} f(\alpha^{i_1}) \\ \alpha^{i_2} f(\alpha^{i_2}) \\ \cdots \\ \alpha^{i_{2t}} f(\alpha^{i_{2t}}) \end{bmatrix}$$

where $f(x) := f_0 + f_1 x + f_2 x^2 + \cdots + f_{2t-1}x^{2t-1}$. But these identities show that the polynomial $f(x)$ has $2t$ distinct roots in $\mathbf{F}$. Since $f(x)$ is of degree $2t - 1$, it can have $2t$ distinct roots only if it is identically 0, by the fundamental theorem of algebra. Indeed, the addition and multiplication have the same property as for real numbers. Thus $\mathbf{f}M = \mathbf{0}$ only if $\mathbf{f} = \mathbf{0}$, which shows that M is nonsingular.

Consider a codeword $\mathbf{c}$ that has at most $2t$ nonzero symbols. The identities $H\mathbf{c}^T = \mathbf{0}$ then imply that $M\mathbf{c}^T = \mathbf{0}$ for some $(2t) \times (2t)$ submatrix of M obtained by selecting the columns of H that correspond to the nonzero symbols of $\mathbf{c}$. Since M is nonsingular, it follows that $\mathbf{c} = 0$. Consequently, any nonzero codeword must have at least $2t + 1$ nonzero symbols.

Finally, assume that $\mathbf{c}_1$ and $\mathbf{c}_2$ are two distinct codewords. Then $\mathbf{c} := \mathbf{c}_1 - \mathbf{c}_2$ is a nonzero codeword (it is a multiple of $\mathbf{g}$ since $\mathbf{c}_1$ and $\mathbf{c}_2$ are). Consequently, $\mathbf{c}$ has at least $2t + 1$ nonzero symbols, which implies that $\mathbf{c}_1$ and $\mathbf{c}_2$ differ in at least $2t + 1$ symbols. This argument shows that the minimum distance between two codewords is at least $2t + 1$.

Reed-Solomon Codes

These codes, developed by Reed and Solomon in 1960, are a subclass of the nonbinary BCH codes that have powerful error-correcting properties. We explain that these codes can correct bursts of errors.

We define the (n, k) Reed-Solomon code $RS(n, k)$ where $n = 2^b - 1$ for some integer b and $1 \le k \le n = k + 2t$ for some integer t. The words that the source—the sourcewords—sends have k symbols. Each symbol is an element of $\mathbf{GF}(2^b)$.

The $RS(n, k)$ code is the set of words $\mathbf{c} = (c_0, c_1, \ldots, c_{n-1})$ with n symbols in $\mathbf{GF}(2^b)$ such that

$$c(x) = a(x)g(x) \tag{6.5}$$

where

$$g(x) = (x + \alpha)(x + \alpha^2)(x + \alpha^3) \cdots (x + \alpha^{2t}) = g_0 + g_1 x + g_2 x^2 + \cdots + g_{2t-1} x^{2t-1} + x^{2t}.$$

Consider a sourceword $u_{k-1} u_{k-2} \cdots u_2 u_1 u_0$ with symbols in $\mathbf{GF}(2^b)$. We associate the polynomial $u(x) = u_{k-1} x^{k-1} + u_{k-2} x^{k-2} + \cdots + u_2 x^2 + u_1 x + u_0$. To encode this sourceword, we perform the long division of $u(x)x^{2t}$ by $g(x)$. That is, we calculate

$$u(x)x^{2t} = a(x)g(x) + r(x).$$

The codeword is $a(x)g(x)$. This encoding procedure is identical to the calculation of the other codes we explained earlier, except that the operations are in $\mathbf{GF}(2^b)$ instead of modulo 2 without carry. The circuit that performs this calculation is the same as that shown in Figure 6.16 except that the addition and multiplications are now in $\mathbf{GF}(2^b)$. Simple logic circuits can be designed to perform the operations in that field.

The remarkable property of these codes is that their minimum distance is $2t + 1$, as we show next. First note that if $\mathbf{c} = (c_0, c_1, \ldots, c_{n-1})$ is an $RS(n, t)$ codeword, then

$$H\mathbf{c}^T = \mathbf{0} \qquad \text{where } H := \begin{bmatrix} \alpha_0 & \alpha_1 & \cdots & \alpha_{n-1} \\ \alpha_0^2 & \alpha_1^2 & \cdots & \alpha_{n-1}^2 \\ \alpha_0^3 & \alpha_1^3 & \cdots & \alpha_{n-1}^3 \\ \cdot & \cdot & \cdots & \\ \alpha_0^{2t} & \alpha_1^{2t} & \cdots & \alpha_{n-1}^{2t} \end{bmatrix} \tag{6.6}$$

with $\alpha_i := \alpha^i$.

Indeed,

$$H\mathbf{c}^T = \begin{bmatrix} \alpha_0 & \alpha_1 & \cdots & \alpha_{n-1} \\ \alpha_0^2 & \alpha_1^2 & \cdots & \alpha_{n-1}^2 \\ \alpha_0^3 & \alpha_1^3 & \cdots & \alpha_{n-1}^3 \\ \cdot & \cdot & \cdots & \\ \alpha_0^{2t} & \alpha_1^{2t} & \cdots & \alpha_{n-1}^{2t} \end{bmatrix} \begin{bmatrix} c_0 \\ c_1 \\ \cdots \\ c_{n-1} \end{bmatrix} = \begin{bmatrix} c(\alpha) \\ c(\alpha^2) \\ \cdots \\ c(\alpha^{2t}) \end{bmatrix} = \mathbf{0}$$

where the last equality follows from $c(x) = a(x)g(x)$ and $g(\alpha^i) = 0$ for $i = 1, \ldots, 2t$.

We claim that these equations are linearly independent and impose $2t$ constraints on the n symbols of $\mathbf{c}$ which has therefore $k = n - 2t$ free symbols. To prove the claim, consider any $(2t) \times (2t)$ submatrix M of H obtained by selecting the columns $i_1, i_2, \ldots, i_{2t}$ of H. This submatrix is nonsingular. Indeed, assume that $\mathbf{f} = (f_0, f_1, \ldots, f_{2t-1}) \in \mathbf{F}^{2t}$ is such that $\mathbf{f}M = \mathbf{0}$. Then

$$\mathbf{0} = \mathbf{f}M = \begin{bmatrix} \alpha^{i_1} f(\alpha^{i_1}) \\ \alpha^{i_2} f(\alpha^{i_2}) \\ \cdots \\ \alpha^{i_{2t}} f(\alpha^{i_{2t}}) \end{bmatrix}$$

where $f(x) := f_0 + f_1 x + f_2 x^2 + \cdots + f_{2t-1} x^{2t-1}$. But these identities show that the polynomial $f(x)$ has $2t$ distinct roots in **F**. Since $f(x)$ is of degree $2t - 1$, it can have $2t$ distinct roots only if it is identically 0, by the fundamental theorem of algebra. Indeed, the addition and multiplication have the same property as for real numbers. Thus $\mathbf{f}M = \mathbf{0}$ only if $\mathbf{f} = \mathbf{0}$, which shows that M is nonsingular.

Consider a codeword $\mathbf{c}$ that has at most $2t$ nonzero symbols. The identities $H\mathbf{c}^T = \mathbf{0}$ then imply that $M\mathbf{c}^T = \mathbf{0}$ for some $(2t) \times (2t)$ submatrix of M obtained by selecting the columns of H that correspond to the nonzero symbols of $\mathbf{c}$. Since M is nonsingular, it follows that $\mathbf{c} = 0$. Consequently, any nonzero codeword must have at least $2t + 1$ nonzero symbols.

Finally, assume that $\mathbf{c}_1$ and $\mathbf{c}_2$ are two distinct codewords. Then $\mathbf{c} := \mathbf{c}_1 - \mathbf{c}_2$ is a nonzero codeword (it is a multiple of $\mathbf{g}$ since $\mathbf{c}_1$ and $\mathbf{c}_2$ are). Consequently, $\mathbf{c}$ has at least $2t + 1$ nonzero symbols, which implies that $\mathbf{c}_1$ and $\mathbf{c}_2$ differ in at least $2t + 1$ symbols. This argument shows that the minimum distance between two codewords is at least $2t + 1$.

6.8.3   *Convolutional Codes*

The codes we explained so far transform a sourceword of k bits into a codeword of n bits with $n > k$. Accordingly there are 2^k codewords in a set of 2^n binary words. A good code is one where the codewords are as far from each other as possible. Moreover, a good code should be easy to use: coding and decoding should be efficient. These codes are called block codes because they group the source bits into blocks of k bits that they encode and decode independently of one another. This form of coding is similar to isolated word recognition in handwriting or speech recognition: one word at a time.

Continuous word recognition uses the information about previous words to decode the next one. For instance, assume that the receiver has decoded the words "once upon a ?i??" and tries to decode the next word. Remembering the previous three words helps the decoder because successive words are not independent.

In this section we explain that convolutional codes introduce a dependency among successive bits that are transmitted and that the decoder exploits this dependency to correct errors. We explain the concepts in one simple example to eliminate the need for heavy notation.

An Example

We start with a simple example that illustrates the main features of convolutional codes. Let $u_0, u_1, u_2, \ldots$ be the successive bits that the source produces. At time n, bit u_n enters the encoder. The decoder then calculates two bits: c_{2n} and c_{2n+1}. These two bits are calculated as linear functions of u_n, u_{n-1}, and u_{n-2}. For instance, $c_{2n} = u_n + u_{n-2}$ and $c_{2n+1} = u_n + u_{n-1} + u_{n-2}$.

The *rate* of this encoder is $\frac{1}{2}$. This means that the rate of the source bits is half of the rate that the encoder produces. In other words, the encoder encodes $\frac{1}{2}$ a source bit per transmitted bit.

The encoder is easy to build. It consists of a shift register that stores the last 3 bits of the source and of three logic functions that add some of these three bits modulo 2.

State Diagram

To understand how to decode this code and to determine the distance properties of the code, one summarizes the operations of the encoder by a *state diagram*. The state of the encoder at time n is $z_n = (u_{n-1}, u_{n-2})$. Knowing z_n and u_n we can calculate the outputs $y_n := (c_{2n}, c_{2n+1})$ and the next state z_{n+1}. That is, we can write

$$z_{n+1} = f(z_n, u_n) \qquad \text{and} \qquad y_n = g(z_n, u_n). \tag{6.7}$$

Figure 6.18 shows the state diagram that summarizes the relations (6.7). For instance, when $z_n = (u_{n-1}, u_{n-2}) = (0, 1)$ and $u_n = 1$, one has $z_{n+1} = (u_n, u_{n-1}) = (1, 0)$ and $y_n = (c_{2n}, c_{2n+1}) = (u_n + u_{n-2}, u_n + u_{n-1} + u_{n-2}) = (1 + 1, 1 + 0 + 1) = (0, 0)$.

With this diagram we can quickly determine the output string that corresponds to a particular source string. We always start with the register in state $(0, 0)$. For instance, if the input string is 0101110100, we find that the code string is 001101111110101100 and the encoder is back to state $(0, 0)$. As this example shows, we can use this code to encode messages with a finite length by resetting the encoder with two inputs equal to 0. Thus, this code can be used as a block code.

Minimum Distance

We want to calculate the minimum distance d_{min} between two valid sequences of code bit strings $\mathbf{c} = \{c_n, n \geq 0\}$. As for block codes, the larger this minimum distance, the less likely it is that the noise transforms one transmitted code string into another one and mistakes the receiver. We always start with the initial state $z = (0, 0)$. To calculate d_{min} we first note that the code is linear. That is, if the input string $\mathbf{u}$ produces the code string $\mathbf{c}$ and if $\mathbf{u}'$ produces $\mathbf{c}'$, then $\mathbf{u} + \mathbf{u}'$ produces $\mathbf{c} + \mathbf{c}'$. Indeed, equations (6.7) are linear. Consequently, the minimum distance between two valid codes $\mathbf{c}$ and $\mathbf{c}'$ is the minimum distance between a nonzero valid code $\mathbf{c}$ and the string $\mathbf{0}$ of all 0s. This distance is the number of 1s in the code string. Thus, we want to find the valid nonzero code string with the smallest number of 1s.

Let's try by hand to find a valid code string with few 1s. Consider the state diagram of Figure 6.18. We start from state $(0, 0)$. We must leave that state to produce a nonzero code string. The encoder is supposed to run for a long time. To produce a code string with few 1s

FIGURE 6.18

Convolutional encoder (left) and its state diagram (right).

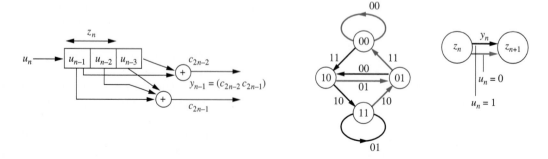

we must find a sequence of states in the state diagram that goes back to the state $(0, 0)$ and outputs to few 1s. A simple look shows that the path $(0, 0) \rightarrow (1, 0) \rightarrow (0, 1) \rightarrow (0, 0)$ is the path from $(0, 0)$ to itself with the fewest output bits equal to 1. The number of 1s produced by that path is 5. Thus, we have found that $d_{\min} = 5$ for this code.

For a more complex convolutional code, we can find the minimum distance by using a simple algorithm that we explain in our example. Denote by $d(z)$ the minimum number of 1s of the output string for a path that goes from the state z to state $(0, 0)$. For instance, $d(0, 0) = 0$. Also, $d(1, 0) = \min\{1 + d(1, 1), 1 + d(0, 1)\}$. We can write a similar equation for every state. These equations are identical to the Bellman-Ford equations, which is not surprising since we are trying to find a shortest path in the graph. By solving these fixed-point equations we can determine the minimum distance of a convolutional code.

Optimum Decoding

We conclude our quick look at convolutional codes by discussing the optimum decoding of such a code.

Consider once again Figure 6.18. Assume that the receiver gets the code string $\mathbf{r} :=$ 10100011 and that it knows the encoder went from state $(0, 0)$ back to state $(0, 0)$. The receiver must determine the input string that produces this code string.

The transition diagram shows that transmission errors must have occurred. For instance, the first two bits of the code string were either 00 or 11, not 10. We assume that bit errors are independent and occur each with probability $\epsilon < 0.5$. We don't want to assume a prior likelihood of input strings. Instead we will try to find the input string $\mathbf{u} = (u_0, u_1, u_2, u_3)$ that maximizes the probability that the transmission errors have produced the received string $\mathbf{r}$. For instance, the input string $\mathbf{u} = 0100$ produces the code string $\mathbf{c} = 00110111$. Three bit errors transform $\mathbf{c}$ into $\mathbf{r}$. As another example, the input string $\mathbf{u}' = 1100$ produces the code string $\mathbf{c}' = 11101011$. It now takes only 2 bit errors to transform $\mathbf{c}'$ into $\mathbf{r}$. Accordingly, we decide that the input string $\mathbf{u}'$ is a more plausible cause of $\mathbf{r}$ than $\mathbf{u}$. Looking at other possibilities we conclude that the most plausible input string, based on the received string $\mathbf{r}$, is $\mathbf{u}'$.

This decision is called a *maximum likelihood detection.* This maximum likelihood detection is the input string that maximizes the probability of the received string. That is, if we denote by

$$P[\mathbf{r}|\mathbf{u}]$$

the probability of receiving the string $\mathbf{r}$ given that the input string is $\mathbf{u}$, then the maximum likelihood detection is the string $\mathbf{u}$ that maximizes $P[\mathbf{r}|\mathbf{u}]$.

It is not difficult to make systematic our above search of the maximum likelihood input string. To do this, we construct the diagram of Figure 6.19. This diagram shows the four states of the encoder and the possible sequences of states starting from $(0, 0)$. The left diagram also shows the received string $\mathbf{r}$. On every possible transition, we have indicated the number of different bits between the two code bits and the received bits. For instance, the transition from state $(1, 0)$ at time 1 to state $(0, 1)$ at time 2 produces the two code bits 01 (see Figure 6.18). The received bits at that time are 10. Accordingly, that transition is marked with the number 2 since 10 and 01 differ in two bits. The maximum likelihood decoding of $\mathbf{u}$ given $\mathbf{r}$ is the string $\mathbf{u}$ that produces the path in the left diagram with the

FIGURE 6.19

Illustration of the maximum likelihood decoding.

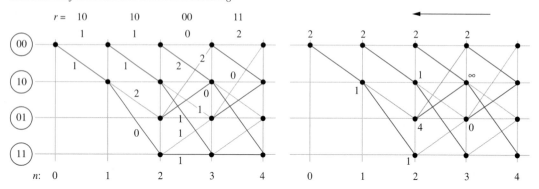

smallest total length. The length of the path is the sum of the numbers on its segments. By inspection of the left diagram, we see that the shortest path goes through the states $(0, 0) \rightarrow (1, 0) \rightarrow (1, 1) \rightarrow (0, 1) \rightarrow (0, 0)$ and its length is 2. This path corresponds to the input $\mathbf{u} = 1100$, as we have already observed in the previous paragraph. To find this path, we use a shortest path algorithm such as Dijkstra's or Bellman-Ford. The Bellman-Ford algorithm applied to this situation is called the Viterbi algorithm.

In practice, one is interested in long strings and in more complex encoders. Finding the shortest path that produces the maximum likelihood decoder then requires clever tree pruning methods to keep the complexity reasonable. Please refer to the literature for a discussion of these important implementation issues.

6.8.4   Turbo Codes

Turbo codes are a new class of codes that exhibit remarkable performance over very noisy channels. Although we do not have the space to discuss these codes in detail, we think that you should at least be aware of their existence. We explain the structure of turbo codes and we discuss their decoding.

General Structure

Roughly speaking, a turbo code is a combination of a number of block codes applied to different permutations of the source word. Recall that a block code encodes a k-bit sourceword $\mathbf{u}$ into an n-bit codeword $\mathbf{c}$. If the block code is linear, we can write $\mathbf{c} = \mathbf{u}G$ for some $k \times n$ matrix G. (We view binary words as row vectors.) The matrix G is an (n, k) encoder.

For $j = 1, \ldots, J$, let G_j be an (n_j, k) encoder and Π_j a $k \times k$ permutation matrix. The turbo code maps the k-bit sourceword into the codeword

$$\mathbf{c} = \mathbf{u}[\Pi_1 G_1 | \Pi_2 G_2 | \cdots | \Pi_J G_J].$$

Figure 6.20 illustrates the encoding procedure. In the simplest case, the same block code is used twice: once on $\mathbf{u}$ and a second time on a permutation $\mathbf{u}\Pi$ of $\mathbf{u}$. In this case,

$$\mathbf{c} = \mathbf{u}[G | \Pi G].$$

Why should this particular combination of codes work well? The intuition that although the noise corrupts the reconstruction of the permutations $\mathbf{u}\Pi_j$ of $\mathbf{u}$, one should be able to find the most likely $\mathbf{u}$ consistent with these corrupted reconstructions. The various permutations are introduced so that a noise with a particular pattern will have different effects on the reconstructions of $\mathbf{u}$.

Decoding

The turbo decoding algorithm is iterative and based on the individual codes. That is, the decoding algorithm does not look at the turbo code as a big block code. Rather, it exploits the structure of the code as providing different noisy images of the same codeword. Thus, the algorithm enforces a decomposition of the decoding algorithm, which greatly reduces its complexity. We explain the procedure when $J = 2$.

Figure 6.21 shows the turbo encoder with $J = 2$, the noisy channel modeled by the noise strings $\mathbf{n}_1$ and $\mathbf{n}_2$ and the decoder. The decoder consists of two decoders that collaborate. Each decoder D_j observes $\mathbf{y}_j$ equal to the codeword $\mathbf{c_j}$ corrupted by the noise vector $\mathbf{n}_j$. For $t \geq 0$, at step $2t + 1$, decoder D_2 uses $\mathbf{y}_2$ and a vector $\mathbf{z}_1(2t)$ it gets from D_1 to produce the vector $\mathbf{z}_2(2t+1)$ and some estimate $\mathbf{u}_2(2t+1)$. At the next step, decoder D_1 performs similar operations. The two decoders collaborate in that way, hopefully refining their estimates as they go on.

At step $2t$, decoder 1 tells decoder 2 that it believes that $P(u_i = 1)/P(u_i = 0)$ is equal to component i of the vector $\mathbf{z}_1(2t)$. Based on that information, and on the vector

FIGURE 6.20

The general structure of a turbo encoder.

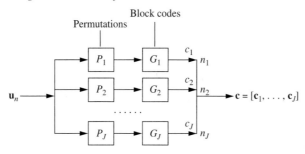

FIGURE 6.21

Turbo encoder and decoder.

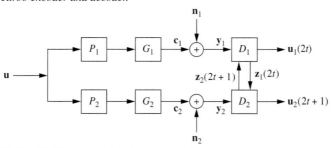

$\mathbf{y}_2$, decoder 2 calculates its estimates of $\alpha_{2,i}(2t + 1) = P(u_i = 1|\mathbf{y}_2)/P(u_i = 0|\mathbf{y}_2)$ and provides these estimates as the vector $\mathbf{z}_2(2t + 1)$ to decoder 1. Decoder 2 also produces a temporary guess of the sourceword $\mathbf{u}$ by deciding that u_i was one if $\alpha_{2,i}(2t + 1) > 1$ and that it is 0 otherwise. Decoder 1 then performs a similar step, and the procedure continues.

Ideally one would hope that the successive estimates of the decoders converge to

$$\alpha_i := \frac{P(u_i = 1|\mathbf{y}_1, \mathbf{y}_2)}{P(u_i = 0|\mathbf{y}_1, \mathbf{y}_2)}.$$

If that were the case, then the decoders would decode $\mathbf{u}$ by deciding that u_i was probably equal to 1 if $\alpha_i \geq 1$ and that is was probably equal to 0 otherwise. This decision would correspond to selecting the most likely value of u_i given the output of the channel.

Unfortunately, simple examples show that the turbo decoding algorithm that we described may not converge, or that it may converge to the wrong values. Nevertheless, for reasons that are not well understood yet, these codes and the decoding algorithm perform well in practical applications.

6.9 Complement 2: Correctness of ABP

In this section we show that ABP is correct when the packets are not numbered sequentially but are instead numbered modulo some integer $M + 1$. We will show that $M = 1$ when the channels between the sender and the receiver are FIFO. This case is the alternating bit protocol. We study non-FIFO channels in Section 6.10.

We first assume that the sender marks packet n and its copies with sequence number n, for $n \geq 1$. Similarly, the receiver marks the acknowledgment of a packet with sequence number n also with sequence number n. We call this numbering the *explicit numbering*.

Initially, the receiver is waiting for packet 0. The sender transmits packet 0. If it does not get the acknowledgment of packet 0 within T, the sender transmits another copy of the same packet, again with sequence number 0. Eventually, the receiver gets packet 0 correctly and eventually the sender also gets the acknowledgment of packet 0 correctly. In the meantime, the receiver may get a number of copies of packet 0 which it acknowledges but discards. The sender then transmits packet 1 and starts waiting for the acknowledgment with sequence number 1 and retransmits copies of packet 1 whenever the acknowledgment of packet 1 fails to arrive within T. In the meantime, the sender may get additional acknowledgments of packet 0, which it ignores. The procedure then continues as for packet 0 for the subsequent packets. From this description, it is believable that the protocol is correct when the sender and receiver use the explicit numbering.

To show that the alternating bit sequence numbering suffices, we examine a feature of the protocol we just described. Consider the system whose inputs are the packet sequence numbers sent by the sender and whose outputs are the acknowledgment sequence numbers received by the sender. This system is first in, first out (FIFO) but can lose items. Let us call this system the *packet-to-acknowledgment system,* which we abbreviate as PTA. See Figure 6.22.

We introduce a proof method that we also use for the other retransmission protocols. We prove by induction that the sender, receiver, and the PTA always satisfy some property that guarantees that the reduced numbering is sufficient. Our induction hypothesis $H(L)$

FIGURE 6.22

The channel from packets to acknowledgments and the induction hypothesis.

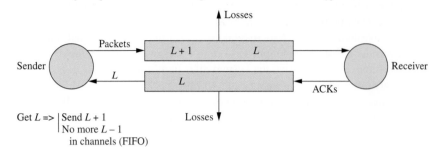

Get L => | Send $L + 1$
           | No more $L - 1$
             in channels (FIFO)

for some $L \geq 0$ is the following. If the sender last got the acknowledgment with sequence number L, then the sender can transmit only packet $L + 1$ and the PTA contains at most items L and $L + 1$.

Note that this induction hypothesis is true initially with $L = 0$ if we consider item 0 as being a non-item. Assume that $H(L)$ is true and that the sender last got an acknowledgment with sequence number L. The sender transmits packet $L + 1$ repeatedly every T seconds until it gets an acknowledgment $L + 1$. Since the PTA is eventually successful, the sender gets that acknowledgment eventually. At that time, the PTA, which contained at most items L and $L + 1$, cannot contain any item L anymore, by the FIFO assumption: the item $L + 1$ that just left the PTA when the acknowledgment $L + 1$ reached the sender entered the PTA after every L item and therefore leaves the PTA after every L item. Now that the sender has received acknowledgment $L + 1$ it can transmit only packet $L + 2$ and the PTA can therefore contain at most items $L + 1$ and $L + 2$: the items $L + 1$ that were still in the PTA when the sender got its first acknowledgment $L + 1$ and the items $L + 2$ that the sender injected in the PTA. Hence, $H(L)$ implies $H(L + 1)$ and moreover $H(0)$ is true. Consequently, $H(L)$ is true for all $L \geq 0$.

It follows from $H(n + 1)$ that if the sender last transmitted item $n + 1$, it can get only acknowledgments n of the previous packet and $n + 1$ of the last packet. Similarly, $H(n)$ implies that if the receiver last got packet n correctly, it can only get other copies of packet n or a new packet $n + 1$. Consequently, the numbering can be simplified to that of ABP which suffices to distinguish the two possibilities for the receiver and for the sender. Hence, ABP is correct. Note also that the proof that $H(L)$ implies $H(L + 1)$ also shows that the protocol makes *progress* when transmitting packets: If the sender gets acknowledgment L and has another packet to transmit with sequence number $L + 1$, it gets acknowledgment $L + 1$ after a finite time.

6.10 Complement 3: Correctness in a Non-FIFO Network

6.10.1 ABP

We saw in Section 6.9 that ABP is correct if the packets and ACKs are numbered sequentially. The sequential numbering requires arbitrarily large sequence numbers if there is no bound on the number of packets of a connection. It is intuitively obvious that the sequence numbers can be limited to $\{1, 2, \ldots, K\}$ if the network cannot contain more than $K - 1$ packets or ACKs of one connection: the sender and receiver can then recycle old sequence numbers. We prove that result next, again using the same method of proof by induction.

We define the degree of overtaking of an arbitrary channel that can lose items as follows. Consider a system with input $12 \cdots n \cdots$. We say that the degree of overtaking is less than D if item n cannot leave the system after item $n + D$ for all $n \geq 1$. We then define the degree of overtaking as being equal to D if it is less than $D + 1$ but is not less than D. In particular, the degree of overtaking is 0 if the channel is FIFO. If there is a bound on the time to live of packets in the communication network, then we can derive a bound on the degree of overtaking of the PTA. For instance, if any packet can spend at most T seconds in a network and if the maximum transmission rate of S and R is G packets per second, and if R acknowledges the packets in order, the degree of overtaking of the PTA is at most $2T \times G$.

Consider again ABP with explicit numbering. Instead of assuming that the PTA is FIFO, we assume that its *degree of overtaking* is less than M. Thus, for FIFO channels, we have $M = 1$.

The induction hypothesis $H(L)$ for an arbitrary $L \geq 0$ is that if the sender last got acknowledgment L, then it can transmit only packet $L + 1$ and the PTA contains at most items $L - M + 1, L - M + 2, \ldots, L + 1$. (Items n with $n \leq 0$ are nonitems.) See Figure 6.23. Assume that $H(L)$ is true and that the sender last got acknowledgment L. The sender keeps on transmitting copies of packet $L + 1$ until it eventually gets acknowledgment $L + 1$ and can then transmit only packet $L + 2$. Because of our assumption, the PTA can now contain only items $L - M + 2, L - M + 3, \ldots, L + 2$. Indeed, since the degree of overtaking of the PTA is less than M, item k for $k \leq L - M + 1$ cannot leave the PTA after item $L - M + 1 + M = L + 1$. Since transmissions are eventually successful, it follows that when item $L + 1$ leaves the PTA, no packet k with $k \leq L - M + 1$ can remain in the PTA.

Therefore $H(L)$ implies $H(L+1)$. Since $H(0)$ is true, we conclude that $H(L)$ holds for all $L \geq 1$. This property implies that we can reduce the explicit numbering to a numbering modulo $M + 1$. Indeed, with this numbering, if the sender last got an acknowledgment with sequence number $A \in \{0, 1, \ldots, M\}$, it is sending the next packet with sequence number $K(A) := A + 1$ if $A < M$ and with sequence number $K(A) := 0$ if $A = M$, until it gets the acknowledgment with sequence number $K(A)$. When it gets that acknowledgment, the sender can be sure that this is the acknowledgment of the last packet it transmitted. Indeed,

FIGURE 6.23

The non-FIFO channel from packets to acknowledgments and the induction hypothesis.

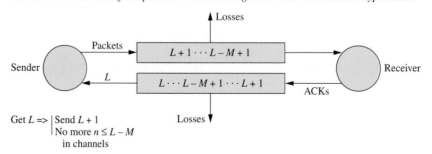

the previous acknowledgments of earlier packets with the same sequence number had left the PTA before the sender transmitted the last packet. Similarly, one can verify that the receiver cannot be confused by the reduced numbering.

6.10.2 GBN

With numbering modulo K ($K \geq W + 1$), the rules are identical except that the sequence numbers are now modulo K and that if the timer S started when it sent a packet with sequence number n, then S starts the retransmissions of W packets from the last packet it sent with that sequence number. Also, if R last did not discard a packet with sequence number n, then the next packet it does not discard is the first one with the next sequence number in the numbering modulo K (i.e., $n + 1$ if $n < K$ and 0 if $n = K$).

Correctness

We assume that the PTA is FIFO. (See the section on correctness of the ABP for the definition of PTA.) The proof that GO BACK N is correct when explicit numbering is used is similar to that of ABP with explicit numbering. We let the reader adapt the argument.

Consider GO BACK N with explicit numbering. Assume that the sender last received acknowledgment n. As with ABP, we can prove by induction that the system between the sender and the receiver contains at most items from the set $\{n, n+1, \ldots, n+W-1\}$ and that the sender will not send any more packets with sequence number $m \leq n$. We let the reader carry out the induction step. It follows from this property that at any time the system from the sender to the receiver contains at most items from the set $\{n, n+1, \ldots, n+W\}$ for some n. As with ABP, this fact implies that the GO BACK N numbering prevents confusions. Hence, the GO BACK N protocol is correct. Note that GO BACK N for $W = 1$ is ABP.

Negative acknowledgments (NACKs) can be used to speed up the GO BACK N protocol. A NACK is transmitted as soon as the receiver realizes that it received an incorrect packet. This occurs when the receiver notices a gap in the numbering of the correct packets it receives. The sender can then retransmit a copy of the corrupted packet immediately without having to wait for the timeout, as would be the case without NACKs.

GO BACK N is correct with window size W and with numbering modulo $W + M$ if the degree of overtaking of the PTA is less than M. The proof of this fact is almost identical to that of the correctness of ABP under similar assumptions. We let the reader adapt the argument.

6.11 Complement 4: Congestion Control in Internet

The congestion and flow control mechanisms of TCP adjust the window size W of the retransmission protocol.

Flow control means adjusting the rate of the source transmissions to avoid overwhelming the destination. The destination host advertises a maximum acceptable window size W_{max} in the acknowledgements that it sends to the source host. The source limits the window size it uses to W_{max}. For instance, if the destination runs out of buffer space, it can set $W_{max} = 0$, which stops the source transmissions when it gets that message.

Congestion control means preventing router buffers from getting congested. The destination may not be aware of such congestion. The source detects congestion either by noticing that the acknowledgments of its packets are excessively delayed or that acknowledgments fail to arrive and that packets must have been lost. When the source believes that the link is congested, it reduces the window size it uses (even if the receiver does not reduce W_{max}). Otherwise, the source keeps increasing the window size. The window adjustment algorithm aims at sharing the network capacity *fairly* among the active connections.

In this section we study congestion control. We start our discussion by examining the objectives and mechanisms of congestion control. We then explore some mechanism in detail.

6.11.1   Objectives and Mechanisms

The objective of congestion control is to share the network resources efficiently and fairly. Efficiently means that the links are used close to their capacity and that the backlogs in the buffers are small. Fairly is more difficult to define. One definition is that a fair allocation maximizes the minimum average rate per connection.

Mechanisms for congestion control include reservation and pricing, priority, rate control, and window control. The mechanisms depend on the information they use: The mechanisms can be based on delays (measured round-trip times), on losses of packets (duplicated acknowledgments or time-outs), and on information from routers about congestion.

Consider N connections that happen to share one 10-Mbps link. An efficient and fair mechanism allocates $10/N$ Mbps per connection. If these connections are video conferences that need 0.2 Mbps to work satisfactorily, it is necessary to limit N to 50. To limit N, the network must keep track of committed resources and implement some form of call admission control. An alternative approach, which is still being pursued by some network experts, is to hope that the transmission rate of the links increases fast and that progress in video compression is so remarkable that the network is able to keep up with a rapidly increasing demand without requiring admission control and resource reservation.

6.11.2   Delay/Window Mechanism (Vegas)

We first explain a simple congestion control mechanism that adjusts the window of a retransmission protocol based on round-trip times. Figure 6.24 illustrates the setup where a single connection uses a bottleneck router with transmission rate C packets per second and with a round-trip delay equal to D seconds when there is no queue in the buffer. The source uses a window with size W packets and we denote the queue size by x packets. The measured round-trip time is T seconds. That is, T is the delay of a typical acknowledgment. The

FIGURE 6.24

A single connection through one bottleneck router. D is the round-trip delay when there is no backlog, i.e., when $x = 0$. T is the measured round-trip delay, W is the window size, and R is the transmission rate.

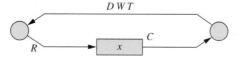

source transmits at an average rate of R packets per second. Note that with the definitions of the figure, about RD packets or acknowledgments are stored in the pipe from the source to the destination and back. Indeed, the source injects packets at rate R and the pipe has a delay of D. Accordingly, the window size W must be equal to RD plus the packets stored in the router. Hence, $W = RD + x$, or

$$x = W - RD. \tag{6.8}$$

To evaluate D, the source calculates $\min\{T\}$, the minimum of the round-trip time it has observed. From this, we conclude that the source can use a simple algorithm to adjust x to a small target value, say $x = 2$: If $x \geq 3$, then reduce W; If $x \leq 1$, then increase W.

Let's try to see how this algorithm would behave if many connections were to implement it. This situation is shown in Figure 6.25.

We assume that the N sources use the following algorithm, called Vegas. For $k = 1, \ldots, N$, source k adjusts W_k according to following rules:

Algorithm (Vegas):

$$\text{If } W_k - R_k D_k \leq 1, \text{ then increase } W_k;$$
$$\text{If } W_k - R_k D_k \geq 3, \text{ then decrease } W_k.$$

The justification for this algorithm is that $W_k - R_k D_k$ is an estimate of x_k, the backlog of connection k inside the router. Accordingly, if the algorithm converges and results in x_k being close to 2, then all the connections will have an equal share of the router and have R_k close to C/N. Moreover, the total backlog will be rather small and the algorithm achieves a fair and efficient sharing of the resources.

6.11.3   *Loss/Window Mechanisms (Tahoe, Reno)*

We go back to the case of one connection shown in Figure 6.26 and we consider mechanisms that adjust the window based on observed losses of packets. The source notices that a packet has been lost when the timer for the ACK of that packets runs out (time-out) or when duplicated ACKs arrive at the source (assuming that the ACK number is that of the next byte or next packet expected, as in TCP).

When the source detects a loss, it thinks that the loss is due to congestion and decreases the window W of its retransmission protocol. In the absence of losses, the source increases its window size to "fill up the pipe." Let's try a naive protocol first:

Algorithm (Naive). Source adjusts W according to the following rules:

While no loss, increase W at the rate of u units per second;
When loss, set $W = 1$.

FIGURE 6.25	**FIGURE 6.26**
N connections share a single bottleneck router.	*A single connection through a router.*

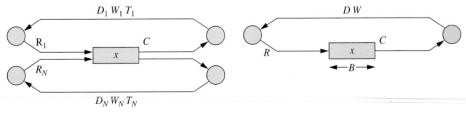

For W small, $R = W/D$ and $x = 0$. When W reaches CD, x starts growing and R saturates to C. Thus, as before, $x = W - RD$ and $R = C$ when $x > 0$. A packet is lost when x reaches B, i.e., when W reaches $CD + B$. With our adjustment mechanism, we expect W to grow from 1 to $CD + B + uD$ when we detect the loss (with a delay of D) and then the window drops to 1. Figure 6.27 shows the evolution of the window size with the naive algorithm. This is not very good: the average window size is about half of the desired value (assuming u not too large), we lose many packets that have to be retransmitted.

How do we increase the window size at a constant rate u? One way to do this is as follows. Assume that the window size is W. The source increases the window size by $1/W$ every time it gets an ACK. (The window size is measured in bytes and one unit here is some number of bytes, say 64 Kbytes. Accordingly, if $W = 16$, $1/W$ is 4 kbytes.) During one round-trip time D, the source gets back about W ACKs and increases the window W times by about $1/W$ each time, so that it increases the window size by one unit every D seconds. Thus, the window size increases linearly with rate $u = 1/D$. Note that the rate depends on D; this fact will come back to haunt us later.

We can use three simple ideas to improve the naive algorithm. We look at these ideas one by one.

First improvement (Slow Start): Increase window size fast initially and increase more slowly when we approach the value that was being used when the loss occurred. Thus, if the loss occurs when the window size is W, we set $V = W/2$, reset $W = 1$, increase W fast until W reaches V, then increase W linearly as before (this last precaution is called congestion avoidance). How do we increase the window size fast? Here is a simple method: The source increases the window size W by 1 every time it gets an ACK. As a result, during one round-trip time D the source gets W ACKs and increases the window size by W. Accordingly, W doubles every D seconds and W grows exponentially: $W(t) = 2^{t/D}$, roughly. The naive algorithm with this first improvement is called slow start with congestion avoidance.

Second Improvement (Fast Retransmit): When the source detects a third duplicated ACK, it retransmits the lost packet and decreases the window size from W to W/2. Recall that in TCP the destination host numbers an ACK by the sequence number of the byte that the destination expects next. Consequently, if a packet is lost, the source receives a number of ACKs with the same sequence number. These ACKs are called duplicated ACKs. The source wait for a third duplicated ACK to make sure that the packet was not simply delayed. We discuss below the motivation for this multiplicative decrease of W (by 50 percent). The motivation for not going down to 1 is clear by looking at Figure 3. The name fast retransmit refers to the fact that the source does not wait for a time out to occur before it retransmits.

FIGURE 6.27

Evolution of the window size with the "naive" algorithm.

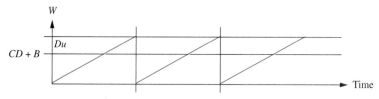

Third Improvement (Fast Recovery): The objective of this modification is to better control the backlog inside the pipe than by simply resetting the window size. The steps in Figure 6.28 explain the mechanism. The source gets a third duplicated ACK when it was using the window size $W = 2K$. The source retransmits the lost packet L and resets the window to $W = K + 3$. Whenever it gets an ACK, the source increases the window W by 1. When W reaches the value $2K + 1$, the source can transmit new packets again. Eventually, the source gets the ACK of the retransmitted packet L. This ACK signals the end of the fast recovery period. The source then resets W to $W = K$ and resumes the normal operations of the congestion avoidance mechanism. The advantage of this scheme is that it keeps the pipe busy with K packets. Instead of flushing the pipe, the fast recovery reduces the backlog from $2K$ to K. In the history (now, folklore) of TCP fine-tuning, the original implementation, Tahoe, incorporated slow start and congestion avoidance. The next version, Reno, included fast retransmit and fast recovery.

6.11.4 Additive Increase—Multiplicative Decrease

What happens when a number of connections adjust their transmission rate? How can we hope that the adjustments converge to a fair allocation of the transmission rate? We discuss

FIGURE 6.28

Details of the fast recovery mechanism.

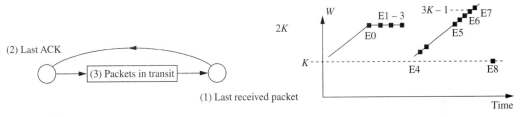

(2)	(3) *Packets in Transit*	(1)	*W*	*Event*
$L-1$	$L+2K-1, \ldots, L+3, L+2, L+1, L$	$L-1$	$2K$	[E0] L was lost
$L-1$	$L+2K-1, \ldots, L+3, L+2$	$L+1$	$2K$	[E1] 1st dup ACK: Don't change W
$L-1$	$L+2K-1, \ldots, L+3$	$L+2$	$2K$	[E2] 2nd dup ACK: Don't change W
$L-1$	$L, L+2K-1, \ldots, L+4$	$L+3$	$K+3$	[E3] 3rd dup ACK: Retransmit L, reset W
$L-1$	$L, L+2K-1, \ldots, L+5$	$L+4$	$K+4$	[E4] $W+1$ for each ACK (even dup)
.	...	...	...	Continue
$L-1$	$L+2K, L, L+2K-1, \ldots, L+K$	$L+K-1$	$2K+1$	[E5] W large enough => send new packet
$L-1$	$L+2K+1, L, L+2K-1, \ldots, L+K+1$	$L+K$	$2K+2$	Continue increasing W, sending new packets
.	...	...	...	
$L-1$	$L+3K-2, \ldots, L$	$L+2K-1$	$3K-1$	[E6] Continue
$L+2K-1$	$L+3K-1, \ldots, L+2K$	L	$3K$	[E7] End of fast recovery
$L+2K$	$L+3K, \ldots, L+2K+1$	$L+2K$	K	[E8] Reset $W = K$ and resume congestion avoidance

a simple argument that shows that one particular mechanism, called additive increase and multiplicative decrease, has a chance to converge to such a fair allocation. Figure 6.29 summarizes the evolution of the transmission rates of two connections that share one bottleneck.

The figure shows that the two sources increase their rate by the same amount when the router is not congested, i.e., when the sum of the rates is less than C. Such an increase corresponds to the arrows. When the sum of the rates exceeds C, the sources find out about the congestion and reduce their rates by a common factor. Such a decrease corresponds to a black arrow that points toward the origin $(0, 0)$. Iterating this process, we see that the two rates converge to $C/2$. To try to make all the sources aware of the congestion, instead of only the source that happened to lose a packet, some routers implement a random early discard of packets when they become congested. Such routers are called RED. Of course, when implementing such an algorithm, the sources do not update their rates by the same amount. When there is no congestion, we learned that the rate of increase depends on the round-trip delay. If we take this fact into account, we find that additive increase and multiplicative decrease does not converge to a fair allocation.

Another shortcoming of this mechanism is that if the rate is adjusted by modifying the window size of a retransmission protocol, then some care is required to avoid making the mechanism even more unfair. To see this effect, assume that one modifies the algorithm by applying the additive increase and multiplicative decrease to the window sizes instead of to the rates. Arguing by analogy, we can expect the algorithm to adjust the windows so that they become close to each other. However, the same window size corresponds to a smaller rate for a connection with a longer round-trip delay. This effect combines with the previous one and results in a highly unfair sharing of the common link.

6.11.5   Incompatibility of Reno and Vegas

Now for a bit of controversy. We have seen that Vegas converges (at least in the simplistic case of a single bottleneck) to a fair allocation. However, most implementations of TCP use Reno. What happens if a source implements Vegas and others implement Reno? Consider Figure 6.30.

FIGURE 6.29

Convergence of the additive increase—multiplicative decrease algorithm.

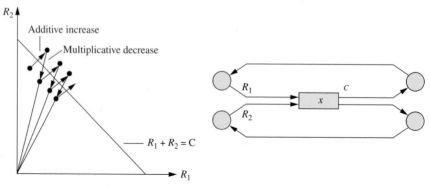

As we explained, Vegas tries to maintain x_2 close to 2. On the other hand, Reno pushes x all the way to overflow, then backs off more or less cleverly. In any case, Reno ends up with x_1 substantially larger than x_2. Since the rate of a connection is proportional to the fraction of its packets in the router, we see that Vegas loses in this competition. The situation is much worse if a connection deliberately does not back off when the router is congested. For instance, you can implement a connection over UDP and ignore all delays in ACKs. All the Reno and Vegas connection would back off to make room for you.

6.11.6   Rate-Based Control: ABR in ATM

So far we have looked at window-based congestion control. We now examine one example of rate-based control. Such a form of control is implemented in a service called "Available Bit Rate" (ABR) in asynchronous transfer mode networks.

Figure 6.31 illustrates a very simple example of one for of ABR. The switch selects a target value B^* for its buffer occupancy and some rate $C^* > C$. In addition, the switch monitors the number N of ABR connections that are active. (Many connections may be set up but dormant.) The sources have to limit their transmission rate (with a leaky bucket) to the rate E (for explicit) advertised by the switch. If all the sources follow this algorithm and are eager to transmit when they are active, then the total rate into the buffer is C^* when $x < B^*$ and is 0 when $x > B^*$. Accordingly, the rate of increase of x is given by

$$\frac{dx(t)}{dt} = \begin{cases} C^* - C, & \text{if } x(t-d) < B^* \\ -C, & \text{if } x(t-d) > B^* \end{cases} \tag{6.9}$$

where d is the round trip delay between the sources and the router (which we assume the

FIGURE 6.30

Two connections share a buffer. One connection uses Reno and the other uses Vegas.

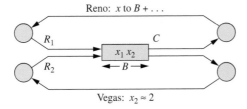

FIGURE 6.31

Explicit rate control in ABR.

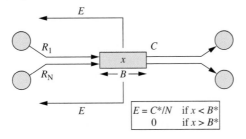

same for all sources). If we take into account the inequal round-trip propagation delays and we assume that these delays are all bounded by T, we see that the sources might, in the worst case, transmit at rate C^* for some time T even after x has crossed the value B^*. When this happens, the largest possible value of x is $B^* + (C^* - C)T$. Similarly, the lowest value of x if all the sources are always eager is $B^* - CT$. We can calculate the values of C^* and B^* that lead to efficient use of the switch ($x > 0$ always) and no overflow ($x < B$ always).

This scheme is fair because all the sources use the same explicit rate.

When there are multiple switches, they each advertise an explicit rate and the sources have to respect the minimum value they received from the switches.

6.11.7   Detecting Late Acknowledgments

The source detects late acknowledgments as follows. The source maintains an estimate τ of the average delay of recent acknowledgments and an estimate D of the average absolute value of recent errors of this estimate. The source decides that an acknowledgment is late if its delay T (measured from the time the packet was sent until the acknowledgment came back) exceeds $R + 4D$.

The values of τ and D are obtained by first order low pass filters of the successive values of T and $|T - \tau|$, respectively. That is,

$$\tau = a\tau + (1 - a)T \text{ and } D = bD + (1 - b)|T - \tau|. \tag{6.10}$$

A typical implementation uses $a = 7/8$ and $b = 3/4$, which makes the calculations simple in integer arithmetics.

Let us explain how the relations (6.10) calculate average values of recent quantities. We write the first relation as

$$\tau_n = a\tau_{n-1} + (1 - a)T_n \tag{6.11}$$

where T_n is the delay of the nth acknowledgment that arrives at the source (for $n \geq 1$) and τ_n is the nth update of the estimate τ. One chooses $\tau_0 = 0$. The above equalities imply that

$$\tau_n = (1 - a) \sum_{m=1}^{n} a^{n-m} T_m, \text{ for } n \geq 1, \tag{6.12}$$

as you can check by induction on n. The equality (6.12) shows that τ_n is an exponentially weighted average of the $T_n, T_{n-1}, T_{n-2}, \ldots, T_1$. The weights are more important for recent values of the T_m than for older ones. Note also that if the times T_n happen to converge to some value T, then so does τ_n. The same considerations apply to the second relations in (6.10).

Note that the source should not update the estimate when it receives the acknowledgment of a retransmitted packet. Indeed, the source cannot distinguish a late acknowledgment of the original transmission from the acknowledgments of the retransmitted packet and a wrong guess would throw off the delay estimate. Since the source stops updating its estimate, it might happen that the estimate is always too small and that the source keeps on retransmitting. To prevent this situation from persisting, the source increases its timeout value geometrically when it receives the acknowledgment of a retransmitted packet.

6.12 Complement 5: Efficiency of Protocols in the Presence of Errors

Not surprisingly, transmission errors reduce the efficiency of retransmission protocols. To quantify this effect, we study the efficiency of GO BACK N when packets are corrupted independently of one another.

For simplicity, we assume that the window size of GO BACK N has the smallest value that results in an efficiency of 100 percent in the absence of error. That is, we assume that

$$T = S = W \times \tau. \tag{6.13}$$

This assumption means that the window size is just large enough that the source gets the acknowledgment of the first packet when it has transmitted W packets. Also, the duration of the timeout T is exactly matched to the round-trip delay S. Note that in practice the round-trip delay is not the same for all the acknowledgments so that our assumptions are somewhat unrealistic.

The left part of Figure 6.32 shows a sequence of events with GO BACK N under these assumptions. As the figure indicates, an error in packet k or in its ACK eventually leads the sender to waste an amount of time equal to $T = S$ before restarting the transmissions from packet k. If there is no error during the transmission of packet k or of its ACK, then the sender takes a time equal to τ before transmitting the next packet. As a consequence, the evolution of GO BACK N can be represented as in the right part of Figure 6.32. The circle marked n is *state n;* it designates a stage in the execution of the GO BACK N protocol. Specifically, state n represents the start of the transmission of a sequence of n packets. Similarly, state $n - 1$ represents the start of the transmission of a sequence of $n - 1$ packets.

We denote by p the probability that both the packet and its ACK arrive correctly. The horizontal arrow from state n to state $n - 1$ signifies that with probability p the first of the n packets (here, packet k) and its ACK are correctly transmitted, and the sender starts transmitting the first of the remaining $n - 1$ packets after a time equal to τ. The circular arrow from state n to itself signifies that, with probability $1 - p$, the first of the n packets or its ACK is not correctly transmitted; the sender restarts the transmission of the n packets after having wasted an amount of time equal $T = S$. One finds that the average time $E(X)$ per packet, i.e., to move from state n to state $n - 1$, is such that

$$E(X) = p\tau + (1 - p)\{S + E(X)\}.$$

FIGURE 6.32

GO BACK N with transmission errors. The right part of the figure shows the evolution of the protocol "going back" with probability $1 - p$ and progressing with probability p.

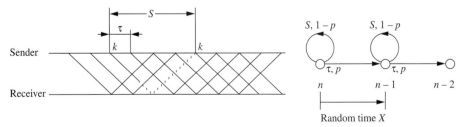

Solving this equation for $E(X)$, we find that

$$E(X) = \tau + \frac{(1-p)S}{p}.$$

Since the useful transmission time per packet is τ, the transmission time, we conclude that the efficiency η_{GBN} of GO BACK N is given by

$$\eta_{\text{GBN}} = \frac{\tau}{E(X)} = \frac{\tau}{\tau + \frac{(1-p)S}{p}}.$$

Hence,

$$\eta_{\text{GBN}} = \frac{1}{1 + \frac{1-p}{p}\frac{S}{\tau}} \quad \text{when } S = T = W\tau. \tag{6.14}$$

For instance, if $S = 20\tau$, one finds $\eta_{\text{GBN}} = 31\%$ for $p = 0.9$ and $\eta_{\text{GBN}} = 83\%$ for $p = 0.99$. For comparison, using formula (6.2), assuming $T = \tau$ in ABP, we find $\eta_{\text{ABP}} \approx 4.9\%$ for $p = 0.9$ and $\eta_{\text{ABP}} \approx 5\%$ for $p = 0.99$.

Summary

- The data link layer converts a bit pipe into a packet link. The layer implements two tasks to implement the packet transmission service: framing and error control.
- Framing enables the receiver to locate packets into the bit stream and adds control information that the sender and receiver need to monitor the transmissions.
- Error control uses either error correction or error detection codes. When error detection is used, the data link layer arranges for the sender to retransmit erroneous packets. Error correction is used when retransmissions are not possible or not practical. Error correction codes add enough redundant information in the packets to enable the receiver to correct the transmission errors.
- Retransmission protocols keep track of the packets that need to be retransmitted. These protocols use acknowledgments and timers. More complex protocols are more efficient when the round-trip delay is large compared to a packet transmission time. The stop-and-wait protocol is the simplest protocol that operates correctly if one has a bound on the acknowledgment delays. ABP works correctly without that information. GO BACK N and SRP are more efficient retransmission protocols than ABP.
- The correctness of protocols can be verified formally by using finite-state machine descriptions of the protocols. An algorithm explores all the possible evolutions of the protocols and detects if there is an evolution that is not correct. The correctness of some protocols (e.g., GO BACK N) can be proved by induction arguments.
- The efficiency of a retransmission protocol is defined as the fraction of the line transmission rate that is actually used to transmit packets. This efficiency can be analyzed from the parameters of the link and the protocols.
- The CRC, a widely used error detection code, appends a group of bits to a packet. By checking the CRC, the receiver can detect most likely transmission errors.

- The number of errors that an error control code can detect or correct depends on the minimum distance between two codewords. The BCH and RS codes are constructed by algebraic methods to have large minimum distances. The convolutional codes have a particularly simple maximum likelihood decoding algorithm. The turbo codes exhibit remarkable performance over very noisy channels.
- Retransmission protocols form the basis of flow control and congestion control algorithms: By reducing the window size of a retransmission protocol when it detects congestion, the source can adapt its transmission rate to the available capacity of the network. Designing fair congestion control algorithms that utilize the network capacity efficiently is a challenging problem that is not satisfactorily solved to date.

Problems

Problems with a * are somewhat more challenging. Those with a ^c are based on material in a complement. Problems with a ^1 are borrowed from the first edition.

1. Consider a satellite connection with a one-way propagation time of 0.27 s from the sending ground station to the receiving ground station (via the satellite) and a transmission rate of 1 Mbps with packets of 1000 bits. What is the minimum window size of GO BACK N that results in an efficiency of 100 percent?

^c2. A half-duplex link transmits a packet correctly with probability 0.7 in each direction. You use ABP. How many times do you expect to send each packet, on average?

3. Consider the system shown in Figure 6.33. The packets and acknowledgments have 1000 bits. Each wireless link is half-duplex.
   a. Assume that the packet transmission error rate is negligible. Calculate the maximum throughput from *A* to *B* when each wireless link uses ABP.
   b. Repeat the calculation assuming now that ABP is used from end to end between *A* and *B*.

^c4. Repeat Problem 3 assuming that the wireless links each have a PER = 0.1 in each direction.

5. Consider the network shown in Figure 6.34. The links are 10-km fibers, full-duplex at 45 Mbps. We examine transmissions of 53-byte ATM cells by store-and-forward. Node *A* sends a packets to *B*, using GO BACK N from end to end. The acknowledgments are also 53 bytes long. We assume that the processing time is 3 μs for every cell in each node along the path and 10 μs in the end node (*B* for data and *A*

FIGURE 6.33

A network path with two wireless and one fiber link.

FIGURE 6.34

Network for Problem 5.

for acknowledgments). Calculate the smallest window size for GO BACK N that achieves 100 percent efficiency in the absence of errors. (*Hint*: Draw a timing diagram.)

6. Figure 6.8 shows the state machine of SWP. Draw the corresponding state machine for ABP.

*7. Consider ABP, as in Section 6.4, but model the acknowledgment delay S and the packet transmission time τ as random variables. Assume that the packet error rate is negligible. Analyze the efficiency of the protocol.

^{c}8. Calculate the CRC that corresponds to the message $\mathbf{u} = 1001001011$ and the generator $\mathbf{g} = 10011$.

^{c}9. In this problem we examine $\mathbf{GF}(2^3)$. Let $f(x) = 1 + x + x^3$. Show that $f(x)$ is a prime polynomial. Derive the addition and multiplication table of $\mathbf{GF}(2^3)$ based on this polynomial. Next repeat the problem with $f'(x) = 1 + x^2 + x^3$. In what sense are the two constructions of $\mathbf{GF}(2^3)$ based of $f(x)$ and $f'(x)$ equivalent?

^{c}10. Construct the BCH(7, 2) code. What is its minimum distance?

^{c}11. Construct the RS(7,3) code.

^{c}12. A convolutional code is defined by $c_{2n-2} = u_{n-1} + u_{n-4}$ and $c_{2n-1} = u_{n-1} + u_{n-3} + u_{n-2}$. What is the rate of this code? Draw its state diagram and explain the maximum likelihood decoding of the code.

^{c,*}13. Discuss the correctness of SRP for non-FIFO channels.

^{c}14. In Section 6.11 we claim that the additive increase–multiplicative decrease applied to window sizes is unfair to connections with larger round-trip delays. Provide the analysis that supports that claim.

References

Chapter 2 in Bertsekas (1992) was our main source for this chapter. That chapter presents a useful discussion of framing and of retransmission protocols, including an original discussion of the overhead imposed by framing. Tanenbaum (1996) describes a number of examples of the data link layer. Our discussion of error control codes borrows heavily from Lin (1983). Turbo codes are explained in McEliece (1995). Har 'El (1990) explains a software system (COSPAN) for protocol verification. Protocol verification is studied in Holtzmann (1991) and Aggrawal (1990). For the flow control in TCP/IP, see Jacobson (1988) and Jain (1989). The study of the correctness of protocols with non-FIFO channels was done in collaboration with Jean-Yves Le Boudec. Section 6.11 was written with J. Mo.

Bibliography

Aggrawal, S., Courcoubetis, C., and Wolper, P., "Adding liveness properties to coupled finite state machines," *ACM TOPLAS,* April 1990.

Bertsekas, D., and Gallager, R., *Data Networks,* Prentice-Hall, 1992.

Har' El and Kurshan, R. P., "Software for analytical development of communications protocol," *AT&T Tech. J.,* pp. 45–59, January/February, 1990.

Holtzmann, G., *Design and Validation of Computer Protocols,* Prentice-Hall, 1991.

Jacobson, V., "Congestion avoidance and control," *Proc. SIGCOMM '88 Symposium,* pp. 314–329, August 1988.

Jain, R., "A delay-based approach for congestion avoidance in interconnected heterogeneous computer networks," *ACM Computer Communication Review,* 19, pp. 56–71, October 1989.

Lin, S., and Costello, D. J., *Error Control Coding,* Prentice-Hall, 1983.

McEliece, R. J., Rodemich, E. R., and Cheng, J.-F., "The Turbo decision algorithm," *33rd Allerton Conference on Communication, Control, and Computing,* October 1995.

Tanenbaum, A. S., *Computer Networks,* 3rd ed., Prentice-Hall, 1996.

CHAPTER 7 Physical Layer

In this chapter we examine the technologies of communication links: optical links, copper lines, and radio waves. Some understanding of these systems is part of a basic technological literacy and helps one to appreciate the characteristics of networks. For readers not interested in details, we summarize the key ideas and results in Section 7.1. The subsequent sections are more detailed and may be skimmed by non–electrical engineering majors. Sections 7.2, 7.3, and 7.4 examine optical links, copper lines, and radio links, respectively. Each section starts with an overview of the technology.

The complements provide more details about some aspects of implementations and of communication theory. These complements can be read independently of one another.

In the first section we explain how links transport bits and the characteristics of the links having the most impact on their performance.

7.1 Communication Links and Their Characteristics

In this book we consider only communication links that use electromagnetic waves. We ignore acoustic links even though they are useful in some specialized applications.

7.1.1 Digital Link

A digital link is illustrated in Figure 7.1. The link delivers bits by first converting them into signals that propagate through a channel. The receiver converts the signals back into bits. Seen as a network element, the link is characterized by its bit rate, bit error rate, and distance.

The signals propagate through the medium as electromagnetic waves. These waves may be guided by an optical fiber, a pair of wires, or a cable, or the waves may propagate in free space as radio waves or as optical waves.

The graphs in Figure 7.2 show the maximum length as a function of the bit rate for communication links built with a coaxial cable, a wire pair, and three types of optical fiber:

Digital link. The main components of a digital communication link are a modulator, the channel, and a demodulator. The figure also sketches the signals.

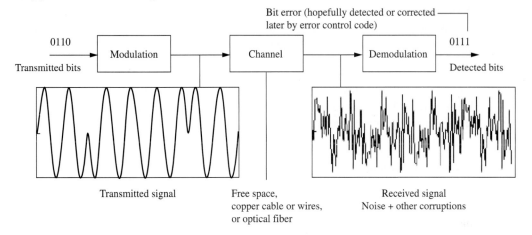

Achievable distance as a function of the bit rate for optical links, wire pairs, and cable. (The graphs are approximate.)

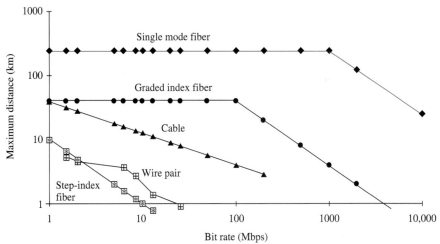

single mode, graded index (GRIN), and step index. Section 7.2 describes these fibers. The graphs are drawn for typical operating conditions. For instance, you can see from the figure that if the bit rate is 5 Mbps, then the maximum length is about 20 km for a coaxial cable, 4 km for a wire pair, and 2 km for a step-index fiber; it is 40 km for a GRIN fiber and 200 km for a single-mode fiber. (These numbers are approximate.)

You should be aware that some characteristics other than the bit rate and distance are important when choosing a link technology. For instance, for high-speed access to Internet, the choice between cable modem and high-speed modem over telephone lines depends on a number of factors. Although cable has a larger bit rate than a wire pair, the cable of the CATV plant is shared by up to a few hundred users whereas the telephone line is not shared. A shared channel requires a MAC protocol and security, and is more prone to interference.

We now examine the propagation of the electromagnetic waves and the limitations on the transmission rate and distance of links. These sections explain how the graphs in Figure 7.2 are obtained.

We then discuss the conversion of bits into signals (called *modulation*). In particular, we discuss the tradeoff between the range of frequencies that the signals occupy and the bit error rate. This tradeoff is important because applications differ in the scarcity of the frequencies and in the susceptibility to errors.

7.1.2   *Frequency and Propagation*

We consider links where electromagnetic waves transport the bits. Two qualities of electromagnetic waves make them suitable for transporting information in a network. First, they *propagate,* enabling them to move from one place to another, as from a transmitter to a receiver. Second, they contain energy that can be used to carry messages. We know, from quantum mechanics, that this energy should be viewed as being carried by *photons.* One can think of a photon as a minute burst of electromagnetic energy. An electromagnetic wave is a stream of photons. The energy of a single photon is so minuscule that an ordinary lightbulb emits about 10^{20} photons every second. In this section we look at the propagation of electromagnetic waves to see how they transport bits between the nodes of a network. We explain that electromagnetic waves are guided by using their different propagation properties in different media. Waves are modified, distorted, and weakened during their propagation. The telecommunications engineer takes these modifications into account when designing the transmission equipment. We discuss how the wave modifications limit the rate at which a link can transmit messages reliably.

The propagation of electromagnetic waves is described by the equations postulated in 1863 by James Clark Maxwell, an ingenious Scottish physicist and mathematician. Maxwell understood that the phenomenon of propagation is caused by the interactions of an oscillating electrical field and an oscillating magnetic field "pushing" one another through empty space or some other medium. These ideas developed from Michael Faraday's work. Faraday had shown earlier that a changing magnetic field generates an electrical field; this effect is at work in a car alternator. Maxwell argued that in an electromagnetic wave a changing magnetic field induces a changing electrical field, which in turn generates a changing magnetic field, and so on, thus causing the wave to propagate. The left part of Figure 7.3 illustrates the interaction of the varying electrical field E and magnetic field M.

Light is a familiar example of propagating electromagnetic waves. *Radio waves* are electromagnetic waves, too, but they differ from light waves by their *frequency,*, which is lower by a few orders of magnitude. The frequency is the number of oscillations of the electric and magnetic fields in one second. Frequency is measured in hertz; one hertz is written as 1 Hz, and it corresponds to one full oscillation per second. (See the center part of Figure 7.3.) Thus, the electric and magnetic fields of an electromagnetic wave with a frequency of 10^8 Hz make 10^8 oscillations per second. In the case of visible light, different frequencies are perceived by the eyes and the brain as different colors. The

FIGURE 7.3

Electromagnetic waves: Their propagation (left), frequency (center), and typical frequency ranges (right).

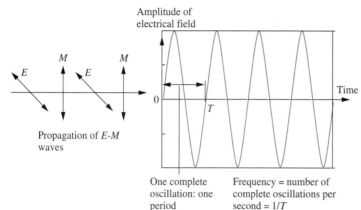

Amplitude of electrical field

Frequency ranges:

Twisted pairs: 0–a few 100 kHz
Coaxial cable: a few 100 kHz–1 GHz
Waveguide: 1 GHz–a few hundred GHz
Optical fiber: $\approx$ 100 THz (10^{14} Hz)
AM radio: $\approx$ 1 MHz
VHF TV: 20 MHz–80 MHz
FM radio: 88 MHz–108 MHz
UHF TV: 300 MHz–600 MHz
Cellular phones: 850 MHz, 1.2 GHz, . . .
Satellite: 1 GHz–100 GHz

Propagation of E-M waves

One complete oscillation: one period

Frequency = number of complete oscillations per second = 1/T

propagation speed of an electromagnetic wave in a vacuum or in air is approximately equal to $c := 3 \times 10^8$ meters per second. Thus, light from the moon, which is about 400,000 km from Earth, reaches us in slightly more than 1 second.

Frequencies of typical electromagnetic waves are between 88 MHz (1 MHz = 1 megahertz = 10^6 Hz) and 110 MHz for FM stations, from 30 MHz to 300 MHz for TV stations. Satellites transmit microwaves (above 1 GHz = 1 gigahertz = 10^9 Hz). Visible light covers the frequencies from 4×10^{14} Hz (red light) to 7×10^{14} Hz (blue light). See the right part of Figure 7.3 for typical ranges of frequencies transported by cables and wires and used in various applications.

An electromagnetic wave can be described by its *wavelength* instead of by its frequency. The wavelength is defined as the speed of propagation in a vacuum (c) divided by the frequency. For instance, the wavelength of red light is the speed of light (c) divided by the frequency (4×10^{14} Hz) and is, therefore, equal to 0.75 μm. (1 μm = 1 micrometer = 10^{-6}m.) Thus, the *wavelength* of visible light ranges from about 0.43 μm for blue light to 0.7 μm for red light. The wavelengths 0.8 μm, 1.3 μm, and 1.5 μm are in the *infrared range;* i.e., they correspond to frequencies less than that of red light; those are typical wavelengths in optical fiber communication links.

Communication networks transmit information over electromagnetic waves that propagate in air, with radios, or in vacuum, with satellites. Communication networks also use transmission media that guide the propagation of electromagnetic waves, such as optical fibers and copper lines and cables. The electrons in the wires or the cable of a copper line interact with the electromagnetic wave and guide it.

After having discussed the general propagation of electromagnetic waves, we examine the sources of limitations to the transmission of bits by electromagnetic waves.

7.1.3   Limitations

There are four basic phenomena that limit the rate and distance of bit transmissions by electromagnetic waves: attenuation, distortion, dispersion, and noise. We examine each of these phenomena, illustrated in Figure 7.4, and describe the limitations they introduce.

FIGURE 7.4

Four basic phenomena that limit the transmission of bits by electromagnetic waves.

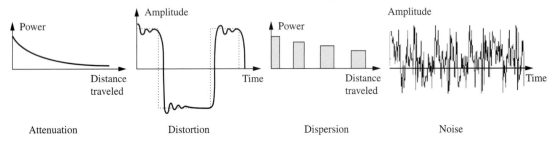

Attenuation   Distortion   Dispersion   Noise

Attenuation

Attenuation is the loss of energy of the wave as it propagates. When a wave propagates in a fiber or a transmission line, some of its energy is absorbed and converted into heat. Attenuation is illustrated in the left part of Figure 7.4. An electromagnetic wave that propagates in a vacuum or in the air spreads out as a cone or as a sphere. Consequently, the energy of the wave per unit area decreases as the wave propagates. Accordingly, when the wave reaches an antenna or an optical detector, that device captures an energy that has decreased with the propagation distance. Electromagnetic waves may also be reflected, refracted, or absorbed by physical objects in their path. The absorption is called *shadowing* because of the analogy with visible light. The antenna may capture the superposition of electromagnetic waves sent by the same transmitter that have been reflected or diffracted differently. Such superposition modifies the power the antenna receives.

After propagating over some distance, the guided or unguided electromagnetic wave does not have enough power for a receiver to detect it reliably. Thus, attenuation limits the usable propagation distance. The maximum usable distance depends on the rate at which the propagation reduces the power and on the amount of power that the wave can lose before becoming too weak for the receiver. The attenuation in a transmission line or fiber depends on the frequency of the electromagnetic wave. The attenuation in radio transmissions depends on the shadowing and reflections.

Distortion

We call a superposition of electromagnetic waves a *signal.* Remarkably, a signal with a single frequency keeps its frequency as it propagates, although the propagation attenuates and delays the signal. When a signal is composed of multiple frequencies, its propagation in a transmission link or fiber attenuates and delays differently the different frequencies. As a result, the shape of the received signal differs from that of the transmitted signal. Communication engineers call this shape modification *distortion.* Figure 7.4 illustrates the distortion of a signal. When it leaves the transmitter, the amplitude of the signal alternates between two opposite values. After transmitting through a cable, this amplitude has a more complicated shape. Cable or wire pair transmission links use a device called an *equalizer* to compensate for these different attenuations at different frequencies. The equalizer amplifies differently the various frequencies to make up for the different attenuations.

Dispersion

A burst of electromagnetic energy that a transmitter sends in wires, a cable, or an optical fiber spreads as it propagates. Communication engineers call this phenomenon *dispersion.* Consequently, bursts sent in rapid succession tend to merge as they propagate and become harder to distinguish. Figure 7.4 sketches the spreading out of bursts as they propagate. If the bursts correspond to bits, these bits become difficult to detect after some distance, all the more so when the bursts are close together. Thus, dispersion limits the usable propagation distance to a maximum value that depends on the size of the bursts the transmitter sent, i.e., on the bit rate. We explain in Section 7.2 that dispersion limits the product $R \times L$ of the bit rate R times the distance L. The limit on $R \times L$ depends on how fast bursts spread as determined by the physics of the transmission medium.

Noise

A fourth phenomenon affects the transmission of bits: *noise.* Noise is an unpredictable variation in the signal that reaches the receiver. The physical causes of noise include the thermal agitation of electrons in conductors, uncertainty in the number of photons that a light source generates, and electromagnetic waves that sources other than the transmitter produce and the receiver picks up. Figure 7.4 illustrates the noise that corrupts a received signal.

Without noise it would be possible to transmit bits arbitrarily fast, at least as fast as the circuits can operate. To appreciate this fact, consider a transmission line, say a 1-km-long telephone line. Assume that when one applies a voltage of V volts across one end of the line, after a fixed time, say 1 ms, the voltage across the other end of the line stabilizes to the value $\alpha \times V$. Here, α captures the attenuation. If there were no noise, one could determine α by selecting $V = 1$ and measuring $\alpha \times V$. Thereafter, one could transmit an arbitrarily long string of bits $b_1 b_2 \cdots b_n$ by selecting $V = b_1 2^{-1} + b_2 2^{-2} + b_3 2^{-3} + \cdots + b_n 2^{-n}$. The receiver would then measure αV, calculate V by dividing by the known value of α, and recover the bits $b_1 b_2 \cdots b_n$ by writing out the binary expansion of V. Using this procedure, we could transmit n bits in two milliseconds: one millisecond to measure α and the other to send the bits. Since n was arbitrary, this argument shows that it would be possible to transmit bits arbitrarily fast over a noiseless transmission line.

The noise prevents us from measuring the voltage V accurately and therefore limits the number of bits in this binary expansion that can be recovered reliably. It seems therefore plausible that the larger the noise, the fewer the number of bits we could transmit in a given time. Accordingly, we suspect that the transmission rate decreases if the line is more noisy.

Also, if we can vary the voltage V faster, then we can send bits faster. The speed at which we can change the voltage and hope that these variations are reflected accurately (except for the noise) at the other end of the line is limited by the *bandwidth* of the transmission line. This bandwidth specifies the range of frequencies that the line can transmit. (See Figure 7.3 for typical values.)

The above arguments indicate that the rate at which one can transmit bits over a transmission line increases with the bandwidth of the line and decreases with the noise.

In 1948, Claude Shannon made these intuitive arguments precise. He explained that noise introduces a fundamental limit on the rate at which a communication channel can transmit bits reliably. Communication engineers call this limit the *capacity* of the channel.

For instance, if the capacity of a channel is equal to 30 kbps, then it is possible to design a transmitter and a receiver that transmit 29,999 bits per second over the channel with an error rate smaller than 10^{-9}, i.e., with fewer than 1 bit out of 10^9 being incorrectly received. Such a channel cannot transmit reliably faster than 30,000 bps. We explain the main concepts of Shannon's theory in Section 7.5.

A good transmission link makes few errors. For instance, the *bit error rate* of a typical optical fiber link is 10^{-12}. Such a link corrupts 1 bit out of 10^{12} bits, on average. If the transmission rate of the link is 155 Mbps, then one incorrect bit arrives every $10^{12}/155 \times 10^6$ s, i.e., about every 2 hours, on average.

Copper lines (wire pairs and coaxial cables) have larger bit error rates: 10^{-7} is typical. A transmission link that sends packets of N bits each with a bit error rate equal to BER corrupts some fraction of the packets. That fraction is the *packet error rate* PER of the link. Assume that the bit errors are *independent* (see Appendix A). The packet error rate PER is the probability that the N bits of one packet are not all received correctly and is equal to

$$ \text{PER} = 1 - (1 - \text{BER})^N . $$

You can verify that

$$ \text{PER} \approx N \times \text{BER} \qquad \text{if} \qquad N \times \text{BER} \ll 1. $$

For instance, if $N = 10^5$ and BER $= 10^{-7}$, then PER $\approx 10^{-2}$.

Whereas the bit error rate of a coaxial cable or optical fiber link can be made very small, the situation is very different in wireless links. In those communication links, the bit error rate can be as large as 10^{-3}. Moreover, the error rate fluctuates widely over time. The wireless receiver may find itself in a region where the power it receives is too low to recover the bits successfully. This effect that reduces the signal power is called *fading*. Fading is caused by the superposition of different reflected fractions of the transmitted electromagnetic wave that annihilate one another and by the shadowing that objects in the propagation path produce.

We can now revisit Figure 7.2 and explain it in the light of the preceding discussion. The attenuation limits the usable length. The maximum usable length imposed by the attenuation is indicated in Figure 7.2 by a horizontal line for optical fiber links. For a coaxial cable, the attenuation increases rapidly with the frequency and, consequently, with the bit rate. There is practical limit to the power of an electronic transmitter. As a result, the maximum length of a cable link decreases rapidly with the bit rate, as shown on the figure.

In a fiber link, the dispersion limits the product $R \times L$ of the bit rate times the length. This limitation corresponds to a limit on $\log (R \times L) = \log (R) + \log (L)$. Thus, the dispersion limit translates into a line with slope -1 in the figure since its axes have logarithmic scales. That dispersion limit is smallest for step-index fibers and largest for single-mode fibers, as we explain in the next section. As a result, the dispersion limits for these three types of fibers correspond to three distinct graphs, as the figure shows.

We have examined the four fundamental sources of limitations of transmission links. Next we explore the practical methods that the links use to transmit the bits. That is, we discuss the conversion between bits and signals.

7.1.4   Converting between Bits and Signals

The transmitter converts bits into signals. This conversion is called *modulation*. The receiver performs the reverse conversion: the *demodulation* from signals to bits. This conversion takes place in a *modem* (for modulator-demodulator) when the channel is a transmission line. When the transmission is optical or over radio waves, the conversions are made by a *transmitter* and a *receiver.*

Modulation schemes differ in their bit error rate and the range of frequencies in the signal they produce. As a rule, modulation schemes that occupy a smaller range of frequencies have a larger bit error rate. Thus, different modulation schemes are appropriate in different situations, depending on the scarcity of the frequencies and the noise level.

We describe a few representative modulation procedures for transmitting a group of bits. We indicate the technologies that use each method: copper lines (L) such as wire pairs and coaxial cables, optical fibers (F), and radio links (R).

Asynchronous Baseband Transmission (L)

Figure 7.5 illustrates the operations of an asynchronous baseband transmission system over a copper line. The transmitter and the receiver each have a clock with approximately the same rate R (Hz). The transmitter first groups the bits into short words of K bits. (For instance, $K = 8$.) Initially, the transmitter sets the voltage at its end of the transmission line to 0 volt. To send the first word, the transmitter, using its clock, sets the voltage successively to V volts during $T = 1/R$ seconds for each bit 0 in the word and to $-V$ volts for each bit 1. (For instance, $V = 5$.) Engineers call this representation of bits *bipolar modulation*. The transmitter then resets the voltage to 0 volt for at least some short duration before continuing with the subsequent words.

FIGURE 7.5

Asynchronous transmission over a transmission line. The transmitter and receiver have clocks with approximately the same frequency. The receiver starts its timing when the first bit arrives. The difference in clock rates might cause the receiver to operate incorrectly.

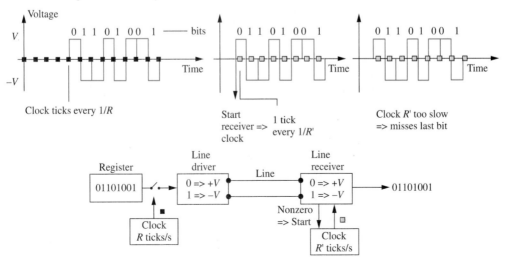

The receiver detects that a word is arriving when it notices that the line voltage jumps away from its initial zero value. At that time, the receiver starts its clock and uses it to determine when it should measure the line voltage to recover the successive bits. The receiver then gets ready to receive another word: it waits until the line voltage once again jumps away from zero.

Communication engineers call this method *asynchronous baseband transmission.* Asynchronous refers to the fact that the transmitter and receiver are not synchronized to each other. Transmissions of words of K bits can occur at arbitrary times. Baseband means that the fluctuations of the transmitted signal follow the changes of the successive bits and do not occur at higher frequencies.

The timing of the asynchronous transmission method limits the size K of the words and the bit rate R to small values. Indeed, since the receiver and transmitter clocks never have exactly the same rate, the timing of the receiver becomes less accurate as it keeps on measuring bits. Figure 7.5 shows that a small difference in the clock rates makes the receiver operate incorrectly if the bit string is too long.

The serial line (RS-232-C) between your computer and modem or printer uses this transmission method. (See Section 7.9 for details.)

Asynchronous Optical Transmission (F)

Some optical links use a procedure similar to asynchronous baseband transmission. The transmitter sends words of K bits by switching the light source ON for each bit 1 and OFF for each bit 0, in each case for T seconds. This procedure is called ON-OFF keying (OOK). The transmitter adds a leading 1 to each word. The receiver uses this start bit to detect the arrival of a word.

The infrared remote control of your TV set uses this transmission procedure.

Asynchronous Broadband Transmission (L, R)

This method groups bits into words and times the bits as in the previous methods. When using *frequency shift keying* (FSK), the transmitter sends a 0 by transmitting a sine wave of frequency f_0 and a 1 by transmitting a sine wave of frequency f_1, each during T seconds. With *binary phase shift keying* (BPSK), the transmitter sends a 0 by transmitting a sine wave of frequency f_0 and a 1 by transmitting a sine wave of the same frequency but whose phase differs by 180°. (See Figure 7.6.)

The communication engineer selects frequencies f_0 and f_1 that the transmission line transmits well. For instance, a telephone line is designed to transmit the frequencies between 300 Hz and 4000 Hz, which cover the main range of the human voice. A suitable choice for such a line is $f_0 = 1070$ Hz and $f_1 = 1270$ Hz.

Assume that a modem is sending a bit stream with rate 150 bits per seconds using BPSK with a frequency $f_0 = 600$ Hz. The resulting signal is a sequence of sine wave bursts with frequency f_0 and phases that vary as the bits being transmitted alternate between 0 and 180°. This signal is shown in Figure 7.7. The energy density diagram in the right part of Figure 7.7 shows how much energy the signal contains at the various frequencies. That is, if one filters the signal so that only the frequencies between f Hz and $f + 1$ Hz are retained (as we do in audio with a graphic analyzer), then the energy of the filtered signal is proportional to the energy density of the signal at the frequency f.

FIGURE 7.6

FSK and BPSK. To transmit bits using FSK, the transmitter sends a short sine wave burst per bit. The frequency of the burst depends on the bit 0 or 1 being transmitted. Using BPSK, the sine wave bursts have the same frequency for bits 0 and 1 but their phases are different.

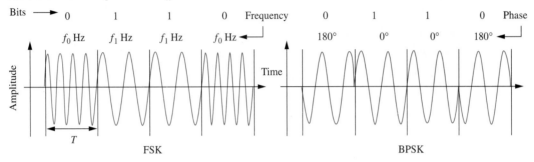

FIGURE 7.7

A bit stream at 150 bps sent by BPSK with sine waves at 600 Hz. The right part of the figure shows the frequencies where the signal has energy. Most of the energy is in the range [600 − 75 Hz, 600 + 75 Hz].

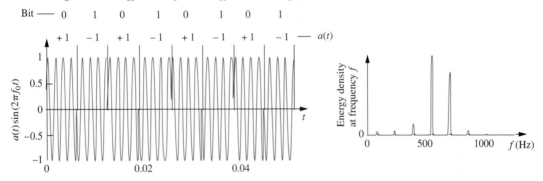

The energy of that signal covers a range of frequencies around f_0. In this example, one can show that most of the energy in the signal is in the range [525 Hz, 675 Hz]. Thus, the width of this range of frequencies, called the *bandwidth* of the signal is equal to 150 Hz. We conclude that, for this specific modulation scheme BPSK, the bandwidth in hertz is approximately equal to the bit rate in bits per second.

Other modulation schemes use a bandwidth in hertz that is some small multiple (typically between 0.15 and 0.5) of the bit rate in bits per second. The smaller the multiple, the more susceptible the signal is to be degraded by noise. That is, robustness against noise is achieved at the cost of some bandwidth. The precise tradeoff between bandwidth and noise robustness is characterized by *rate distortion theory,* which falls beyond the scope of this text.

Radio links use procedures similar to BPSK and FSK, but with higher frequencies. An antenna radiates and receives electromagnetic waves effectively if the length of the antenna is comparable to the wavelength of the electromagnetic wave. The wavelength of an electromagnetic wave with frequency f Hz is equal to c/f, where $c = 3 \times 10^8$ m/s. Thus, a 1-foot antenna is suitable for frequencies around 1 GHz and would therefore be appropriate for f_1 and f_0 around that frequency.

We explain other modulation schemes in Section 7.3.

Synchronous Baseband Transmission (L)

Synchronous transmissions use a different timing mechanism than asynchronous methods. To keep the receiver synchronized to the incoming bits, even after thousands of bits, the synchronous methods use a *self-synchronizing code,* i.e., a code that contains the timing information in addition to the transmitted bits.

The Manchester code is a self-synchronizing code that many copper lines use. This code represents a bit 1 by setting the line voltage high for $T/2$ seconds and low for the following $T/2$ seconds. The code represents a bit 0 by reversing the order of the two values. (See Figure 7.8.)

When the transmitter sends a sequence of bits using the Manchester code, the line voltage makes a transition in every epoch of T seconds. The receiver uses these transitions to remain synchronized to the incoming bits. Specifically, the receiver uses a phase-locked loop, as we explain in Appendix C. To start the synchronization, the transmitter sends a set of extra bits, called a *synchronization preamble,* before the bits to be transmitted. An example of a synchronization preamble is $0101010 \cdots 010111$. If the receiver gets synchronized during the preamble, then it can determine the beginning of the bits by detecting the last bit of the preamble. Most local area networks, including Ethernet at 10 Mbps and token ring, use the Manchester code.

By introducing one extra transition for every bit, the Manchester code doubles the frequencies in the signal. When used at a high transmission rate, say 100 Mbps, over a twisted pair of wires, these higher frequencies make the line radiate an unacceptable level of power in the FM band. This unacceptable radiation power is why 100-Mbps Ethernets and FDDI (fiber distributed data interface) networks use other self-synchronizing codes over twisted wire pairs: MLT-3 and NRZI. Figure 7.8 illustrates those codes.

The NRZI (non-return to zero with inversion) signal makes a transition for every bit 1. When using this signal, the transmitter needs to make sure that the bit stream contains enough 1s. Since the transmission line must work for arbitrary bit strings, including those that contain few 1s, NRZI is used together with a *line code* that inserts 1s in the original bit stream at the transmitter and removes them at the receiver. The 4B/5B code that we explain on the next page (in the section on synchronous optical transmissions) is an example of codes that insert the needed 1s.

FIGURE 7.8

Some widely used self-synchronizing baseband modulation schemes.

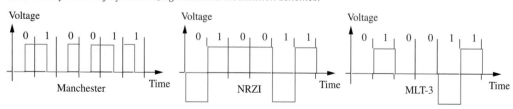

The MLT-3 (multilevel-ternary) signal transmits 0s as 0 volt and 1s alternatively as $+V$ and $-V$. As for NRZI, this scheme requires a line code that inserts 1s.

You will note that these codes generate a signal that changes less rapidly than the Manchester code. To achieve the 100-Mbps transmission rate over twisted pairs, Ethernet and FDDI use two or three twisted pairs in parallel.

Synchronous Broadband Transmission (L, R)

Synchronous broadband transmissions use the modulation methods we described for asynchronous broadband transmission. The sine waves contain the timing information, so these signals are self-synchronizing.

Synchronous Optical Transmission (F)

A synchronous transmission over an optical fiber also uses a self-synchronizing code. Some optical links use the Manchester code to produce the voltage that drives the light source.

To limit the rate of transitions of the optical signal, high-bit-rate optical links use another self-synchronizing code. One such code, called 4B/5B, first groups the bits to be transmitted into words of 4 bits. The code then represents each of the 16 possible 4-bit words by a 5-bit word. The sixteen 5-bit words of the 4B/5B code are selected to contain enough transitions between 0s and 1s. The transmitter then sends the successive 5-bit words using OOK. The resulting light signal contains enough transitions to keep the receiver synchronized. A link may keep on sending a reserved 5-bit word, called the *idle symbol,* to indicate the absence of data bits and to keep the receiver clock synchronized.

The optical links of FDDI use the 4B/5B encoding. Other links use a similar code called 6B/8B.

In the next sections, we describe how optical, radio, and wired communication links work. That description helps us to understand the characteristics of the links.

7.2 Optical Links

FDDI uses optical links to connect computers separated by more than 100 m. Some Ethernet connections are over optical links. High speed (e.g., 155 Mbps) and long ($\geq$ 100 m) links of ATM networks are optical. Finally, all long-distance telephone links are optical. Some systems use free-space infrared transmissions to set up wireless local area networks or to communicate between a camcorder and a TV set or between computer devices such as a computer, printer, keyboard, and mouse. In this section we focus on links with optical fibers.

7.2.1 Overview

A guided optical link consists of a transmitter equipped with an optical source, an optical fiber, and an optical detector attached to a receiver. The transmitter converts the bits into an optical signal, for instance using OOK as we explained in the previous section. The optical signal propagates in the fiber and hits the detector. The detector and the receiver convert the optical signal back into bits. We discuss the characteristics and sources of limitations of optical links.

End-to-End Characteristics

Optical fiber communication links transmit at a very large bit rate over a long distance. Some links can transmit at rates larger than 10 Gbps over more than 100 km. Moreover, bit streams that transmitters send using different colors of light can coexist in the same fiber. The bit error rate of these links can be kept below 10^{-12}. The cost per km and per Mbps of these links is much lower than with any other technology. Optical fiber links make the modern high-performance communication networks possible.

The large bit rate and long distances are possible because of the small dispersion and low attenuation in optical fibers. We examine those phenomena next.

Dispersion Limit

We explain why dispersion limits the product $R \times L$ of the bit rate R times the usable length L of fiber. Imagine a pulse of light that propagates in a fiber. The laws of propagation imply that different fractions of the energy in that pulse travel at different speeds along the fiber. Let us say that the slowest fraction of the energy takes D seconds more to cover 1 km of fiber than the fastest fraction of that energy. If the pulse travels across L km of fiber, then the slowest fraction of energy reaches the end $D \times L$ seconds later than the fastest fraction. Consequently, if the transmitter injects a pulse with a duration of T seconds, then that pulse has a duration of $T + D \times L$ seconds after L km of fiber.

Consider now a sequence of bits 1010101010101 $\cdots$ modulated by OOK into a light wave. Say that the bits are sent at rate R bps, so that each bit 1 corresponds to a pulse with duration $T = 1/R$ seconds and each 0 to the absence of light, also for T seconds. After L km of fiber, the pulses that represent the 1s have a duration equal to $T + D \times L$. Accordingly, the gaps between these pulses are equal to $T - D \times L$ after L km. The bit stream is difficult to detect if these gaps are a small fraction of T, say if they are less than $0.50 \times T$. Thus, the maximum value of L is such that $D \times L \leq T/2 = 1/2R$. We can rewrite this condition as $R \times L \leq 1/2D$. Recapitulating our discussion, we have found that dispersion imposes the limit

$$R \times L \leq \frac{1}{2D}. \tag{7.1}$$

Recall that, in this expression, D is the difference in travel times of the slowest and fastest fractions of the energy of the light after 1 km of fiber. We call D the *dispersion rate* of the fiber. The value of D depends on the type of fiber, as we explain later.

Three Types of Fiber

An optical fiber is a long cylinder made of plastic or glass. Three types of fibers are used in networks: step index, graded index, and single mode. These three fibers are sketched in Figure 7.9.

Note the different shapes of the rays of light in the fibers. As we explain in the next section, these different shapes result in different values of the dispersion rate D of these fibers. Consequently, these fibers have different dispersion limits, as we saw in (7.1). The limits are as follows:

$$R \times L \leq \begin{cases} \text{10 Mbps} \times \text{km for a step-index fiber} \\ \text{1 Gbps} \times \text{km for a graded-index fiber} \\ \text{200 Gbps} \times \text{km for a single-mode fiber} \end{cases} \tag{7.2}$$

We should note that some single-mode fibers are specially designed to have an even smaller dispersion rate. These *dispersion-compensated fibers* achieve a dispersion limit that can exceed 1000 Gbps × km.

Figure 7.10 shows the attenuation of an all-glass fiber as a function of the wavelength. An attenuation of A dB/km means that the power $P(L)$ after L km is equal to the power $P(0)$ injected in the fiber multiplied by a factor $10^{-(AL)/10}$. That is,

$$P(L) = P(0) \times 10^{-\frac{AL}{10}} \quad \text{and} \quad 10\log_{10}\left\{\frac{P(0)}{P(L)}\right\} = AL \text{ dB}.$$

Consequently, the attenuation after L km is equal to $A \times L$ dB, or A dB/km.

Note two ranges—or windows—of wavelengths where the attenuation is particularly low. These windows cover about 0.1 μm, one around 1.3 μm and the other around 1.55 μm. The attenuation in these windows is about 0.4 dB/km and 0.2 dB/km, respectively. These

FIGURE 7.9

Three types of fiber: step index, graded index (GRIN), and single mode (SMF).

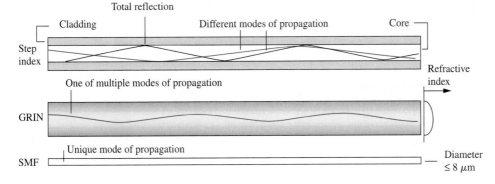

FIGURE 7.10

Attenuation in an all-glass fiber.

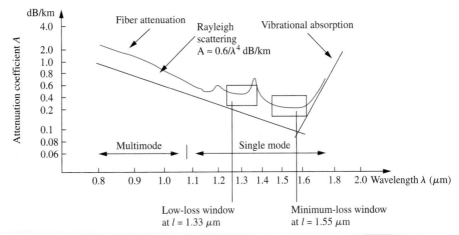

windows cover a range of frequencies with a width equal to a few terahertz. (1 terahertz = 10^{12} Hz.) An optical fiber has a bandwidth so large that it could transmit simultaneously all the telephone conversations that happen at any given time in the world. However, the rates of optical links are limited by dispersion and by the maximum rate of optical transmitters and receivers.

The propagation speed in these fibers is c/η where $c = 3 \times 10^8$ m/s and $\eta \approx 1.5$ is the refractive index of the fiber. Thus, the propagation speed is approximately 2×10^8 m/s, which corresponds to a propagation time equal to 5 μs per kilometer of fiber.

7.2.2   Propagation in Fibers

We explain how light propagates in step-index, GRIN, and single-mode fibers. Our discussion helps us understand the result in (7.2). We discuss the three fiber types separately, starting with the step-index fiber.

Step-Index Fiber

A step-index optical fiber is a cylindrical *core* of glass or plastic with refractive index η that is surrounded by a tube of glass or plastic with a slightly smaller refractive index ν. (See Figure 7.9.)

In a step-index fiber, rays that make an angle larger than some critical angle θ with the axis of the fiber gets refracted into the outer material and escape the fiber. The critical angle θ is such that $\cos \theta = \nu/\eta$. The other rays propagate through a series of total reflections, as shown in Figure 7.9. Different angles of propagation are called *modes* and, accordingly, the step-index fiber is called a *multimode* fiber. Depending on their angle, rays of different modes cover different distances for each kilometer of fiber that they go through. All these rays travel at velocity c/η in the core with refractive index η. The different distances that different modes cover result in different travel times, which corresponds to some dispersion rate D, as we explained above in "Dispersion Limit."

To calculate D, note that rays parallel to the axis of the fiber have a travel time equal to $T_1 = 1/(c/\eta) = \eta/c$ per kilometer of fiber. Rays with the maximum reflected angle θ cover the maximum distance equal to $1/\cos \theta$ km to go through 1 km of fiber. The travel time of these rays is then $T_2 = T_1/\cos \theta = T_1 \times (\eta/\nu)$. Hence, $D = T_2 - T_1 = (\eta/c)(\eta/\nu - 1)$. For a typical fiber, $\nu = 0.99 \times \eta$. Thus, $D \approx (\nu/c)(\eta/\nu - 1) = (\eta - \nu)/c$. Using (7.1) we then find that the dispersion limit is given by

$$R \times L \leq \frac{c}{2(\eta - \nu)}. \tag{7.3}$$

With $\eta - \nu = 0.01$, this bound gives the first result indicated in (7.2).

GRIN Fiber

In addition to the step-index fibers, communication systems also use *graded-index* (GRIN) fibers. The refractive index of the core of a GRIN fiber decreases with the distance from the fiber axis. In such a fiber, the rays are subject to continuous refractions and propagate along oscillatory paths as illustrated in Figure 7.9. Thus, a GRIN fiber is also a *multimode* fiber. The modes are different oscillating paths.

The rays that travel longer distances go through regions of the fiber where the refraction index is smaller and, therefore, where the propagation speed is larger. Thus, remarkably, longer distances are compensated by larger average speeds. As a result, the propagation time of the longer rays is similar to that of rays which travel closer to the center of the fiber. As a consequence, the modal dispersion of GRIN fibers is substantially smaller than it is in step-index fibers. The analysis of GRIN fibers shows that (7.3) becomes

$$R \times L \leq \frac{2c\eta_1}{(\eta_1 - \eta_2)^2} \qquad \text{for GRIN fibers.} \qquad (7.4)$$

In this expression, η_1 is the refractive index at the center of the fiber and η_2 is the refractive index at the periphery of the fiber. Typical values give $R \times L \leq 1$ Gbps $\times$ km, i.e., 10^9 bps $\times$ km.

Single-Mode Fibers
We learned that multiple modes limit the transmission rate in a fiber. The GRIN fibers reduce but do not eliminate the modal dispersion. Is it possible to construct a fiber having a single transmission mode? The answer is yes.

If the radius of the core of a step-index fiber is small enough (about 8 μm), then only one mode can propagate. The reason is that the successive reflections of a ray interfere with each other and such interference destroys some modes. This phenomenon is somewhat similar to the phenomenon which makes a pipe organ vibrate only at specific frequencies. The analysis of the solutions of Maxwell's equations shows that only a finite set of modes propagates along the fiber, instead of the continuum of angles from 0 to θ that Figure 7.9 deceivingly suggests. When the radius is small enough, only the mode that is parallel to the fiber exists and such a fiber is called a *single-mode fiber*. Because single-mode fibers have no multimodal pulse spreading, they have a small dispersion rate D.

Note, however, that all fibers—including single-mode fibers—are subject to *material dispersion*. This dispersion is caused by the differences of the refractive index at different wavelengths, which results in different velocities of different wavelengths. A light source emits light that covers some frequency band characterized by a *spectral width*. The spectral width σ of a light source is the width of the interval of frequencies where the power of light is at least equal to half its peak value. Using this definition, one can show that the material dispersion imposes a limit given by

$$R \times L \leq \frac{1}{4\Phi\sigma} \qquad (7.5)$$

where

$$\Phi = \frac{d^2\eta(\lambda_0)}{d\lambda^2} \qquad (7.6)$$

is the second derivative of the refractive index with respect to the wavelength, evaluated at λ_0, the average of the wavelength of the source. The value of Φ is 10^{-12} s/km $\times$ nm for a silica fiber at $\lambda_0 = 1.3$ μm. The value of σ is about 1 nm for a GaAsP laser and about 3 nm for a laser diode. (GaAsP designates a specific semiconductor crystalline alloy.) We discuss laser diodes and other transmitters below. These values correspond to the limits $R \times L \leq 250$ Gbps $\times$ km and $R \times L \leq 80$ Gbps $\times$ km, respectively.

We now turn our attention to the light sources.

7.2.3   *Light Sources*

Optical links use two types of light source: light-emitting diodes (LEDs) and laser diodes (LDs). The main characteristics of these sources are the intensity and wavelengths of the light they produce and the maximum rate at which they can be modulated. A laser diode can be modulated faster than an LED and has a smaller spectral width. A smaller spectral width produces a smaller dispersion and enables many LDs to share the 100-nm window around 1.55 μm of an optical fiber.

An LED is a *PN semiconductor junction.* An LED connected to a *constant* current source produces a light beam with an intensity proportional to the current. The physical phenomenon at work is that a fixed fraction of the injected electrons induce electron-hole recombinations; when such a recombination occurs, the energy lost by the electron is emitted as a photon. Typical LEDs generate a few milliwatts [1 milliwatt (mW) = 10^{-3} watt] of optical power when about 50 mA (1 mA = 10^{-3} ampere) are injected. If the injected current changes rapidly, then the optical power emitted by the diode drops with the rate of change. Physically, the electron-hole pairs take some time to recombine and they cannot do so efficiently if the current changes too fast. This effect limits the frequency at which one can *modulate,* i.e., modify, the intensity of the light beam. This frequency limit is indicated by the *cutoff* frequency at which the power is reduced to 70 percent of its peak value. Typical values of the cutoff frequency of LEDs range from 1 kHz to 100 MHz.

LEDs can be used for transmitting strings of bits by OOK or analog signals that take a continuous set of possible values. LEDs are used in consumer electronic devices ranging from remote controls to wireless headphones; in those applications, the LEDs emit infrared light. Infrared LEDs are also used in local area communication networks. LEDs that emit visible light are also used as inexpensive and long-lasting indicators.

Essentially, a laser diode (LD) is a PN junction in an optical cavity. That is, the junction is terminated by two parallel semireflecting faces *A* and *B;* the distance between the faces is some multiple of half a wavelength generated by electron-hole recombinations. The junction is forward-biased and emits photons. Some photons "hit" free electrons that then recombine and emit coherent photons, by *stimulated* emission. That is, when an incident photon leads an electron to lose an amount of energy equal to that of the photon, then a second photon is emitted with the same frequency and phase as the first photon. These stimulated emissions form a chain reaction. Some of the emitted light waves are attenuated by out-of-phase interference with reflections from the faces *A* and *B*. That is, the interference with the reflections acts as a filter which attenuates light waves unless they have a specific phase and wavelength. The chain reaction is self-sustaining if the gain due to stimulated emissions is larger than the loss due to the imperfect reflections (and to absorption). This self-sustaining reaction occurs when enough electron-hole pairs recombine in the junction and, therefore, when the forward bias is sufficient.

The cutoff frequency of an LD is a few orders of magnitude larger than that of an LED. Some LDs can be modulated at up to 11 Gbps. LDs are temperature-sensitive. The optical power for a given value of the injected current depends significantly on the temperature because heat increases the generation of free electrons and free holes. This dependency can be controlled with sophisticated circuitry that monitors the LD temperature and adjusts the injected current correspondingly. LDs are used in compact disc players and in long-distance optical communication systems.

7.2.4   *Light Detectors*

The receiver of an optical link contains a light detector that converts the light into an electrical signal. The light detector is followed by a preamplifier that amplifies the weak current that comes out of the detector. The output of the preamplifier enters a circuit that measures the amount of current that the light detector produces during the successive bit durations. This circuit contains the PLL that we discussed earlier.

The light detector is a PN junction, a PIN diode, or an APD (avalanche photo-diode). The main characteristics of a light detector are the range of wavelengths it responds to, its quantum efficiency, and the rate of variation of intensity of light it can track. The quantum efficiency is the fraction of incident photons that the light detector converts into electrons.

A *PN photodiode* is a reverse-biased PN junction. The reverse bias increases the electrical field in the junction and prevents charges from moving across it. When the junction is illuminated, some photons are absorbed by electrons in the p region close to the junction; these electrons become free when the energy of the photons is adequate. The freed electrons are pushed across the junction by the large electrical field and have a good chance of making it through the device and of contributing an external current. Thus, a PN photodiode acts as a light-dependent current source.

The rate at which the PN-photodiode detects bits depends on how fast it can follow variations in the intensity of light. The long travel time of freed electrons across the device limits that rate to about 1 MHz.

The second type of optical receiver is the *PIN diode,* the most commonly used photodetector in communication networks. The PIN diode has three layers which are, respectively, p-doped, intrinsic (i.e., not doped), and n-doped, hence the name PIN. The intrinsic layer is large; since it is not doped, its resistivity is high. As a result, when the diode is reverse-biased, there is a large electrical field across the intrinsic region. Most photons are absorbed in the intrinsic region and free charges that race across the device. This fast travel results in a rate of detection that can be as high as 20 GHz.

The third type of optical detector is the *avalanche photodiode.* An APD is a PIN diode with a very large junction electrical field. The operating mechanism is roughly that the reverse bias of the diode is so large that freed charges race through the diode so fast that they free additional charges through "impacts." This effect results in an avalanche that makes the APD much more sensitive than a PIN diode: a single photon may, through the avalanche effect, free as many as 100 charges.

7.2.5   *Free-Space Infrared*

Infrared communication is also used in free space. Applications include connecting a computer to a peripheral such as a printer or a keyboard and interconnecting computers.

IrDA, the Infra-Red Data Association, is an independent industry association formed in June 1993 with the objective of defining standards for low-cost infrared interconnections. The standards defined so far are for the transmission rates 2400 bps, 115.2 kbps, 1.152 Mbps, and 4 Mbps.

The IrDA transmitters use the wavelength 0.86 μm with a 1-meter range. The bit encoding is OOK at the low bit rates and *pulse position modulation* (PPM) at 4 Mbps. PPM encodes a group of bits by adjusting the timing of a pulse of light. For instance, PPM4 groups the bits 2 by 2 and transmits the bits by sending the pulse in one of four time slots that make up a frame. PPM achieves a lower BER than OOK at large bit rates because the transmitter can use a larger peak value of the light intensity—which is then easier to detect—for a given average emitted power.

7.3 Copper Lines

In this section we explore copper lines such as wire pairs and coaxial cables. The telephone network uses twisted pairs for the subscriber loop between the customer telephone set and the local telephone switch. When the telephone company wires an office building for telephones, it usually installs redundant twisted pairs in anticipation of future needs. Consequently, most office buildings have unused twisted pairs already installed. These spare twisted pairs can be used for wiring local area networks, thereby eliminating the need for a costly dedicated wiring of the building.

7.3.1 Overview

Copper lines guide electromagnetic waves through the interaction of a varying electrical field and magnetic field that are produced by and also induce varying currents in the wires.

A transmission line is characterized by the range of frequencies—the bandwidth—that it can transmit with a low attenuation. The attenuation in a transmission line increases with the frequency because electrons, being charged particles, repulse each other. At high frequencies, the electrons are pushed at the periphery of the conducting material that makes up the transmission line. Consequently, the effective section of conductor that the current goes through is very small and the apparent resistivity of the transmission line is high.

A wire pair radiates electromagnetic fields when traversed by a varying current. Conversely, an external electromagnetic field induces a varying current in a wire pair. These effects produce *crosstalk* between adjacent pairs: a signal in one pair leaks into the other pair. The radiation may also disturb the operations of radio receivers or of other electronic equipment. To prevent such disturbances, the Federal Communication Commission regulates the acceptable level of radiation at different frequencies. The FCC limitations force the communication engineers who design transmitters for wire pairs to employ clever encoding and modulation methods that limit the amount of energy of the signal that the transmitter injects into the twisted pairs at different frequencies.

The radiation of a wire pair is reduced by twisting the pair. Indeed, the fields generated by two successive loops of a twisted pair are of opposite signs and tend to cancel one another. Similarly, a twisted pair is less sensitive to crosstalk than an untwisted wire pair. In practical terms, a single twisted pair can transmit at up to about 50 Mbps over 100 m and up to 4 Mbps over 1500 m.

A coaxial cable, because of its shield that surrounds the center conductor, radiates and picks up very little energy. The coaxial cable has a larger bandwidth than a twisted pair (about 400 MHz over 2 km). With an efficient modulation scheme—using 0.5 Hz/bps—the coaxial cable (or "coax") can transport about 800 Mbps over 2 km. This rate is enough for 500 digital TV channels compressed at 1.5 Mbps each.

7.3.2   *Modulation*

The following scheme, called *quadrature phase shift keying* (QPSK), produces a signal that covers a smaller range of frequencies than PSK and FSK and is used by high-speed telephone and cable modems. The transmitter groups the bits 3 by 3. For each of the 8 possible groups of 3 bits, the transmitter selects a pair of numbers (a_n, b_n) equally spaced on the unit circle, as shown in Figure 7.11. The transmitter then sends the signal $a_n \sin (2\pi f_0 t) + b_n \cos (2\pi f_0 t)$ for T seconds. The figure shows the signals that correspond to the 3-bit words $000, 001, 011$. For instance, to transmit the three bits 011, the transmitter transmits the signal $-0.7 \cos (2\pi f_0 t) + 0.7 \sin (2\pi f_0 t)$ for T seconds. Figure 7.12 shows the signal the transmitter sends to transmit the bit string $000'001'011'000$.

The coefficients of $\cos (2\pi f_0 t)$ and $\sin (2\pi f_0 t)$ change only once every 3 bits. This rate is the symbol rate of the modulation scheme; a symbol represents 3 bits in this example. By analogy with our discussion of BPSK, one can expect the bandwidth of the signal to be approximately equal to the symbol rate.

Instead of using 8 points (a_n, b_n) on the circle, we could choose 2^k points anywhere on and inside the circle. Each of these coefficients is then associated with of word of k bits. This modulation method is called *quadrature amplitude modulation* (QAM). With this choice, the symbol rate is equal to the bit rate divided by k, say R/k. The signal bandwidth is then approximately equal to R/k Hz.

FIGURE 7.11

QPSK. The transmitter groups the bits 3 by 3. To transmit a 3-bit group, the transmitter sends $a_n \cos (2\pi f_0 t) + b_n \sin (2\pi f_0 t)$ for T seconds, where the coefficients (a_n, b_n) correspond to the group.

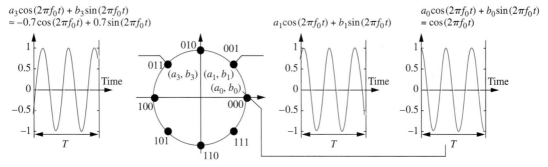

FIGURE 7.12

Transmission of a bit string using QPSK.

$a(t)\cos(2\pi f_0 t) + b(t)\sin(2\pi f_0 t)$

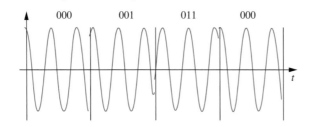

To detect a BPSK, QPSK, or a QAM signal, the receiver measures the coefficients a and b to recover the symbol (a, b) and the group of bits that are encoded by this symbol. By choosing k larger, the bandwidth R/k gets smaller. However, the points (a_n, b_n) that represent the groups of bits become closer to one another as k increases, and this closeness makes it more difficult for the receiver to guess the correct value. This discussion shows that there is a tradeoff between the bandwidth of the signal and the sensitivity to noise. Thus, if a transmission line has little noise but can only transmit a narrow range of frequencies, then QAM with a large k is the suitable choice. Conversely, if the line is noisy but transmits a wide range of frequencies, then BPSK is a better choice.

Fast modems use additional methods to further increase their spectral efficiency. Two such methods, called *minimum shift keying* (MSK) and *Gaussian minimum shift keying* (GMSK), replace the sharp transitions of the coefficient of the sine wave of BPSK by smoother transitions. That is, instead of changing from $+\sin(2\pi f_0 t)$ when transmitting a bit 1 to $-\sin(2\pi f_0 t)$ when transmitting a bit 0, MSK and GMSK change the coefficient of the sine wave from $+1$ to -1 gradually. This smoother change reduces the bandwidth of the signal.

Multiplexing

A transmission link with bandwidth $[f_0, f_1]$ can transmit different signals that cover nonoverlapping ranges of frequencies in the range $[f_0, f_1]$. The receiver can select one of the signal by filtering out the others. Thus, a coaxial cable with bandwidth [300 MHz, 1 GHz] can transmit 70 TV programs with 6-MHz-wide spectra and a number of digital signals using QPSK. Some cable TV companies are implementing such systems to offer telephone services, access to the Internet, and video on demand.

Telephone companies are also using broadband transmission methods to deliver high-speed data services in addition to the regular telephone service on a regular copper subscriber loop. This combination—called ADSL for *asymmetric digital subscriber loop*—competes with the cable TV solutions. (See Section 7.10.)

A *wireless cable*—with its oxymoronic name—replaces a coaxial cable by radio transmitters to provide community antenna TV (CATV). When the transmitter covers only a small radius, the wireless cable also provides a return channel from the subscribers premises. That return channel is used for telephone service or for control messages of a video-on-demand service.

Modems

Wired and wireless modems (modulator-demodulators) use the transmission method described above, sometime using slightly different modulations to encode the bits. Here are a few representative modems:

- *9.6 kbps (V.32):* This modem uses QAM with a 32-point constellation.
- *14.4 kbps (V.32 bis):* This modem uses QAM with a 64-point constellation.
- *28.8 kbps and 33.6 (V.34):* This modem uses QAM with a 960-point constellation at 28.8 kbps and a 1664-point constellation at 33.6 kbps. The modem adjusts both the symbol rate and the carrier frequency.
- *56 kbps:* This modem provides an asymmetric service (56 kbps downstream and 30 kbps upstream). (See the next page for some details.)

- *Compression MNP5 and V.42 bis:* To reduce the number of bits they must transmit, some modems first compress the bit stream before transmitting it and decompress it at the destination. The MNP5 compression uses run length encoding: Instead of sending a string of K zeros, the modem sends the number K. Since the number K can be encoded with approximately $\log_2 K$ bits, this compression scheme can be very effective when K tends to be large. V.42 bis uses the Lempel-Ziv compression algorithm that we explain in Chapter 8.
- *Cable modems:* Fast modems can take advantage of the large bandwidth of coaxial cables to transmit data very fast on a CATV network. These modems typically use a modulation scheme such as QPSK with a fast rate of signal changes.

56-kbps Modems.   These recent modems use an approach different from QAM. To understand this approach, let us examine how bits are sent by the central office of the telephone company to the subscriber. When the central office gets a bit stream to deliver to the subscriber, it assumes the successive bytes v_n are analog voice samples s_n that were first compressed as $\mu(s_n)$ before being quantized (digitized) as v_n. The function $\mu(.)$ is a function that compresses large values that are rare. This compression results in smaller quantization errors for the frequent small-valued samples and reduces the quantization noise in the digitized samples v_n.

Accordingly, the line interface electronics in the central office decompresses the samples by computing $\mu^{-1}(v_n)$. The electronics then produces an analog signal $s(t) = \sum_n g(t - n\tau)\mu^{-1}(v_n)$ where $g(.)$ interpolates between the values $\mu^{-1}(v_n)$ placed every $\tau = 125\,\mu$s.

When the bytes v_n are data, the demodulator in the 56-kbps modem must recover these bytes from $s(t)$. The optimal decoder for these bytes must be designed by assuming that the noise in the channel is Gaussian. Remarkably, this approach doubles the effective rate of the modem over the rate that can be achieved by using QAM. The magic happens because this approach exploits the precise shape of the compression.

As the above description shows, this approach works only if there is a single analog-to-digital conversion in the link. Thus, you can use the 56-kbps modem to download data from a remote server that is attached to the network with an ISDN line or another digital service. This modem does not work (at 56 kbps) if the server itself is sending the data over a voiceband modem.

7.3.3   CATV and Video-on-Demand Systems

The cable television system evolved in four major steps. Initially, the transmission was analog over a tree-like coaxial cable plant. Amplifiers compensate the loss of power due to the attenuation of the signal as it propagates. Since amplifiers also amplify the noise, the noise limits the range of the distribution system. For the NTSC (National Television System Committee) standard, different TV channels are modulated in different frequency bands that are 4.5-MHz wide.

The second step is the replacement of sections between the CATV head-end station and segments of the coaxial cable by optical fibers. This hybrid fiber–coax plant has an improved signal quality because fibers have a smaller attenuation than coaxial cable. The cost of the optical links is shared by the users attached to the same coaxial cable segment.

The third step is the digital transmission of compressed video. The compression uses the MPEG1 standard at 1.5 Mbps. The decompression is performed by the user's set-top box. With a suitable QPSK modulation of the 1.5-Mbps bit stream, a TV channel then occupies only 600 kHz instead of 4.5 MHz. Consequently, the coaxial cable can carry about 500 TV channels. If 500 users are attached to the same coaxial segment, each can have a dedicated channel.

The fourth step is the two-way transmission of control messages between the users' equipment and the CATV station. This transmission is packetized and takes place on dedicated frequency bands. Amplifiers in the upstream direction, from the users to the CATV station, are needed to transmit the messages. With this fourth step, the users can request video programs or pieces of information. Thus, 500 users can each request different video programs from the CATV station. The upstream channels can also be used to provide local access to the Internet or to other service providers. The IEEE 802.14 is defining standards for transmission and multiple access over cables.

7.4 Radio Links

Radio links are used when mobility is required or to avoid the cost of setting up a wired link.

7.4.1 Overview

An antenna radiates and receives an electromagnetic wave efficiently if its length is about half the wavelength of the electromagnetic wave. Thus, to radiate an electromagnetic wave at 10 kHz, the antenna should have a length close to half of 3×10^5 km/s/10^4 Hz = 30 km, which is not practical. If the frequency of the wave is 100 MHz, then the length required is about 1.5 m, which is the length of the antenna of an FM receiver. For a frequency of 1 GHz, the length becomes 0.15 m, which is the size of a direct satellite broadcasting TV antenna.

If the signal to be transmitted has a low frequency, such as a music program with a frequency range from 0 to 10 kHz, then the transmitter first modifies the frequencies in the signal before sending it to the antenna. To perform this modification, the transmitter generates a sine wave at a high frequency f_0 Hz—called the *carrier*—and it uses the signal to be transmitted to modify the amplitude, the frequency, or the phase of that carrier. This operation—called *modulation*—produces a signal with a frequency close to f_0 whose amplitude, frequency, or phase variations represent the signal to be transmitted. The antenna has a length suitable for transmitting the carrier. When the receiver gets the modulated carrier, it converts the variations in frequency or amplitude or phase into the transmitted signal.

By using carriers at slightly different frequencies, the transmitter can send different signals on the same antenna. These modulated carriers have different frequencies and the receiver can select which one it listens to. For instance, all the broadcast radio and TV programs share the air waves and the radio or TV set can select which one it receives. Thus, modulation serves two purposes: (1) modifying the frequency of the signal to make the antenna efficient and (2) separating different signals in frequency.

7.4.2   *Propagation*

The analysis of propagation shows that the power per unit area first decays with the square of the distance and then decays with its fourth power. Detailed measurements reveal that reflections on buildings, vegetation, and other objects make the actual power decay less regular than the theory predicts. Computer programs enable communication engineers to model the environment where the radio operates and to predict the strength of the electromagnetic wave as a function of the location. Such computer tools are useful to plan the location of transmitters. However, the only sure way of determining the actual range of coverage of an antenna is to measure the power of the signal received at various locations.

7.4.3   *Cellular Networks*

Cellular networks are widely used to provide mobile telephone services. These systems started in the 1970s with analog transmissions where voice was frequency-modulated to provide a number of telephone channels. The region is divided into cells, and cells that are far enough apart can use the same set of frequencies. Thus, if seven cells repeat periodically to cover the area, each using a different set of N frequencies, then $7N$ frequencies are used for the whole area. The density of active users that the system can accommodate is then equal to N/S users per square kilometer where S is the surface area of one cell.

These analog systems are progressively being complemented or replaced with digital systems. Digital systems take advantage of voice compression to reduce the range of frequencies that a voice channel occupies. In addition, digital transmission can be made more robust to noise.

One key mechanism of a digital cellular system is the sharing of the frequency band allocated by the regulating agencies to the system. One method uses frequency-division multiple access (FDMA) combined with time-division multiple access (TDMA). When this method is used, the frequency band is first divided into a set of channels. Each channel is then decomposed into a number of voice circuits by time division. That is, a channel transports a bit stream that is arranged into frames and each frame is divided into K slots. The slots are allocated to the different voice circuits. In another method, code-division multiple access (CDMA), the signals of the different users all occupy the complete frequency band. However, these signals are obtained by multiplying the users bit streams by distinct pseudo-random sequences that take the values $+1$ and -1 and change rapidly. To recover the bit stream of a particular user, the receiver multiplies the signal it receives by the pseudo-random sequence that characterizes that user. The signals of other users, when multiplied by this particular sequence, appear to the receiver as a fast-changing noise that can be averaged out.

7.5   Complement 1: Shannon Capacity

The capacity can be calculated from the characteristics of the link—such as the attenuation and delay at different frequencies—and of the noise. Shannon showed that the capacity of channels with dominant thermal noise is given by the following formula:

$$C = B \log_2 \left\{ 1 + \frac{S}{N} \right\}. \tag{7.7}$$

In this expression, B is the width of the range of frequencies that the channel transmits in Hz and S/N is the *signal-to-noise ratio* equal to the power S of the signal divided by the power N of the noise.

Engineers usually specify the signal-to-noise ratio in *decibels* (dB). To express a signal-to-noise ratio S/N in decibels, one calculates

$$S/N(\text{dB}) := 10 \log_{10} \left(\frac{S}{N} \right).$$

For instance, assume that $S/N = 100$. That is, the power S of the signal is 100 times larger than the power N of the noise. This signal-to-noise ratio corresponds to $10 \log_{10}(S/N) = 10 \log_{10}(100) = 20$ dB. Thus, a signal-to-noise ratio of 20 decibels, or 20 dB, means that the signal power is 100 times larger than the power of the noise.

More generally, if the signal-to-noise ratio is equal to A dB, then

$$\frac{S}{N} = 10^{A/10}.$$

Indeed, if $S/N = 10^{A/10}$, then we calculate that signal-to-noise ratio by calculating $10 \log_{10}(S/N) = 10 \log_{10}(10^{A/10}) = 10 \times (A/10) = A$ dB, since $\log_{10}(10^x) = x$ for any real number x.

As an illustration of these formulas, consider a telephone line which transmits the frequencies from 300 Hz to 3400 Hz with a signal-to-noise ratio of 35 dB. The capacity of the line is given by

$$C = \{3400 - 300\} \times \log_2\{1 + 10^{35/10}\} = 3100 \times \log_2\{1 + 3162\} = 36{,}044 \text{ bps}.$$

If the transmission line is allowed to transmit a wider range of frequencies, then its capacity is much larger. In practice, the achievable transmission rate on a typical telephone line ranges from 1.5 Mbps for up to 5.5 km to about 52 Mbps for up to 300 m.

We study Shannon's theory in more details in Appendix A, Section A.5.

7.6 Complement 2: Sampling and Quantization

In this section we discuss the principles that underlie the digitization of information, i.e., its conversion into bits. We will explain these principles in the cases of voice signals and video signals. You will see that there are two steps in the digitization of voice and video, and, more generally, of a time-varying analog signal: *sampling* and *quantization*. Sampling is the periodic measurement of the signal every T seconds. These periodic measurements are called *samples*. Quantization is the approximation of the possible values of the samples by a finite set of values. The advantage of quantization is that we can number the finitely many approximate values with binary numbers. Each sample is then replaced by the binary number that designates its approximating value. As a result, the succession of periodic samples is replaced by a succession of binary numbers, i.e., by a bit stream. Thus, the signal has been digitized. We will explain these steps in more detail below.

Let us first consider the digitization of a voice signal. A microphone produces an analog signal $v(t)$ which represents the pressure of the air to which it is exposed at time t. As time goes by, the air pressure and $v(t)$ change. The variations in the pressure are perceived by the ears and the brain as sounds. If the microphone picks up the sound of voice, then most of the energy of the signal is in the frequency range from 300 to 4000 Hz. Assume that the signal does not contain any energy at frequencies higher than 4000 Hz. *Nyquist's sampling theorem* (see Appendix C) establishes that $v(t)$ can be recovered *exactly* from the samples $\{v(nT), n = 0, \pm 1, \pm 2, \ldots\}$ provided that $T < 1/8000$ s, i.e., provided that the signal is sampled at least 8000 times per second. In other words, there is only one signal $v(t)$ with maximum frequency 4000 Hz which takes the value $v(nT)$ at time nT for all n. In general, Nyquist's theorem states that a signal with maximum frequency f_{max} can be recovered exactly from samples that are measured more frequently than $2 \times f_{max}$ every second. That is, the *sampling frequency* must be larger than twice the maximum frequency in the signal. That theorem confirms the intuitive idea that the faster a signal changes, the faster it must be measured to notice the changes.

Sampling is the first step in the digitization of a voice signal. The second step is to *quantize* the samples. That is, the samples which can take a continuous range of values are approximated by values in a finite set. For instance, if a sample $v(nT)$ can take arbitrary values but is very likely to be between -1 and 1, then that sample can be approximated by the closest point in $\mathbf{A} := \{\pm\delta, \pm3\delta, \pm5\delta, \ldots, \pm(2^N - 1)\delta\}$ with $\delta = 2^{-N}$, which is in the interval $(-1, +1)$. **Figure 7.13** illustrates that approximation with $N = 3$. There are 2^N points in the set $\mathbf{A}$, and they can be identified with N bits. We write the approximation of $v(nT)$ as $v_s(nT)$. The *sampling error* $e(nT) := |v(nT) - v_s(nT)|$ is less than 2^{-N}

FIGURE 7.13

Sampling and quantization. The signal $v(t)$ is sampled periodically. The samples are identified with circles. The samples are then approximated by the closest value marked by a small horizontal line. These approximation values are designated by binary words. As a result, $v(t)$ is encoded into a sequence of binary words.

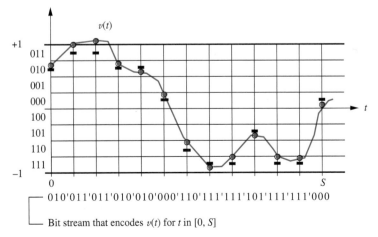

whenever $|v(nT)| \leq 1$. One can view the sampling error as adding a *sampling noise* to the signal. The power of that noise is the average value of $e^2(nT)$. Assuming that $v(nT)$ is equally likely to take any value in $(-1, \ +1)$, one can calculate that the average value of $e^2(nT)$ is approximately equal to

$$N_s := \frac{1}{3} \times 2^{-2N}.$$

Under the same assumptions, the average power of the signal is the average value of $|v(t)|^2$, which is equal to $P_v := \frac{1}{3}$.

It follows that the *signal-to-noise ratio* P_v/N_s is equal to

$$P_v/N_s = 2^{2N}.$$

This signal-to-noise ratio can be expressed in decibels:

$$10 \times \log_{10} \frac{P_v}{N_s} = 10 \times \log_{10} 2^{2N} \approx 6 \times N \text{ dB}.$$

(We used the fact that $\log_{10} x \approx 0.3 \times \log_2 x$.)

Let us look at two examples of sampling and quantization. The first example is the digital transmission of voice in the telephone network. The sampling rate is 8000 Hz, and 8 bits are used per sample. The sampling rate corresponds to a maximum frequency of 4000 Hz. The voice signal is filtered so that its spectrum is [300 Hz, 3400 Hz]. The 8 bits per sample introduce a quantization noise with a signal-to-noise ratio equal to $6 \times 8 = 48$ dB. The second example is a compact disc. The sampling rate is about 41 kHz, which corresponds to a maximum frequency of 20 kHz. Each sample is encoded into 16 bits. The signal-to-noise ratio is, therefore, about equal to 96 dB. Thus, to send a voice signal as a bit stream, one must send 8 bits 8000 times per second. This corresponds to a bit rate of 64 kbps. The bit rate in a compact disc is about $41 \times 10^3 \times 16$ for each of the two channels (left and right), i.e., the total rate is 1.3 Mbps. Reed-Solomon error correction bits are added to that bit stream.

The digitization of a video signal follows similar rules. For instance, the maximum frequency in the NTSC video signal is about 5 MHz. Using 8 bits per sample and a sampling frequency of 11 MHz leads to a bit rate equal to 88 Mbps. Alternatively, if the camera uses a grid of photosensors, one can quantize each sensor value. Say that the camera uses $300 \times 400 = 12 \times 10^4$ sensors for each of the three fundamental colors red, green, and blue. Using 8 bits per sample requires $8 \times 3 \times 12 \times 10^4 \approx 2.88 \times 10^6$ bits per image. If 30 images are sent every second, the needed bit rate is equal to 86.4 Mbps.

The number of bits needed to transmit a given piece of information can be reduced by a technique called *information compression*. This technique will be explained in Chapter 8.

7.7 Complement 3: SONET

The telephone networks are increasingly using SONET (synchronous optical network) technology, called SDH (synchronous digital hierarchy) in Europe. With this technology, the transmitters are synchronized to a common master clock. This synchronization permits the byte interleaving of different digital streams. Given the importance of this technology, we describe the architecture of SONET and its framing structure.

7.7.1   SONET Architecture

Figure 7.14 shows SONET. This network consists of switches, multiplexers, and regenerators connected by optical fibers. This network provides bit ways between users. SONET bit ways are used to transport telephone calls. The bit ways are also used to connect Internet routers or ATM switches.

In SONET terminology, the link that the network makes available to users is a *path.* The path carries user bits between two access points. The access points may be attached to ATM switches, to Internet routers, or to telephone switches. The path packages the user bits in frames called *synchronous payload envelopes* (SPEs). Each SPE contains a path overhead field that the access points use to monitor the connection. Thus, an SPE has two parts: the path *overhead* and the *payload* that contains the user bits. The user bits may encode telephone conversations, video or audio signals, or ATM cells.

The path itself is transported across a set of *lines* between its source and its destination. A line is the transmission facility that transports bits between two multiplexers. A line transports line frames that consist of some overhead monitoring information plus the line payload bits. The line payload bits transport SPEs. Thus, we may think of a line as a conveyer belt that carries SPEs.

Finally, a line may consist of a number of *sections.* A section is the part of a line between devices where a light-to-electronics conversion occurs. These devices include multiplexers and also regenerators that clean up and amplify the signals. A section transports section frames that contain some overhead information plus the section payload bits which transport line frames. The transmission itself occurs on optical fibers.

FIGURE 7.14

SONET. The figure shows the three-layer hierarchy of the synchronous optical network: sections make up lines that make up paths.

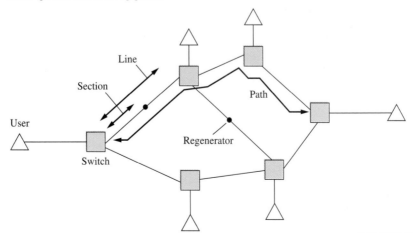

7.7.2   *Frames*

Figure 7.15 sketches the frames that SONET transports on its paths, lines, and sections. We illustrate the frames transported by lines at 155 Mbps. The frame, transmitted row by row, contains 9×270 bytes and it repeats 8000 times per second. The total bit rate is therefore equal to $8 \times 9 \times 270 \times 8000 = 155.52$ Mbps. The first factor (8) in the product converts bytes into bits. Out of the $9 \times 270 = 2430$ bytes of the frame, $4 \times 9 = 36$ are section overhead bytes, $5 \times 9 = 45$ bytes are LINE overhead bytes, and $1 \times 9 = 9$ bytes are the path overhead bytes. Thus, the user data rate is equal to $8 \times 260 \times 9 \times 8000 = 149.76$ Mbps.

We stated that all the transmitters in SONET are synchronized to a common clock. Now imagine that you are using a SONET path to carry a video signal or packets generated by your workstation or telephone signals produced by non-SONET equipment. To prevent the need to synchronize your equipment with the rest of the network, SONET enables SPEs to *float* with respect to the line frames. This floating is illustrated in Figure 7.15. That is, the SPEs can start at arbitrary times within a line frame. A special pointer in the line overhead indicates where the SPE starts. The receiver uses that pointer to locate and recover the SPE. Moreover, when not synchronized, the user may generate bits slightly faster or more slowly than the line bit rate. If the user rate is slightly larger, then the SPE payload overflows periodically. When an overflow occurs, SONET places the extra user bits in a specific field within the line overhead. Conversely, if the user rate is slower, then the SPE payload underflows whenever there are not enough user bits to fill it up. A specific field in the line overhead indicates the underflow and specifies the number of empty bytes inside the SPE payload.

FIGURE 7.15

SONET frame over 155 Mbps. The frame is read from left to right and from top to bottom. OH represents overhead.

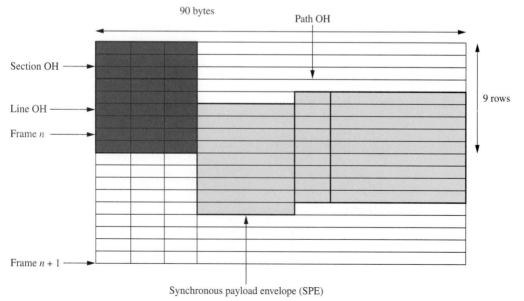

Synchronous payload envelope (SPE)

FIGURE 7.16

Overhead bytes in SONET frame.

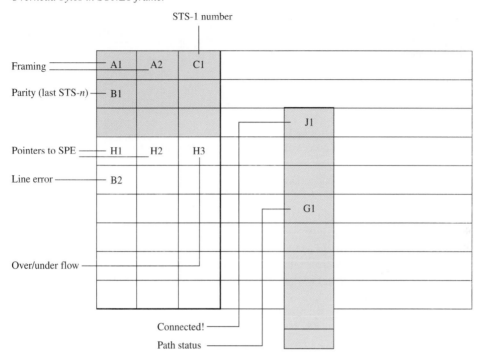

The SONET and SDH standards specify the meaning of the overhead bytes shown in Figure 7.16. Some of these bytes are error detection codes. Some section overhead bytes identify the constituting signals that are multiplexed. The SDH and SONET standards—probably needlessly—differ in the meaning of the overhead bytes, which complicates the interfacing of these two standards.

The 155-Mbps signal that we described is called STS-3, for synchronous transmission signal of level 3. An STS-1 signal at 51.84 Mbps was standardized in the United States but not in Europe. That signal is well suited for transporting the 45-Mbps DS-3 signals that the non-SONET telephone equipment carries.

Different SONET signals are multiplexed by byte interleaving. Thus, four STS-3 signals get combined to form a STS-12 signal. The rate of STS-12 is exactly 4 times that of STS-3, i.e., 622.08 Mbps.

7.8 Complement 4: Power Budget in Optical Link

In this section we explain how to calculate the power flow through an optical link. The objective is to find out how long the link can be, given the characteristics of its components. Figure 7.17 shows the link. We specify the power of light that a source produces in watts

FIGURE 7.17

Optical link used in the power budget calculation.

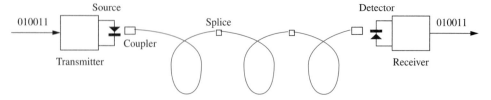

or in dBm. By definition, a power of P_T milliwatts corresponds to

$$10 \log_{10} \left(\frac{P_T \text{ mW}}{1 \text{ mW}} \right) \text{ dBm.}$$

For instance, a power of 1 μW $= 10^{-6}$ watt is equal to -30 dBm, since 10 $\log_{10}(10^{-6}/10^{-3}) = 10 \log_{10}(10^{-3}) = 10 \times (-3) = -30$.

The *receiver sensitivity* [P_R watts or P_R(dBm)] is the power the receiver must receive to detect the bits with a bit error rate equal to 10^{-12}. Consequently, the propagation along the fiber can reduce the power of the light by a fraction P_T/P_R, i.e., by

$$10 \log_{10} \frac{P_T}{P_R} = 10 \log_{10} \frac{P_T/1 \text{ mW}}{P_R/1 \text{ mW}}$$

$$= 10 \log_{10} \frac{P_T}{1 \text{ mW}} - 10 \log_{10} \frac{P_R}{1 \text{ mW}}$$

$$= P_T(\text{dBm}) - P_R(\text{dBm}).$$

This last expression is the attenuation in decibels that is acceptable before the received power falls below the sensitivity of the receiver.

The attenuation in the fiber is A dB/km. Consequently, the maximum length of the fiber, as determined by the attenuation, is

$$L_{\text{att}} = \frac{P_T(\text{dBm}) - P_R(\text{dBm})}{A \text{ dB/km}}.$$

In practice we must also take into account the power loss caused by couplers that attach the source and detector to the fiber and by splices that join fiber segments together. These losses are a few hundredth of a decibel per device. Finally, we should keep a safety margin for the loss of sensitivity caused by noise in the preamplifier and the source detector.

Thus, a more accurate determination of the maximum length of the fiber uses the following formula:

$$L_{\text{att}} = \frac{P_T(\text{dBm}) - P_R(\text{dBm}) - 0.05 \times N - 6}{A \text{ dB/km}}. \tag{7.8}$$

In this expression, N is the number of couplers and splices, 0.05 is the loss per coupler or splice, and 6 is the safety margin for noise. The above formula is called the *power budget formula* for the link.

Once we have determined the maximum length from the attenuation using (7.8), we must take into account the limitations due to the dispersion. These limitations are expressed in (7.2).

We illustrate these calculations on one example: a single-mode fiber at 500 Mbps. The wavelength is 1.55 μm and we use a glass fiber with an attenuation of 0.2 dB/km. The source is a laser diode with a transmitted power of 10 mW and the receiver sensitivity is equal to -50 dBm. We want to calculate the maximum length of the link. To use the power budget formula (7.8) we first convert the transmitted power into dBm: P_T (dBm) $= 10 \log_{10} (10 \, \text{mW})/(1 \, \text{mW}) = 10$ dBm. We need one coupler at the source and one at the detector. The number of splices is the number of fiber segments minus one. The number of fiber segments depends on the total length and on the length of each segment, say 10 km. Assume that we need K segments, i.e., $K + 1$ splices and couplers. We find

$$L_{\text{att}} = \frac{10 - (-50) - 0.05(K + 1) - 6}{0.2} \approx 270 - \frac{K}{4} \quad \text{km.}$$

Next we use the dispersion limit (7.2). For the single-mode fiber we find that $R \times L \leq 200$ Gbps/km. Since $R = 500$ Mbps, we conclude that $L \leq L_{dis} := 400$ km. Thus, the attenuation limits the usable length to about 264 km whereas the dispersion limits the length to 400 km. We conclude that the maximum length of the fiber is 264 km. Using a dispersion-compensated fiber would not increase the usable length at 500 Mbps.

7.9 Complement 5: RS-232-C

The connection between a printer and a computer, a computer and a terminal, or a computer and a modem often conforms to the Electronics Industries Association's (EIA) standard *RS-232-C* (*recommended standard 232-C*). That standard is commonly known as the *serial line interface.* A very similar version of this standard is recommended by the ITU under the name *V.24.* These standards specify the connectors, the assignment of connector contacts (pins) to the various signals, and the sequence of operations involved in a transmission.

The RS-232-C standard is intended for transmission at up to 38,400 bps over short distances (typically less than 15 meters). RS-232-C uses asynchronous baseband transmission (see Section 7.1) with bipolar modulation.

The RS-232-C connectors have 9 or 25 pins. (See Figure 7.18.) A typical connection uses only four to nine connection wires, which are not twisted. The pins are assigned to

FIGURE 7.18

RS-232-C serial line with its connector and the signals.

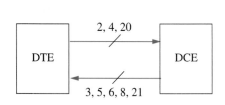

2. Transmit data
4. Request to send
20. DTE ready

3. Received data
5. Clear to send
6. Data set ready
8. Carrier detected by modem
21. Remote call indication

ground, data signals (send and receive), and to control signals. The main control signals are *request to send, clear to send,* and some *timing signals.* The pin assignments depend on the device type: either *DTE* (data terminal equipment, such as a terminal or a computer) or *DCE* (data circuit-terminating equipment, such as a modem or a printer).

RS-232-C is suitable only for short connections since a current may flow between the two grounds of the two connected devices. Such a ground current induces a voltage drop which modifies the voltage differential between the signal and the ground wires. If the connection is too long, then the voltage drop may be so large that bits are received incorrectly. In addition, a large loop formed by the signal and the ground wires may be subjected to electromagnetic interference that might also introduce errors in the transmission.

A typical full-duplex transmission from a DTE (a computer, for example) to a DCE (a printer, for example) using RS-232-C proceeds as follows:

- *DTE to DCE:* The DTE sends data on line 2; it can do so when wire 6 (data set ready) indicates a 0 (i.e., a positive value). Thus, the DCE can stop the transmission by the DTE by dropping the voltage value on the DSR line so as to indicate a 1. If line 5 (clear to send) is connected, the DTE can transmit only when it is positive (this is the case for many computers).

- *DCE to DTE:* The DCE sends data on line 3; it can do so when line 20 indicates a 0. If line 4 is connected, the DCE can transmit only when it is positive.

Two additional wires are usually used when the DCE is a modem: CD (carrier detect) and RI (ring indication). CD signals that the carrier is present and RI that the modem is being called by a remote device.

One notices a form of *handshaking* between the two devices. Handshaking is the name given to a procedure for agreeing on the terms of an information exchange. In RS-232-C, the handshaking is a coordination between the DCE and the DTE to make sure that the transmission starts only when the destination is ready to receive the data.

Computer devices that communicate using asynchronous transmissions contain an integrated circuit called a *UART* (*universal asynchronous receiver and transmitter*). When transmitting, the UART frames the information bits between a start bit and a parity bit (if one is used) and it transmits the bits one after the other at the rate specified by a clock. The UART performs the reverse operations when the device receives bits.

7.10 Complement 6: ADSL

In this section we examine a recent major development in the transmission technology over existing telephone lines. This development is motivated by the market for high-speed access to the Internet for residential customers and by a possible demand for video-on-demand applications. Business users also demand access to a remote LAN. It is estimated that by 1998 about 50 percent of the U.S. households will have a PC and that 85 percent of them will use a modem.

Asymmetric digital subscriber line (ADSL) is a high-speed digital transmission technology developed for the existing telephone twisted pair lines. ADSL transmits at a rate from 500 kbps to 8 Mbps downstream (from the central office of the telephone operator) and from 64 kbps to 1 Mbps upstream over one pair of copper wires. ADSL can provide

References

There are many good texts and tutorials on the physical layer. Henry (1985) is an excellent introduction to optical communications. Palais (1988) is a very readable presentation of the same subject. Couch (1983), Cooper (1986), and Lee (1988) are recommended texts on communication theory. If you want to study this subject in some depth, you can consult Proakis (1995). If you have time for only two references, read the papers by Shannon (1948).

Bibliography

Cooper, G. R., and McGillem, C. D., *Modern Communications and Spread Spectrum,* McGraw-Hill, 1986.

Couch, L. W., *Digital and Analog Communication Systems,* Macmillan, 1983.

Henry, P., "Lightwave primer," *IEEE Journal of Quantum Electronics,* QE-21, no. 12, pp. 1862–1879, 1985.

Lee, E. A., and Messerschmitt, D. G., *Digital Communication,* Kluwer, 1988.

Palais, J. C., *Fiber Optic Communications,* 2nd ed., Prentice-Hall, 1988.

Proakis, J. G., *Digital Communications,* 3rd ed., McGraw-Hill, 1995.

Shannon, C., "A mathematical theory of communication," *Bell System Technical Journal,* 27, pp. 379–423 and 623–656, 1948.

CHAPTER 8 Security and Compression

We have discussed how networks transport files or bit streams between host computers. In this chapter we explain two important services that complement the transport services: security and compression.

Computer systems that are part of a network are potential targets of a number of threats that we examine, together with basic protections, in Section 8.1. Many security mechanisms involve cryptographic methods that we describe in Section 8.2. We examine security systems in Section 8.3.

The transfer of audio, video, even data over a network is greatly facilitated by compression algorithms. Compression reduces the number of bits needed to encode the information. Section 8.4 discusses general concepts that are useful for understanding all compression algorithms. That section also explains algorithms used for data compression. We explain audio compression in Section 8.5 and video compression in Section 8.6. Complements contain details about cryptography and the theory of compression.

8.1 Threats and Protections

We classify the threats faced by systems into threats against computers (or servers), users, and documents. In each case we describe the threats and we discuss protection mechanisms.

8.1.1 Threats against Computers

In Table 8.1 we summarize the threats against computers and the corresponding protections. The threats against computer systems include physical attack, infection by harmful programs, and intrusion by illegitimate users. Harmful programs enter a computer either by a file copied by a computer's legitimate user or via the network without authorization. Harmful programs that copy themselves in other programs are called *viruses*. They are called *worms* if they copy themselves across the network.

To protect a system against such infection, we can use infection detection programs that look for known sequences of instructions in harmful programs or that monitor changes

TABLE 8.1   Main Threats against Computers and Their Protections

Threats	Protections
Physical attack	Physical security (e.g., locks)
Infection	Virus detection
Intrusion	Password control, firewalls

in the size of files. Another method for protecting the computer system is to prevent or restrict the writing to disk or to memory by some programs. Unfortunately, such restriction require a modification of the operating system in single-user systems.

An *intrusion* is an illegitimate access to a computer. Such intrusion typically occurs by guessing a legitimate user's password or by getting access to a password file. Other forms of intrusion include stealing a connection by impersonating a user on another machine, using a debugging trapdoor in a program, and exploiting a programming mistake—called a *hole*—that provides access to user privileges.

To protect against intrusion, users should choose passwords that are difficult to guess and, preferably, change them periodically. Moreover, we should transmit only encrypted passwords over the network. System administrators should be aware of trapdoors and holes and close them.

Firewalls are computers that screen the traffic that enters a computer network. A firewall is a system that monitors and restricts the packets that go through it. A gateway can be used between networks to limit their access. For instance, an organization may install a gateway between its network and the Internet, or between the personnel department's network and the employees' network. Two types of firewall systems can be set up. The first type is based on a packet filter. This filter screens packets typically based on their address. The difficulty with this approach is that it is easy to forge an address. The second type of firewall is an application level gateway. In such a system, the firewalls force packets to go through a gateway. The system can then restrict the applications that go across the gateway. For instance, only email might be authorized for communication with the rest of the world. One objective might be to prevent confidential files from leaking out. However, email can be used for file transfers, so this approach does not eliminate all undesirable transactions.

8.1.2   Threats against Users

Table 8.2 shows the threats against users and protections against them. The sender of a document may need to be *authenticated.* For instance, in electronic money transfers one needs to ascertain the identity of the person who sent the check. Similarly, the person who signs a document must be authenticated in a way that cannot be repudiated. A user who logs onto a computer system to access files may also need to be authenticated.

Authentication is required in some exchanges of information to prevent being fooled by an impersonator. Figure 8.1 shows two impersonation attacks. The first attack is the stealing of a connection by a "thief" who watches the login of a user. The thief then "zaps" the user by sending a barrage of messages to keep the user busy. The thief can then interact

TABLE 8.2   Main Threats against Users and Their Protections

Threats	Protections
Identity violation	Digital signature, encrypted password, watermarking
Privacy violation	Encryption, relays

FIGURE 8.1

Stealing a connection (left) and directory attack (right).

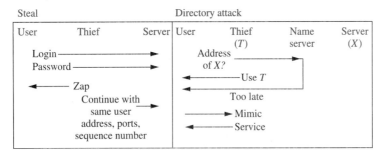

with the server by impersonating the legitimate user. The second attack is the "directory attack." Here, the thief replies to a request for an address by impersonating the name server. The thief can then mimic the server and get information from the user.

We explain in Section 8.3 methods that protect the integrity of data and of entities such as computers or users. Encryption is a family of methods for scrambling the messages so that eavesdroppers cannot read their contents. We discuss encryption in Section 8.2.

Watermarking is a method for including the identity of the author of a document. This method is the electronic equivalent of the watermark on paper currency. Watermarking is used to prevent another user from stealing a document and claiming authorship or ownership.

We say that *privacy* is violated when an eavesdropper discovers the identity of a user. To protect their identity, users may encrypt the messages they send. However, this encryption cannot hide the source of the messages. Users can also send their messages to a relay that resends them to their final destination. The relay may use confidential aliases for the destination addresses so that an eavesdropper cannot trace the (source, destination) pairs of messages.

Threats against Documents

Table 8.3 shows the threats against documents and protections against them. The *integrity* of a document is violated when the document is modified without authorization. Forging a signature, modifying the amount on a check, changing the name of an author are forms of integrity violation. Repudiating the authorship of a document is also possible when the integrity of documents is not guaranteed. The integrity of a document can be verified by

TABLE 8.3   **Main Threats against Documents and Their Protections**

Threats	Protections
Integrity violation	Message authentication code
Confidentiality violation	Encryption

including a message authentication code. *Confidentiality* of a document is violated when an unauthorized person discovers its content.

8.2   Cryptography

Cryptography, from the Greek, meaning "hidden writing," concerns the development of mechanisms that protect the contents of messages or the identity of their authors. We start by discussing the general principles of cryptography. We explain that two types of cryptography systems are used: secret key cryptography and private key cryptography. Cryptographic systems also use hashing functions that we introduce. Encryption and hashing are primitive operations that are used to build security systems that we examine in Section 8.3.

8.2.1   *General Principles*

In abstract terms, an encryption system works as indicated in Figure 8.2. The text that the sender—we call her Alice—must transmit in a secure way to the receiver Bob is the *plaintext P*. Alice converts P into the *ciphertext C*. Bob converts C back into P. Alice uses a function $E(.)$ to convert the plaintext. Bob recovers P from C by computing $D(C)$ where $D(.)$ is the inverse of the function $E(.)$. In this scheme, Alice and Bob must agree on the functions $E(.)$ and $D(.)$.

Eavesdroppers should find it difficult to identify the function D by looking at the messages that Alice sends. Experts have developed functions that have that desirable property. Before we examine them, let us discuss two examples of pair of functions E and D to illustrate how we could encode text.

A Naive Code
The *letter substitution code* replaces every letter with another one. For instance, one could have the correspondence $a \rightarrow u$, $b \rightarrow n$, $c \rightarrow f$, etc. With this code, the word $P = cab$ gets encoded into $C = fun$. The inverse operation converts $C = fun$ back into $P = cab$. This code is easily broken when used for plaintexts in English by using the relative frequencies of the letters. For instance, the letter e is the most frequently used in English. Thus, by finding the most frequent letter in the encoded text C, one knows that it should be decoded as e. One can continue in this way with the second most likely letter t, then the letter o, etc.

FIGURE 8.2

Encryption system. Alice encrypts her text P into C = E(P) before sending it over the network. Bob decrypts C into P = D(C).

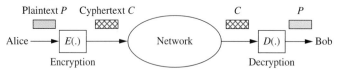

One-Time Pad

We now describe a perfect secret code that is used in top secret military and diplomatic communication.

Alice wants to send a string P of n bits to Bob. Prior to the transmission, Alice and Bob agree on another n-bit sequence R whose bits are chosen independently of one another and are equally likely to be 1 or 0. Alice then sends $C = P \oplus R$ to Bob where the additions are bit-by-bit modulo 2. That is, Alice changes a bit of P whenever the corresponding bit of R is a 1. When Bob gets C, he computes $C \oplus R = P$.

This method is perfect in that the string C is perfectly random: its bits could have been produced by flipping a fair coin n times. Thus, C contains no information about P. Unfortunately, this method is not practical for most applications because R can be used only once for the scheme to be secure.

Communication systems use two types of cryptography: secret key cryptography and public key cryptography. We describe these two systems below.

8.2.2   Secret Key Cryptography

In secret key cryptography, Alice and Bob agree on a pair of secret functions E and D. That is, no one other than Alice and Bob knows which functions they use for their transmission.

The functions should be chosen so that even after observing many ciphertexts $C = E(P)$ it is difficult to reconstruct $E(.)$ and $D(.)$. For some applications, the functions should be difficult to reconstruct even after observing many pairs $\{P, E(P)\}$.

In practice, the functions $E(.)$ and $D(.)$ are a pair of functions $E(.; K)$ and $D(.; K)$. The algorithms to calculate E and D are known, but the *key K* is secret. The key is some n-bit binary word. For such a scheme to be secure, n should be large enough. Otherwise, a brute-force attack which consists in trying out systematically all the 2^n possible keys can succeed within a reasonable time and cost. For instance, $n = 40$ can be broken in a few hours by using a large number of computers on a network to try different keys. Current wisdom is that $n = 56$ bits is marginally secure, that 64 bits may not be secure against powerful organizations, and that $n = 128$ is probably secure for many decades. (Some experts will disagree, obviously.)

We should note that some secret key systems use secret algorithms $E(.; .)$ and $D(.; .)$. These algorithms are implemented in special-purpose integrated circuits that are difficult to reverse-engineer. In this situation, a smaller key may be secure, provided that reverse engineering does not become easier.

Of course, security is a probabilistic notion. An eavesdropper might be particularly lucky, but this event is exceedingly unlikely with a large n even if there are potentially many eavesdroppers interested in your files.

We explore specific secret key systems in Section 8.7.

8.2.3   Public Key Cryptography

In public key cryptography, Bob informs everyone that to send him messages one should use the encryption function $E(.)$, but he keeps the function $D(.)$ private. That is, to send the message P to Bob, Alice computes and transmits $E(P)$. Bob then calculates $D(E(P)) = P$.

In the public key systems we discuss, the functions E and D are inverse of each other. That is $E(D(P)) = D(E(P)) = P$. For such a scheme to be secure, $E(.)$ must be a *trapdoor one-way function*. By definition, a trapdoor one-way function $E(.)$ is one that admits an inverse $D(.)$ and that inverse must be very difficult to construct from the function $E(.)$ itself. Thus, there is little risk that anyone could construct $D(.)$ and start decrypting the messages that Alice sends to Bob.

Note that for a public key system to be usable, Alice needs to be confident that E is the public function of Bob, and not that of someone else pretending to be Bob. That is, one needs to trust the public keys. This certification is done in a hierarchical way.

As a rule, public key cryptography is numerically much more complex than secret key cryptography. That is, the calculations of $C = E(P)$ and of $P = D(C)$ take longer. Accordingly, many systems use secret key cryptography. We discuss the problem of distributing the secret keys in Section 8.7. We explore a specific public key system in Section 8.8.

8.2.4   Hashing

We explain below that cryptographic systems use *hash functions*. A hash function H maps a long text P into a shorter text $H(P)$. For instance, P might be 1000 bits long and $H(P)$ only 160 bits long. Thus, $H(.)$ is a many-to-one function: Many different texts P result in the same $H(P)$.

A hash function must be difficult (computationally) to pseudo-invert. That is, given M it is very difficult to construct a text P' such that $H(P') = M$. Depending of the application, the additional properties may be required:

- *Property 1:* Given both P and $H(P)$ it is difficult to construct $P' \neq P$ such that $H(P') = H(P)$. This condition is stronger than the previous one because knowing both P and $H(P)$ may make it easier to construct P' than if only $H(P)$ is known.
- *Property 2:* It is difficult to construct two texts P and $P' \neq P$ such that $H(P) = H(P')$.
- *Property 3:* Given P but not knowing a string K it is difficult to find $P' \neq P$ such that $H(K \circ P') = H(K \circ P)$. In this notation, $K \circ P$ means the concatenation of two strings K and P, that is, the bits of K followed by the bits of P. (For instance, $10011 \circ 011101 = 10011011101$.)

Our discussion of security systems indicates which properties are required of H.

8.3 Security Systems

In this section we explain security systems that combine encryption and hashing. We classify these systems by the services they provide. We conclude the section by looking at two widely used systems.

8.3.1 Integrity

We explain how cryptography can be used to protect the integrity of the data or of its origin. Imagine Alice sending a message P to Bob. She wants to guarantee that Bob gets P and not some other P'. Moreover, Bob wants to make sure that Alice sent the message, not someone else.

The basic idea of integrity protection systems is the following. To protect the integrity of a text P, Alice calculates $H(P)$ and sends $(P \circ H(P))$ to Bob. (Throughout, H is a hash function.) Note that $H(P)$ must be protected. Otherwise, an intruder Eve can intercept the transmission and send some $(P' \circ H(P'))$ to Alice. Three schemes are used to protect $H(P)$: authentic channel, message authentication code, and digital signature.

Authentic Channel

In the first scheme, Alice sends $H(P)$ to Bob over an *authentic channel..* An authentic channel is defined as a channel whose transmissions cannot be corrupted. For instance, $H(P)$ could be communicated over a secure telephone line (whatever that means). The weakness of this scheme is that, given P, Eve might find some P' such that $H(P') = H(P)$. Eve could then intercept the unsecure transmission of P and replace it with P'. In this application, H should have property 2 described in Section 8.2 under "Hashing."

Message Authentication Code

In the second scheme, $H(P)$ is secured with secret keys. At first, one might think Alice could send P and $E(H(P) \circ K)$ to Bob, using a secret key K. However, this scheme is not secure. For instance, imagine that Alice uses a one-time pad to encrypt $H(P)$. That is, Alice and Bob agree on a secret random string R and Alice sends $Z = R \oplus H(P)$. The intruder Eve sees P and Z and calculates $H(P)$ from P and then computes $H(P) \oplus Z = R$. Once she knows K, Eve can send P' and $H(P') \oplus R$ to Bob who then thinks Alice sent P'.

A suitable strategy is for Alice to send $H(K_2 \circ H(K_1 \circ P))$. That is, Alice first computes $V = H(K_1 \circ P)$, then $H(K_2 \circ V)$. Here, K_1 and K_2 are secret keys shared by Alice and Bob. This scheme fools Eve because she cannot calculate $H(K_1 \circ P')$. Accordingly, for this application, H should have property 2 of "Hashing" in Section 8.2. The string $H(K_2 \circ H(K_1 \circ P))$ is called a *message authentication code.*

Note that this discussion shows that the one-time pad is not secure for integrity applications although it is for confidentiality of the message.

Digital Signature

In this scheme, which uses public key encryption, Alice and Bob use the public function E of Alice whose inverse D is secret to Alice. The basic idea of the scheme is that Alice can sign a document P by sending $C = D(P)$ to Bob. Bob can recover P by calculating $P = E(C)$. (Recall that E and D are the inverse of one another.)

However, as described, the scheme is not suitable. Imagine that an intruder Eve intercepts the channel from Alice to Bob. Eve constructs a message C' and calculates $P' = E(C')$, using Alice's public key. Eve then sends C' to Bob who then calculates $P' = E(C')$ and might think that Alice sent $C' = D(P')$. Because of this possibility, this scheme should not be used: it does not protect the integrity of Alice.

A suitable scheme is for Alice to send $P \circ H(P)$ instead of P. That is, Alice sends $D(P \circ H(P))$ where H is a hash function. This makes it almost impossible for Eve to find another message C' so that $E(C')$ has the form $P' \circ H(P')$ for some text P'. The string $D(P \circ H(P))$ is said to be the *message P signed by* Alice.

Note that the digital signature is suitable even when Alice and Bob do not trust each other since they don't need to share secrets. Accordingly, a digital signature can be used for nonrepudiation of documents, which is critical in electronic commerce.

8.3.2   Key Management

Secret key systems require that Alice and Bob share a secret key K. Alice and Bob can use one of the following strategies to share such a secret key:

1. Hand delivery of the key.
2. Use a secret key to encrypt and distribute other secret keys. Kerberos uses this strategy (see "Kerberos," below).
3. Use a public key to establish secret keys. PGP uses this strategy (see "Pretty Good Privacy," below).
4. Public key agreement scheme. In such a scheme, Alice and Bob end up agreeing on a common key K by exchanging messages that intruders cannot use to compute K. The Diffie-Hellman exchange that we explain next is such a scheme.

Diffie-Hellman Exchange

This strategy named after its inventors. Alice and Bob first agree on a pair of numbers (z, p). These numbers can be made public. Alice then chooses a number a and Bob chooses a number b. Alice computes $\alpha = z^a \bmod p$ and sends it to Bob who computes $\alpha^b \bmod p$. Similarly, Bob computes $\beta = z^b \bmod p$ and sends it to Alice who computes $\beta^a \bmod p$.

We claim that

$$\alpha^b \bmod p = \beta^a \bmod p = z^{ab} \bmod p =: K. \tag{8.1}$$

Moreover, it is difficult to calculate a from α and b from β, so that an eavesdropper who knows (z, p) and observes (α, β) cannot compute K.

To show (8.1), we first note that $\alpha = z^a \bmod p$ so that $\alpha = z^a + mp$ for some integer m. Consequently,

$$\alpha^b = (z^a + mp)^b = z^{ab} + b z^{a(b-1)} mp + \frac{b(b-1)}{2} z^{a(b-2)} (mp)^2 + \cdots + b z^a (mp)^{b-1} + (mp)^b.$$

This equality proves (8.1).

It turns out that the Diffie-Hellman exchange is not robust to a "person-in-the-middle" attack as we explain next. Imagine that an eavesdropper Eve intercepts the Diffie-Hellman exchange between Alice and Bob. Eve can communicate with Alice as if she were Bob and agree with Alice on a common secret key. In the same way, Eve can agree on a secret key with Bob as if she were Alice. At that point, Eve can intercept the messages from Alice to Bob, read them or change them and reencrypt them for Bob. Alice and Bob are unaware of the existence of Eve.

Another weakness of the Diffie-Hellman exchange is that Eve can intercept the message α from Alice to Bob and replace it by 0. Similarly, when the message β comes from Bob to Alice, Eve can replace it with 0. As a result, Alice and Bob end up agreeing on the common key 0 that Eve knows. (Eve could have used 1 or -1 instead of 0.) To prevent this simple attack, the system should verify that the numbers exchanged are not 0, 1, or -1.

A more subtle attack by Eve is as follows. Eve chooses some prime number q that divides $p-1$. She intercepts α from Alice to Bob and replaces it with α^q. She also intercepts β from Bob to Alice and replaces it with β^q. As a result, Alice and Bob end up agreeing on the key K^q instead of the key K. Although Eve does not know K^q, she knows that the key belongs to a smaller subgroup of keys (the exact qth powers) on which the search is considerably easier.

The Diffie-Hellman exchange can be made robust against this attack. Two solutions have been developed: signing the exchange and using safe primes.

When signing the exchange, Alice sends α to Bob and Bob sends β to Alice. Alice then signs (α, β) and sends it to Bob. Bob also signs (α, β) and sends it to Alice. Eve cannot fake the signatures.

By definition, a prime number p is a *safe prime* if $p' := (p-1)/2$ is also prime. In that case, one can show that the only multiplicative subgroups of $\{1, 2, \ldots, p-1\}$ are $\{1\}$, $\{-1, 1\}$, and $\{1, 2, \ldots, p-1\}$. It the suffices in the Diffie-Hellman exchange to check that the numbers exchanged are not 0, 1, or -1.

8.3.3   Identification

The identification problem is to ascertain the identity of some entity such as a computer or a user. In our discussion, Bob wants to ascertain the identity of Alice. We explain five mechanisms: passwords, challenge/response, public key, digital signature, and zero-knowledge proof.

Passwords
The simplest mechanism for identification is to use a password. To certify her identity, Alice has a secret password K and she sends (*Alice, K*) to Bob. Bob maintains a list of names with hashed passwords. This list contains (*Alice, H(K)*) and Bob can thus verify that Alice is who she claims to be. Bob does not keep the list of pairs (*Alice, K*) for obvious reasons. The weakness of the password mechanism is that Bob and the communication line both see the secret password K of Alice.

Challenge/Response
When using the *challenge/response* scheme, Bob sends a string X to Alice. Alice replies with a function $f(X, K)$ where K is a secret key that Bob and Alice share. Bob then checks

the value of $f(X, K)$. If that value is correct, Bob assumes that Alice computed it since no one else knows K. The advantage of the scheme is that the key K is not transmitted over the line. The weakness is that Bob must know K.

Public Key
Let E be the public key of Alice with inverse D known only by Alice. Bob chooses X and sends $E(X)$ to Alice who decrypts $X = D(E(X))$ and sends X back to Bob.

Digital Signature
In this scheme, Bob sends X to Alice who signs it and returns the signed value to Bob.

Zero-Knowledge Proof
A zero-knowledge proof method enables a user to prove that it knows some specific information such as a secret password without revealing any of the information.

One zero-knowledge proof system is based on the notion of *quadratic residue*. Let x, y be positive integers that are relatively prime, i.e., that have no common factor other than 1. Say that y is a quadratic residue of x if there is some w such that $y = w^2 \bmod x$. For instance, 9 is a quadratic residue of 10 since $9 = 7^2 \bmod 10$. It turns out that, if x is hard to factor, then it is difficult to check whether a given y is a quadratic residue of x because it requires factoring x. Also, it is difficult to produce a "square root" w of a given y.

We show how Alice and Bob can use a zero-knowledge proof system to ascertain the identity of Alice.

- *Step 1:* Alice selects w (her password) and calculates $y = w^2 \bmod x$, for some large x. Alice gives y to Bob who then wants to make sure that Alice knows w. Bob cannot compute w from y.
- *Step 2:* Alice chooses some number u that is relatively prime with x and sends $z := u^2 \bmod x$ to Bob.
- *Step 3:* Bob sends back either 0 or 1 to Alice, with equal probabilities. If she gets 0, Alice must provide the value of u. If she gets 1, she must send $v := uw \bmod x$ to Bob. In the former case, Bob can check that $z = u^2 \bmod x$. In the second case, Bob can verify that $zy = v^2 \bmod x$.

An eavesdropper cannot recover w from either u or v alone, and the user cannot make up a value u when asked for a reply since the user does not know whether u or v will be asked. Also, without knowing w, Alice cannot produce a v such that $zy = v^2 \bmod x$ since it is difficult to compute square roots. Thus, Bob believes with probability $\frac{1}{2}$ that Alice knows w.

It is not trivial, but can be shown that repeated exchanges of this type do not reveal information about w.

8.3.4   Replications and Deletions

We have discussed cryptographic systems to protect the integrity and the confidentiality of data and their origin. Such schemes cannot protect a transmission channel against repetitions or deletions of messages. To protect against these corruptions, the system must use other strategies such as serial numbers for the messages.

8.3.5   *Kerberos*

Kerberos, the three-headed dog guarding the entrance of Hades, the home of the dead, gave his name to a secret key network authentication system. A two-headed dog might have been more appropriate, as you will see.

The left part of Figure 8.3 shows Alice confirming her identity to Bob by providing her password. The password is shared between Alice and Bob. This process can be made symmetric by having Bob confirm his identity to Alice. Once Alice and Bob trust each other's identity, they can start a session.

Although this method is used by many systems, it is not very desirable. Each server (Bob) needs to store the passwords of every potential client (Alice). An intruder could then get the list of passwords and break the security of the system. Moreover, every time a new user joins the network, that user's password must be added to all the servers. In addition, after a security breach, all the passwords and password files need to be changed.

These shortcomings are remedied by using a third party to store the passwords. Kerberos is such a third-party authentication system. Using Kerberos, when Alice wants to communicate with Bob, she gets a "ticket" valid for the session with Bob. Before getting tickets, Alice registers with Kerberos by providing her password. This registration is valid for as long as Alice remains logged on.

The procedure, shown in the right part of Figure 8.1, is based on a secret key. Here are the steps of the procedure:

- 1–2: When Alice logs on, she requests a "session key" from the authentication server (AS). This session key remains valid while Alice is logged on.
- 3–4: Before requesting a service from Bob, Alice asks the ticket-granting server (TG) for a "service ticket."
- 5–6: Alice can then ask Bob for the service by showing him the ticket.

Let us examine these steps in more detail.

The AS shares a secret key with each user of Kerberos. For instance, say that $K(Alice)$ is the secret key of Alice. When Alice logs on, her workstation asks (step 1) AS for the session key $S(Alice)$, which AS provides in an encrypted message $E(M, K(Alice))$ (step 2) where

$$M = [S(Alice), T = E(Alice, S(Alice), \textit{Expiration time}; K(AS))].$$

Alice's workstation decrypts the message using $K(Alice)$ and remembers $S(Alice)$ and T.

FIGURE 8.3

Basic authentication (left): Alice confirms her identity to Bob by providing a password. Third-party system (right): Kerberos verifies the passwords and provides a ticket for the session.

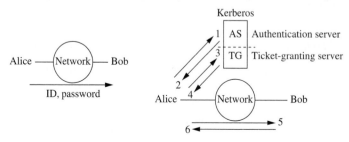

Next, Alice wants to talk to Bob. She sends the following message to TG (step 3):

$$\text{"I want to talk to Bob": } (T, \, Bob).$$

TG decrypts T using the key $K(AS)$ and uses (*Alice, S(Alice), Expiration time*) to verify that Alice is a legitimate user and to obtain the session key $S(Alice)$ of Alice.

TG then provides the service key $S(AB)$ to Alice for the A-B session between Alice and Bob by sending the following message (step 4):

$$E(S(AB); S(Alice)), \, E(Alice, S(AB); K(Bob)).$$

Alice decrypts the first part of the message to get $S(AB)$ and then sends Bob the second part of the message (step 5):

$$\text{Hi Bob, } E(Alice, \, S(AB); \, K(Bob)).$$

When it gets that message, Bob can decrypt it and find (*Alice, S(AB)*) to learn that the session $S(AB)$ will be used for this session between Alice and Bob. Alice and Bob can then talk using the key $S(AB)$.

Note that the keys $S(Alice)$ and $S(AB)$ are time-limited, which reduces the chance that these keys will be discovered.

8.3.6   *Pretty Good Privacy*

Pretty Good Privacy (PGP) is a secure file transfer protocol. The files are transferred using secret key cryptography. The secret keys are distributed using public keys. The public key cryptography is based on RSA. The secret key system uses IDEA. The hashing for authentication uses MD5, a specific hashing function.

The mechanism for certifying the public key is left casual: a key is certified by a trusted friend using the assumption that a friend of a friend is a friend. Thus, if the president of the university certifies the deans, the deans can certify the chairpersons, and so on. The certification is through a certificate that has the form:

$$\text{[John's Public Key is } K \text{—trust level—signed Jim]}.$$

Such a certificate indicates that Jim trusts that the key of John is K. If you trust Jim and believe that his key is secure, then you believe that Jim signed this certificate and you then trust that K is indeed the public key of John. Accordingly, when John sends you a certificate that he signs with his key K:

$$\text{[Gene's Public Key is } K' \text{—trust level—signed John]},$$

you will trust that Gene's public key is K'.

This process can be used to generate lists of public keys with various trust levels. Each user stores certificates of the form

$$\text{[trust—John—Jim—trust—Gene—trust—} \cdots \text{]}$$

that summarize the confidence in the public keys.

When using PGP, Alice sends file F to Bob (with public key B) as

$$[E(K; \, B), \, \text{IDEA(ZIP(File); } K)]$$

and with a signature ($A =$ Alice's public key):

$$[D(K; A), \text{MD5(Message)}].$$

PGP generates keys for you based on user ID, password, and random sequence of characters.

8.4 Foundations of Compression

In this section we explain why compression is possible and we explore some basic algorithms that compression methods use.

8.4.1 *Lossy and Lossless Compression*

Compression is possible because a source output contains *redundant* and/or *barely perceptible information.*

Redundant information is data that does not add information, such as white lines between lines of text on a page, periods of silence in a telephone signal, similar lines in a picture, and similar frames in a video sequence. Redundancy can be reduced by algorithms that then achieve a *lossless compression.* No information is lost by such algorithms even though they reduce the number of bits in the source output.

Barely perceptible information is, as the name indicates, information that does not affect the way we perceive the source output. Examples of barely perceptible information in audio are sounds at frequencies that the human ear does not hear and sounds that are masked by louder sounds. In images, some fine details are difficult to perceive and can be eliminated without much picture degradation. Such information is reduced or eliminated by *lossy compression algorithms.*

8.4.2 *Batch, Stream, Progressive, Multilayer*

Consider a source that produces a file F. We can view a compression algorithm as implementing a function $K(F)$ that produces a new file with fewer bits. The decompression algorithm calculates $X(K(F))$. (We use $K(.)$ for compress and $X(.)$ for expand, to avoid confusion with the cryptography notation.) This reconstruction must be exact in a lossless compression and, hopefully, it is almost exact in a lossy compression. That is, the distance $d(F, X(K(F)))$ between F and $X(K(F))$ is small in a lossy compression. We think of $d(.,.)$ as measuring a subjective perceptual difference, which is of course not easy to quantify.

The above formulation of the compression/decompression process ignores the timing of the operations. Consider the hypothetical compression/decompression of a video signal during a video conference shown in Figure 8.4. The video camera produces a bit stream at 20 Mbps. The compression algorithm converts this stream into a 1.5 Mbps that the decompression converts back into a 20-Mbps stream close to the original. More specifically, the video camera captures successive frames F_n at the rate of 30 frames per second. These frames are converted into a bit stream which is compressed, then decompressed. The receiver uses the decompressed bit stream to reconstruct frames $\hat{F}_n$ that approximate the

FIGURE 8.4

Delay in the compression/decompression of a video stream.

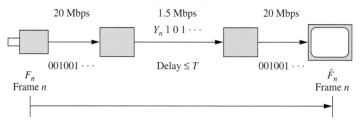

original ones. The delay between the capture of frame F_n and its reconstruction as $\hat{F}_n$ by the receiver is obviously of critical importance in real-time applications.

This delay is important also for stored video. Imagine that the compressed bit stream corresponding to 5 minutes of video is stored in a server. When a user wants to view that video clip, it is desirable for the playback to start as soon as possible. As we discussed in Chapter 4, the delay jitter in the network should be compensated for by a playback buffer. It is also clear that such a stream playback is possible only if the network can deliver the compressed bit stream without building up a backlog.

Similar considerations apply to audio.

In the case of (fixed) pictures, it is desirable to have a progressive reconstruction of the picture. That is, in trying to view an image that is stored as a large file, it is desirable to view quickly a low-resolution version of the image that improves progressively.

Finally, since network connections have widely different bandwidth, it is desirable to be able to request different versions of audio or video clips or streams. These versions correspond to different qualities of the reconstruction, being compressed differently. Multilayer compression algorithms provide complementary versions of a clip, like adding layers of details to a coarse initial version to provide increasingly fine reproductions. Researchers are experimenting with various network protocols that select adaptively the number of layers based on the delays.

8.4.3   Source Coding

We view the source as producing a string of symbols. These symbols may be pixels, sound samples in audio, or characters in a text. By definition, the *entropy* of a source is the minimum average number of bits per source symbol required to describe the source symbols without loss.

Some examples illustrate that concept. First consider a sequence of fair coin flips. Because of their complete unpredictability, it is fairly intuitive that one needs 1 bit per symbol to describe the sequence.

Second, look at a sequence of 1 million 1s. This sequence can be described by the words "1 million 1s." These words require 12 characters, or 96 bits, to be encoded. That is, the sequence of 1 million 1s is encoded with 96 bits which amounts to about 0 bit per symbol.

Third, consider the sequence $\mathbf{Y} = \{X_1, X_1, X_2, X_2, X_3, X_3, \ldots\}$ of twice-repeated fair coin flips. Since it takes one bit per symbol $X_1, X_2, \ldots$, we see that we can describe the sequence $\mathbf{Y}$ with 0.5 bit per symbol.

More generally, we suspect that if there is some correlation between successive bits in a sequence, then we require fewer than 1 bit per symbol to describe the sequence.

Claude Shannon has shown that, under weak assumptions, one can define a quantity H, called the *entropy rate* of the source. This quantity is the minimum average number of bits per source symbol required in lossless compression. We explain the basic ideas of this theory in Section 8.10.

8.4.4   Finding the Minimum Number of Bits

How can one automatically use only the minimum average number of bits per symbol, or at least not much more that this minimum? Practical systems use a combination of clever tricks. Here is a partial list. We revisit these examples in subsequent sections.

- *Models*   Fit model parameters and send the parameters. Examples: fractals, code excited linear predictors.
- *Prediction + error encoding*   Predict next symbol, send error. Examples: video conference compression.
- *Differential encoding*   Send difference with previous symbol. Examples: motion compensation, DPCM.
- *Run length encoding*   Send length of run of repeated symbols. Examples: silence suppression, fax encoding.
- *Dictionary*   Point to, do not repeat a previously seen string. Examples: Lempel-Ziv.
- *Short codes for likely symbols*   Longer for rarer. Examples: Huffman (e.g., JPEG, MPEG).

We examine these basic building blocks of compression algorithms.

8.4.5   Huffman Encoding

The main idea of Huffman encoding is to use short codes for frequent symbols and longer ones for rare symbols. For example, consider a source that produces a string of symbols that take four possible values $\{A, B, C, D\}$ and that, in the long run, these symbols are produced in the following proportions: *A:* 45%, *B:* 45%, *C:* 5%, *D:* 5%. We are not making any independence assumption or using any model for the correlation of successive symbols.

Using the code

$$A: 00, B: 01, C: 10, D: 11,$$

one needs 2 bits per source symbol. However, using the code

$$A: 0, B: 10, C:110, D: 111, \tag{8.2}$$

we need $1(0.45) + 2(0.45) + 3(0.05) + 3(0.05) = 1.65$ bits per symbol. The code that we chose in (8.2) is prefix-free. This means that the first part of no codeword is another codeword. The code *A:* 0, *B:* 00, *C:* 1, *D:* 11 is not prefix-free because the first part 0 of the codeword 00 for *B* is another codeword (for *A*). A prefix-free code has the property that any concatenation of codewords can be decoded without ambiguity. Thus, for the code in (8.2), the string 1011001111001100 can only be decoded as *BCADBACA*. When using a code that is not prefix-free, one must insert special symbols to separate the source symbols, which increases the average number of bits per symbol and ends up transforming the code into a prefix-free code.

The code (8.2) is the prefix-free code that requires the smallest average number of bits per symbol for our example. This code is called the Huffman code. Before explaining the simple procedure that produces the Huffman code, we note that if the symbols are i.i.d. (independent and identically distributed), then we can compute the entropy of the source and we find that we need $H = -0.45 \log 0.45 - \cdots - 0.05 \log 0.05 = 1.47$ bits per symbol. Thus, the Huffman code does not achieve the Shannon bound. Of course, achieving or even approaching that bound is substantially more complicated than constructing the Huffman code. More about this when we study the Lempel-Ziv compression algorithm.

We now explain the general method for constructing the Huffman code. We are given K symbols $\{1, 2, \ldots, K\}$ with the relative proportions in which they appear in a long string of symbols that the source produces: $\{p(1), p(2), \ldots, p(K)\}$. Thus, $p(k)$ is the fraction of symbols that are equal to k. The numbers $p(k)$ are nonnegative and add up to 1.

We construct a tree recursively. At each step, we join the two nodes with the smallest values and we attach the sum of their values to their aggregate. Figure 8.5 shows this procedure for our previous example. We first join the symbols *C* and *D* because their values 0.05 and 0.05 are the two smallest. We join them by creating a subtree whose root is allocated the value sum 0.1 of the values of the two symbols. We then join this root and symbol *B* to create a new root with value 0.55. Finally, we join this new root and symbol *A*. The resulting tree defines a code as follows. The code that corresponds to a symbol is obtained by starting from the top root and traveling down toward the symbol. As we go down, each move to the left corresponds to a 0 and each move to the right corresponds to a 1. For instance, the codeword for symbol *B* is 10 because we first move right (and down), then we move left (and down) to reach *B*.

Figure 8.6 is another example. This algorithm produces a prefix-free code. Indeed, any path down the tree reaches a leaf only at the end of the path and not at any intermediate point along the path. It is not completely straightforward to show that this code has the minimum average length over all prefix-free codes.

FIGURE 8.5

Constructing a Huffman code.

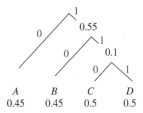

FIGURE 8.6

Another example of a Huffman code.

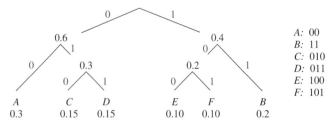

A:	00
B:	11
C:	010
D:	011
E:	100
F:	101

8.4.6   Lempel-Ziv Compression

This algorithm produces a code that achieves the Shannon limit. We should qualify this statement. Consider a source that produces random symbols $\{X_n, n \geq 1\}$. Under the ideal implementation, the average number of bits per symbol that the Lempel-Ziv algorithm requires to encode the first n symbols of the source approaches the entropy rate of the source.

What is remarkable about the Lempel-Ziv algorithm is that it does not require any prior information about the statistics of the source. Versions of this algorithm are implemented in most compression utilities.

Roughly speaking, this algorithm replaces a string by a pointer to the first occurrence of the string, if it has been seen before. For instance, the algorithm encodes the string

$$\text{a bit is a bit is a bit} \tag{8.3}$$

as

$$\text{a bit is } [1, 9][1, 5]. \tag{8.4}$$

The first part of the sentence "a bit is" does not contain pieces that appeared earlier in the sentence. Accordingly, this part must be transmitted by the code. (How else would the receiver know about it?) The next part "a bit is" has been seen previously. Instead of repeating this string, the algorithm replaces it by a pointer to the location of the start of that earlier instance of the string and by the length of that fragment. In our example, the earlier fragment starts in location 1 and has length 9 characters (including spaces). The last piece of the sentence, "a bit" has been seen earlier, starting at location 1 and with a length of 5 characters.

From the code (8.4), there is no ambiguity in reconstructing the sentence (8.3). Moreover, this reconstruction is straightforward. The encoding is more complex. It requires parsing the original sentence to locate fragments that have been seen previously. Encoding algorithms use a "dictionary." The source output is parsed and at each step, the shortest string that is not yet in the dictionary is added to it.

Figure 8.7 illustrates this parsing process. The parsing proceeds from left to right and starts with an empty dictionary. The first bit, 0, is a string that is not in the dictionary. We add that string and indicate that fact by inserting a comma after the first 0. The next shortest string that is not yet in the dictionary is 00. We add it to the list. The process

FIGURE 8.7

Parsing a bit string in the Lempel-Ziv algorithm.

Original string:

0 0 0 1 1 0 0 1 0 1 1 1 0 1 0 0 1 0 0 1 0 0 1 0 1 0 0 1 0 1

Parsing:

0, 0 0, 1, 1 0, 0 1, 0 1 1, 1 0 1, 0 0 1, 0 0 1 0, . . .

continues in this way. To compress the original string, we replace each substring produced by the parsing as follows. We strip the substring of its last bit. This shortened string must have appeared in the dictionary. We replace the substring by a pointer to the location of the shortened string in the dictionary together with the last bit of the substring. For instance, the string 101 is replaced by a pointer 4 to the shortened string 10 together with the last bit 1 of the string. As the process continues, the strings that the parsing produces get longer. A string of length n gets replaced by a pointer to the shortened string. This pointer requires a number of bits approximately equal to the logarithm in base 2 of the number of strings in the dictionary. Remarkably, the analysis shows that the process ends up using an average number of bits per symbol equal to the entropy rate of the source, at least if we have an infinite dictionary to run the algorithm.

8.5   Audio Compression

Audio compression schemes are used in applications that range from regular, cordless, and cellular telephone to audio for digital TV, CD-ROMs, Mini-Disc, Digital Compact Cassette, 3/2 Stereo, Internet Phone, RealAudio. Here is a partial list:

PCM   pulse code modulation

ADPCM   adaptive differential PCM

SBC   sub band coding

VSELP   vector sum excited linear prediction

CELP   code excited linear prediction

MPEG   Motion Picture Experts Group

Figure 8.8 shows the bit rates that these schemes produce. We explain some of these schemes below.

8.5.1   *Differential Pulse Code Modulation (DPCM)*

DPCM transmits differences $X_{n+1} - X_n$ between successive samples X_n instead of transmitting the sample values. Successive audio samples tend to be similar. Accordingly, their differences are small and can be quantized accurately with fewer bits than the samples.

Figure 8.9 illustrates DPCM. The figure shows a sequence of audio samples and their successive differences.

8.5.2   *Adaptive DPCM (ADPCM)*

Adaptive DPCM goes one step beyond DPCM. Instead of simply computing the difference $X_{n+1} - X_n$ between successive samples, ADPCM computes the difference $X_{n+1} - \hat{X}_{n+1}$ between the next sample X_{n+1} and the predicted value $\hat{X}_{n+1}$ of that sample based on samples seen so far. In practice, $\hat{X}_{n+1}$ may be computed from a linear combination of recent sample values. Intuitively, if recent samples are increasing, it may be reasonable to expect that the next one is yet a bit larger. Figure 8.10 shows an example of ADPCM.

8.5.3   *Subband Coding ADPCM*

The idea behind subband coding ADPCM is that the human ear has different sensitivities to different frequencies. It is then sensible to reproduce these different frequency bands

FIGURE 8.8

Bit rates produced by various audio compression algorithms.

Application	Schemes	Bit rates (kbps)
Speech	PCM	64
	ADPCM	40–64
	SBC	16–32
	VSELP-CELP	2.4–8
Audio	PCM	1400
	MPEG	128–384

FIGURE 8.9

Differential pulse code modulation.

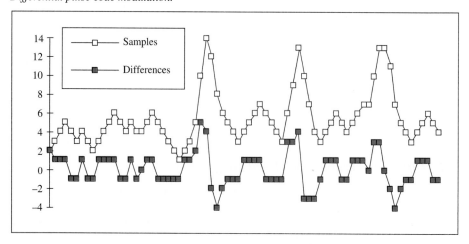

with accuracies that correspond to these sensitivities. The basic block diagram of these algorithms is shown in Figure 8.11.

More complex algorithms combine masking with subband coding. That is, when the volume is much larger in one band than in the other, one may reduce the resolution of that weaker band or suppress it altogether. MPEG audio compression uses that mechanism.

8.5.4   Code Excited Linear Prediction

This method is designed to achieve very low bit rates, at the cost of a significant loss in quality, unfortunately. The block diagram of a CELP system is shown in Figure 8.12.

The speech production process is modeled as a filter that is excited by a collection of different codewords. For a given speech fragment, the algorithm selects the parameters of the filter and the codeword in the codebook that produce the best approximation of the fragment. The algorithm then transmits the filter parameter and the index of the codeword.

The decoder performs the reverse operations. It reproduces the speech fragment by exciting the filter with the codeword.

FIGURE 8.10

Adaptive differential pulse code modulation.

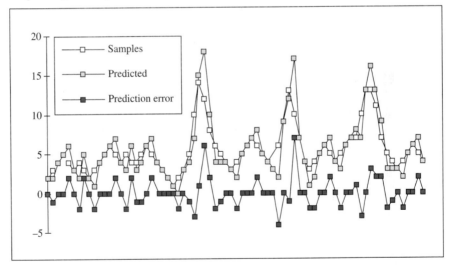

FIGURE 8.11

Subband coding ADPCM.

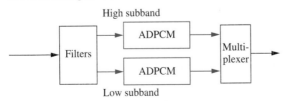

FIGURE 8.12

Block diagram of CELP system.

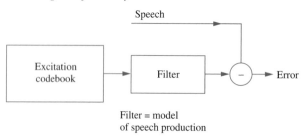

Filter = model
of speech production

8.6 Video Compression

As with audio compression, we are only able to skim the surface of this subject. Video compression is necessary because video clips or signals require too many bits without compression. Fortunately, there is a lot of redundancy and of barely perceptible information in video, so that compression algorithms are very effective.

Without compression, we can estimate the bit rate needed to transmit a video signal as $360 \times 288 \times 8 \times 3 \times 24 = 60$ Mbps. The calculation is picture size times number of bits per color per pixel times number of colors times number of pictures per second. This is for a small image.

HDTV without compression requires about 8000 Mbps. With compression, we get excellent results with 20 Mbps.

8.6.1 Some Algorithms

Here are a few representative video compression algorithms (the numbers in parentheses indicate the applicable steps in the list below):[1]

- JPEG: pictures (1–3)
- H.261 ($p \times 64$): ISDN video conferences
- H.263 (≤ 64): video conferences (1–4)
- MPEG 1: 1.5 Mbps, DBS, CATV (1–4)
- MPEG 2: 10 Mbps and more, DBS, CATV, HDTV, DVC, DVD (1–4)
- MPEG 4: 64 kbps (not ready), model-based

These algorithms combine a number of clever ideas. The image is decomposed into blocks. Then the following steps are applied.

1. Transform image: DCT produces coefficients of the average value and of the higher-frequencies in the image

---

[1] JPEG = Joint Photographic Experts Group, ISDN = integrated services digital networks, DBS = direct broadcast satellites, CATV = community antenna TV, HDTV = high-definition TV, DVC = digital video cassette, DVD = digital video disc.

2. DCT coefficients are correlated. Accordingly, one calculates differences
3. Encode coefficients by Huffman encoding for each frame
4. Motion compensation between frames

We explain these basic ingredients below.

8.6.2 *Discrete Cosine Transform*

An image—more precisely, a block of pixels inside the image—is a matrix $[f(x, y)]$ of numbers that represent the level of light (luminance) of the pixels. (We think of a black and white picture, for simplicity.) The DCT of this block is the two-dimensional Fourier transform $[F(m, n)]$ of the matrix. One can appreciate the meaning of this transform if one knows that we can reconstruct the block from the transform by calculating

$$f(x, y) = \sum_{m,n} F(m, n) \cos (mx) \cos (ny). \tag{8.5}$$

That representation shows that the block is viewed as a superposition of product of sine wave. Figure 8.13 shows three images with their DCT. The left image has a constant luminance. Its DCT has only one nonzero coefficient $C = F(0, 0)$. Indeed, using (8.5) with $F(0, 0) = C$ and the other values of F equal to zero, we find $f(x, y) = C$, so that the luminance is constant across the picture.

The center image has a DCT with two nonzero coefficients: $F(0, 0) = F(1, 0) = C$. Using (8.5) we find $f(x, y) = C + C \cos (x)$. The luminance of the image varies according to a sine wave along the horizontal axis of the image.

The image on the right can be understood similarly. By using more coefficients, one ends up decomposing the image into a sum of gridlike patterns. If the image luminance varies slowly, then only the low-order coefficient F (for small values of m and n) are nonzero.

One key idea is that the eye may not be very sensitive to very fine details, so that we may set to zero the coefficients $F(m, n)$ for large values of m and n that correspond to such fine details.

FIGURE 8.13

Three images and their DCT.

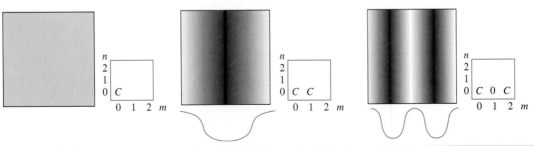

Another important observation is that video images—across a movie—tend to have similar characteristics when transformed by DCT. That is, the coefficients F tend to have comparable magnitudes across images. Consequently, Huffman encoding of these coefficients does not require learning the statistics for video applications. The situation is different for still images where learning the statistics (the fractions of coefficients of different blocks having specific values) improves the performance of the Huffman code considerably.

A third observation is that many DCT coefficients tend to be zero. A good encoding consists in counting the zeros instead of listing them. This method, called run-length encoding, is combined with Huffman encoding. Specifically, the Huffman encoding is applied to the length of zeros followed by the value of the first nonzero DCT coefficient.

8.6.3   Motion Compensation

Substantial gains are achieved in video compression by using the similarity of successive frames. A naive way of doing this would be to compute the difference between two successive frames.

Motion compensation is more subtle. It examines how each block in the image has moved from one frame to the next. Instead of transmitting the next block, one transmits the motion vector that best approximates the motion of the block plus the difference between the translated block and the actual block in the next frame. This idea is illustrated in Figure 8.14.

8.6.4   MPEG

We conclude this section with a very brief look at MPEG, a family of video compression algorithms. MPEG produces a *group of pictures:*

$$I\ B\ B\ B\ P\ B\ B\ B\ P\ B\ B\ B\ P\ B\ B\ B\ P.$$

In this sequence, I designates an "intraframe" that has been compressed by itself, without any reference to other frames. The I frame is encoded using DCT with Huffman encoding

FIGURE 8.14

Motion compensation. A block is moved to find its best match in the next frame. The encoder sends the motion vector and the matching error.

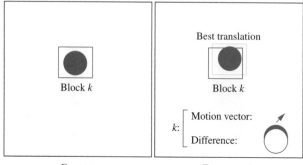

Frame n \qquad Frame $n+1$

of the run lengths of zeros followed by the value of the first nonzero DCT coefficients. These coefficients are listed in the zig-zag order $F(1,0)$, $F(0,1)$, $F(2,0)$, $F(1,1)$, $F(0,2)$, ... to determine the run length of zeros and the first nonzero coefficient. Tables of these values and their Huffman code are standardized. For instance, $(0,0,0,0,1)$ is encoded as $(00110,0)$ where the last zero indicates that the nonzero coefficient is positive. As other examples, $(0,0,0,0,-1)$ is encoded as $(00110,1)$ and $(0,0,0,0,2)$ as $(0000001111,0)$.

The P designates a frame that has been predictive-coded, using motion compensation. The B designates a frame that has been bidirectionally predictive-coded. That is, the prediction is based on past and future I and P frames.

MPEG encodes the motion vectors differentially, because such vectors tend to be similar across a picture (think of a camera pan). Estimating the motion is the time-consuming part of the MPEG algorithm.

Going back to Figure 8.14, we see that estimating the motion consists in finding the translation of a given block in one frame that best approximates the corresponding block in the next frame. Ideally, one would like to explore all the possible translations left and right, up and down, of up to 6 picture elements. However, this amounts to calculating 169 times the difference, say sum of squares, of the blocks.

A number of clever approximate search algorithms have been developed. We give just one example to indicate what can be done instead of an exhaustive search. One algorithm first looks at the nine displacements $(-3,0,+3) \times (-3,0,+3)$. The block is then moved to the best of these nine displacements. In the second step, the algorithm explores the nine displacements $(-2,0,+2) \times (-2,0,+2)$ and moves to the best displacements. Finally, the algorithm explores the nine displacements $(-1,0,+1) \times (-1,0,+1)$. These three steps require calculating only 27 differences, a savings of a factor 6 for a comparable performance.

8.7   Complement 1: Secret Key Cryptography

In this section we discuss the most widely used secret cryptography codes. We then explain how Alice and Bob can agree on the secret functions E and D by communicating over a public channel.

8.7.1   Secret Codes

A few secret codes have withstood the attacks of many clever researchers and are believed to be good codes. We explain these codes. Specifically, we discuss the one-time pad, DES, and IDEA.

DES

One of the standards adopted for encryption uses a combination of substitution and transposition. In 1977, the National Bureau of Standards adopted the *Data Encryption Standard* (DES) as the official encryption standard for the unclassified information of the U.S. government. DES was developed by IBM in cooperation with the National Security Agency. There are inexpensive VLSI (very large scale integration) chips that perform the necessary encryption and decryption.

FIGURE 8.15

Encoding modes for large messages using DES or IDEA. From left to right: cipher feedback, cipher block chaining, and output feedback modes.

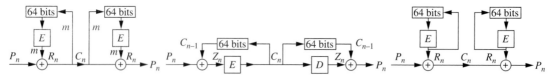

To perform the encryption of a text, DES first divides the text into blocks of 64 bits. Each block of 64 bits is then encoded separately into new blocks, also of 64 bits. Thus, DES is a mapping $C = E_K(P)$ where both C and P are words of 64 bits. The key K is a word of 56 bits.

The first step in the DES encoding of P is to calculate $P_1 = T(P)$, where T is a given bit permutation. The DES algorithm then calculates successively $P_{i+1} = F(P_i, K_i)$ for $i = 1, 2, \ldots, 16$, where the K_i are different subsets of 48 bits of K. The function F is also specified by the algorithm. Finally, the DES algorithm calculates $C = T^{-1}(P_{16})$ where T^{-1} is the inverse of the bit permutation T. The function D_K is calculated by the same algorithm where the order of the keys K_i is reversed.

The encryption E_K has the property that a modification of a single bit in P has about a 50 percent chance of modifying each bit in C and these bit modifications appear independent. It is thus very difficult to reconstruct P from C without knowing the key K.

A successor to DES is being developed and will probably be called AES, for Advanced Encryption Standard.

IDEA

In 1991, Xuejia Lai and James Massey proposed a new secret code called *International Data Encryption Algorithm*. IDEA is similar to DES but is based on a 128-bit key.

Large Messages

Three methods are used for encoding long messages with either DES or IDEA. The methods are shown in Figure 8.15.

The cipher feedback mode has the advantage of being self-synchronizing: no initial value has to be set in the registers. Moreover, an error in a block of m bits has limited error propagation. In most applications, $m = 1$, which avoids the problem of requiring correct alignment of the boundary of these blocks.

The output feedback mode can be thought of as a one-time pad where the pseudo-random sequence is generated by the output feedback.

8.8 Complement 2: Public Key Cryptography

We explain two public key cryptography codes.

8.8.1 RSA

The Rivest, Shamir, Adleman (RSA) public key system is based on the difficulty of finding the prime factors p and q of pq when they are large. The system also uses the following lemma:

Lemma

Let p and q be two prime numbers. Define $n = pq$, $z = (p-1)(q-1)$. Choose e such that d and z have no common divisor other than 1. Then choose d so that $ed = 1 \bmod z$ (i.e., ed is a multiple of z plus 1).

If $P \in \{0, 1, \ldots, n-1\}$ and $C := P^e \bmod n$, then $P = C^d \bmod n$.

For a proof of this lemma, see Section 8.9.

For instance, let $p = 3$, $q = 11$, $n = 33$, $z = 20$, $d = 7$, and $e = 3$. The lemma implies that if $C = P^3 \bmod 33$, then $P = C^7 \bmod 33$ for all $P \in \{0, 1, \ldots, 32\}$. As an example, if $P = 5$, then one verifies that $C = 26$ and that $26^7 \bmod 33 = 5$.

The lemma enables us to construct a public key system by making the pair of numbers (e, n) public. To know the decoding function, an eavesdropper needs the pair (d, n). However, finding d from (e, n) is believed to be a hard problem.

8.9 Complement 3: Proof of RSA Lemma

Here is a proof of the lemma.

Proof of Lemma

Since $C := P^e \bmod n$, as we showed in the verification of (8.1), we know that

$$C^d = P^{ed} \bmod n. \tag{8.6}$$

We show below that

$$P^z = 1 \bmod n. \tag{8.7}$$

This equality implies that (modulo n), $P^{ed} = P^{kz+1} = P$.

It remains to show (8.7). We prove it assuming that P and n have no common factor other than 1. In other words, we assume that P is not a multiple of p nor of q. (The result holds without that assumption.)

To do this, define for $m \geq 1$ the set $F(m)$ of all the integers in $\{1, 2, 3, \ldots, m-1\}$ that are coprime with m. Two numbers a and b are coprime (or relatively prime) if their greatest common divisor $\gcd(a, b)$ is equal to 1.

Using Euclid's algorithm one can check that a and b are coprime if and only if $ax + by = 1$ for some integers x and y. To see this, assume $a < b$ and set $x(1) = 1$, $y(1) = 0$, $x(0) = 0$, $y(0) = 1$ so that $a = q(1)$ and $b = q(0)$ with

$$q(n) = ax(n) + by(n).$$

Next, note that if

$$q(n) = ax(n) + by(n) < q(n-1) := ax(n-1) + by(n-1) = kq(n) + r(n),$$

where k is the quotient of $q(n-1)$ by $q(n)$ and $r(n)$ is the remainder, then $\gcd(q(n-1), q(n)) = \gcd(q(n), r(n))$. Moreover,

$$r(n) = ax(n+1) + by(n+1)$$

with

$$x(n+1) = x(n-1) - kx(n), \; y(n+1) = y(n-1) - ky(n).$$

Continuing in this way, we reach some iteration n with $r(n) = 0$ and it must then be that $q(n) = 1$. Indeed, otherwise $q(n)$ would be the gcd of $q(n-1)$ and $q(n)$ and therefore of $b = q(0)$ and $a = q(1)$. But $q(n) = 1$ shows that $ax + by = 1$ with $x = x(n)$ and $y = y(n)$.

We use that characterization of coprime numbers to show that $F(n)$ is closed under multiplication modulo n. That is, if $a, b \in F(n)$, then $a.b \bmod n \in F(n)$. Indeed $ax + ny = 1 = bx' + ny'$, so that

$$1 = (ax + ny)(bx' + ny') = ab(xx') + n(ybx' + axy' + nyy'),$$

which shows that ab and n are coprime.

Next we show that if $a \in F(n)$, then there is some $b \in F(n)$ so that $ab = 1 \bmod n$. Indeed, a and n are coprime so that $ax + ny = 1$ for some (x, y) so that we can choose $b = x$.

Now, assume that $a, b, c \in F(n)$ with $a \neq b$. We claim that $ac \neq bc \bmod$ n. To see this, let's multiply both sides of the equality by h such that $hc = 1 \bmod n$. We find $ach = bch \bmod n$, so that $a = b \bmod n$.

From the above, it follows that $\{ab, b \in F(n)\} = F(n)$.

Let us denote by ρ the product of all the elements of $F(n)$ and by σ the number of those elements. Then, multiplying all the elements of $\{ab, b \in F(n)\}$ together we should get $a^\sigma \rho \bmod n$, but we should also get ρ since that set is $F(n)$. Hence,

$$a^\sigma \rho = \rho \bmod n.$$

Let δ be such that $\delta\rho = 1 \bmod n$. Multiplying both sides of the above equality by δ we find that

$$a^\sigma = 1.$$

It remains to show that $\sigma = z$ to demonstrate that $a^z = 1$ for any element a of $F(n)$. Remember that $z = (p-1)(q-1)$. It is immediate that $F(p) = \{1, 2, \ldots, p-1\}$ so that the cardinality (number of elements) $|F(p)|$ of $F(p)$ is $p-1$. Similarly, $|F(q)| = q-1$. Recall that $n = pq$. To show $|F(n)| = (p-1)(q-1)$, it suffices to show that each $a \in F(n)$ corresponds to a unique pair (x, y) with $x \in F(p)$ and $y \in F(q)$ such that

$$a = x \bmod p = y \bmod q. \tag{8.8}$$

Assume that (8.8) holds with $x \in F(p)$ and $y \in F(q)$. Then

$$xu + pv = 1, a = x + kp \rightarrow (a - kp)u + pv = 1 \rightarrow (a, p) \text{ coprime.}$$

Similarly, (a, q) coprime. Hence, (a, pq) coprime so that $a \in F(n)$.

Conversely, assume $a \in F(n)$ and that (8.8) holds. Then $au + pqv = 1$, so that by (8.8),

$$a = kp + x \rightarrow (kp + x)u + pqv = 1$$

which implies that (x, p) are coprime. Similarly, (y, q) are coprime. This concludes our proof.

8.10 Complement 4: Source Coding Theory

In this section we outline the source coding theory of Claude Shannon. Shannon showed that it is possible to quantify information. His starting point is that information describes an unpredictable outcome. For instance, the outcome of a fair coin has two equally likely values "head" and "tail." By convention, we agree that it takes one bit of information to describe this outcome.

If the outcome of a random experiment is equally likely to take any value in a set with 2^n elements, then we can identify this outcome with a string of n fair coin flips and we find that this outcome requires n bits to be described.

More generally, if a string of n symbols is equally likely to take any value in a set with 2^{nH} elements, it takes nH bits to describe the n symbols, or H bits per symbol. For example, consider the string (X, Y, Z) produced by a source and assume that the string is equally likely to be any one of the following four strings: $000, 100, 010$, and 001. In this case, we find that it takes 2 bits to describe the three symbols, or $\frac{2}{3}$ bit per symbol.

The key observation of Shannon is that for most sources, the sequence of symbols it produces must be "typical" and there are few typical sequences!

We first explain this observation in the case of i.i.d. (independent and identically distributed) symbols. Let then $\{X_n, n \geq 1\}$ be i.i.d. with

$$P(X_n = k) = p(k), k = 1, 2, \ldots, K$$

where the numbers $p(k)$ are positive and add up to one. Fix $n \gg 1$ and consider $\mathbf{X} = (X_1, X_2, \ldots, X_n)$. Define

$$g(\mathbf{x}) := g(x_1, \ldots, x_n) = \log p(x_1) + \cdots + \log p(x_n), \text{ for } x_i \in \{1, \ldots, K\}, 1 \leq i \leq n.$$

In the above definition, log designates the logarithm in base 2.

Note that, by the law of large numbers, $g(\mathbf{X})/n = [\log p(X_1) + \cdots + \log p(X_n)]/n \approx -H$ where

$$H := -E \log p(\mathbf{X}) = -p(1) \log p(1) - \cdots - p(K) \log P(K). \tag{8.9}$$

Consequently, the random sequence $\mathbf{X}$ is in a set $\mathbf{S}$ of sequences defined as follows:

$$\mathbf{S} := \{\mathbf{x} = (x_1, \ldots, x_n) \text{ such that } g(\mathbf{x}) = -nH\}.$$

But, for any $\mathbf{x}$ in $\mathbf{S}$,

$$P(\mathbf{X} = \mathbf{x}) = p(x_1) \cdots p(x_n) = 2^{g(\mathbf{x})} \approx 2^{-nH}.$$

Consequently, the set **S** has |**S**| elements where

$$|\mathbf{S}| = 2^{nH}.$$

Moreover, **X** is equally likely to take any value in **S.** Accordingly, we conclude that it takes nH bits to describe this random sequence. That is, it takes H bits per symbol to describe the sequence, as we wanted to show.

As a simple example, consider a sequence of flips of an unfair coin such that $P(\text{head}) = 1 - P(\text{tail}) = p$. Using (8.9) we find that for this sequence,

$$H = H(p) := -p \log p - (1 - p) \log (1 - p).$$

This entropy, called the *binary entropy,* is shown in Figure 8.16. As the figure shows, when $p = 0.5$, it takes 1 bit per symbol to describe the outcome of a sequence of coin flips. However, if $p \ll 1$, then it takes much fewer bits to describe the sequence of coin flips because this sequence typically contains very few 1s and there are few such sequences.

This theory extends to non-i.i.d. symbols. In that case, the entropy of the source is H bits per symbol where H is the average entropy of X_{n+1} given the past symbols. For instance, if $\{X_n, n \geq 1\}$ is an irreducible Markov chain on $\{1, 2, \ldots, K\}$ with transition probabilities $P(i, j) = P[X_{n+1} = j | X_n = i]$ and invariant distribution $\pi = \{\pi(i), i = 1, 2, \ldots, K\}$, then

$$H = -\sum_i \pi(i) \sum_j P(i, j) \log P(i, j). \tag{8.10}$$

As an illustration of this formula, consider the Markov chain with the transition diagram shown in Figure 8.17. This source produces random strings with symbols in $\{1, 2, 3\}$. After it outputs the symbol 1, the source outputs the symbol 2 with probability $P(1, 2) = 0.4$ and the symbol 3 with probability $P(1, 3) = 0.6$. Similarly, after it outputs the symbol 2, the source outputs the symbols 1 with probability $P(2, 1) = 0.5$ and the symbol 3 with probability $P(2, 3) = 0.5$. After symbol 3, the source produces the symbol 1 with

FIGURE 8.16

Binary entropy.

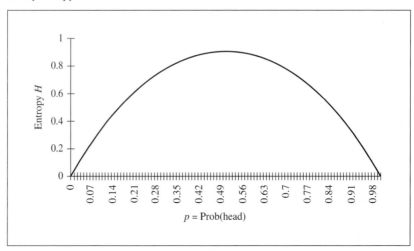

FIGURE 8.17

A Markov source.

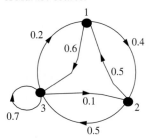

probability $P(3, 1) = 0.2$, the symbol 2 with probability $P(3, 2) = 0.1$, and the symbol $P(3, 3) = 3$ with probability 0.7. The transition probabilities $P(i, j)$ define the matrix **P** given below:

$$\mathbf{P} = \begin{bmatrix} 0 & 0.4 & 0.6 \\ 0.5 & 0 & 0.5 \\ 0.2 & 0.1 & 0.7 \end{bmatrix}.$$

The invariant distribution π is the vector $(\pi(1), \pi(2), \pi(3))$ of nonnegative numbers that are such that

$$\pi\mathbf{P} = \pi \text{ and } \pi(1) + \pi(2) + \pi(3) = 1.$$

Solving these equations one finds that $\pi \approx (0.204, 0.146, 0.650)$. Using formula (8.10) we then compute the value of H and we find that

$$H = 0.204(0.4 \log 0.4 + 0.6 \log 0.6) - 0.146(0.5 \log 0.5 + 0.5 \log 0.5)$$
$$-0.650(0.2 \log 0.2 + 0.1 \log 0.1 + 0.7 \log 0.7)$$
$$\approx 1.1.$$

That is, it takes approximately 1.1 bit per symbol to describe the output of this random source. For example, we would need about 1100 bits to describe the first 1000 symbols produced by the source.

Problems

Problems with a * are somewhat more challenging. Those with a ^{c} are based on material in a complement. Problems with a ^{1} are borrowed from the first edition.

^{1}1. Consider the data encryption scheme. What property should the function F possess in order for the scheme to be decipherable by the receiver? Formulate the decoding algorithm that the receiver should use, assuming that the receiver knows the key K.

^{1}2. Assume that if the key K is known, then it takes 100 ms to decipher a message that has been coded using DES. Suppose that an eavesdropper chooses keys at random and tries to decipher the message. What is the average time required for the correct deciphering? (There are 2^{56} possible keys.)

[1]3. The public key cryptography system is based on the idea that large prime numbers p and q (typically over 10^{100}) will be used. (This is why it is important to know many large prime numbers—it is here that number theorists are needed!) If small prime numbers are used, then deciphering is relatively easy. Here is an example: Suppose a user transmits a message which is an integer between 1 and 10, using $n = 15$ and $e = 3$. The coded message yields the integer 8. What number was transmitted?

[1]4. Show that the public key cryptography system can also be used as an electronic signature. (*Hint:* Check that the encoding function is invertible.)

[1]5. Assume that, in a text, the number of consecutive 0s, X, is equally likely to be any integer between 1 and n. Derive an expression for the compression factor r_n achieved by run length encoding. Compute r_n for $n = 5$ and $n = 10$. Using Stirling's approximation for $n!$ ($n! = \sqrt{2\pi n}e^{-n}n^n$, for large values of n), find an asymptotic expression for r_n. The assumption that X is uniformly distributed is not very reasonable; for instance, in a text one is very likely to find large sequences of 0s due to margins, empty lines, etc. Will that unequal distribution improve r_n?

[1]6. Is it better to use differential interline encoding to transmit a file containing random numbers representing the outcomes of a coin-tossing computer-generated experiment or a file containing the binary encoding of a picture of the sunset over Oahu? Why?

7. Construct the Huffman code for transmitting messages constructed by using the symbols A, B, C, D, and E occurring with frequencies 0.60, 0.30, 0.05, 0.03, and 0.02, respectively. What is the compression factor, defined as the number of bits per symbol in a straightforward binary encoding divided by the average number of bits per symbol of the Huffman encoding?

[c]8. Assume that the entropy rate of the English language is about 0.1 bits per symbol. (This value is hypothetical.) How many bits do we need for a message digest $H(P)$ to be effective for plaintexts of up to 2000 characters?

9. Consider a prefix-free encoding used to encode a string of symbols. Explain why a single bit error in the encoded string can corrupt the decoding of the subsequent symbols.

[c]10. Calculate the entropy rate of the Markov source which produces the symbols 0 and 1 according to the transition probabilities $P(0, 0) = 0.7 = 1 - P(0, 1)$ and $P(1, 0) = 0.4 = 1 - P(1, 1)$.

[c]11. Consider the following simplified model of a digitized audio signal. The successive samples X_n take values in $\{\ldots, -2, -1, 0, 1, 2, \ldots\}$. If $X_n > 0$, then $X_{n+1} = X_n + 1$ with probability 0.4 and $X_{n+1} = X_n - 1$ with probability 0.6. Symmetrically, if $X_n < 0$, then $X_{n+1} = X_n - 1$ with probability 0.4 and $X_{n+1} = X_n + 1$ with probability 0.6. Finally, if $X_n = 0$, then $X_{n+1} = 1$ with probability 0.5 and $X_{n+1} = -1$ with probability 0.5. Estimate the entropy rate of this signal.

References

The material on security was developed in collaboration with Dr. B. Preneel.

An excellent reference on cryptography is Menezes (1997). Kaufman (1995) provides a good introduction to security issues and solutions. The book also explains the cryptography algorithms. In Ahuja (1996) you will find an accessible description of security systems. A discussion of security problems raised by Java can be found in McGraw (1997). For a discussion of message authentication codes and their properties, see Preneel (1995).

Two particularly enjoyable presentations of source coding, as well as other topics, can be found in Feynman (1996) and Rényi (1984). Once again, McGraw (1997) is *the* reference on information theory.

Vetterli (1995) is an excellent presentation of source coding, including multiscale representations using wavelet transforms.

The Lempel-Ziv algorithm is explained in Ziv (1997). The MPEG compression standard is described in LeGall (1991).

Bibliography

Ahuja, V., *Network and Internet Security,* Academic Press, 1996.

Feynman, R. P., *Lectures on Computation,* edited by J. G. Hey and R. W. Allen, Addison Wesley, 1996.

Kaufman, C., Perlman, R., and Speciner, M., *Network Security—Private Communication in a Public World,* Prentice-Hall, 1995.

LeGall, D., "MPEG: A video compression standard for multimedia applications," *Communications of the ACM,* 34, no. 4, pp. 46–58, April 1991.

McGraw G., and Felten, E. W., *Java Security,* John Wiley & Sons, 1997.

Menezes, A. J., van Oorschot, P. C., and Vanstone, S., editors, *Handbook of Applied Cryptography,* CRC Press, 1997.

Preneel, B., and van Oorschot, P. C., "MDx-MAC and building fast MACs from hash functions," *Proc. Crypto '95, LNCS* 963, D. Coppersmith, ed., Springer-Verlag, 1–14, 1995.

Rényi, A., *A Diary on Information Theory,* John Wiley & Sons, 1984.

Shannon, C., "A mathematical theory of communication," *Bell System Technical Journal,* 27, pp. 379–423 and 623–656, 1948.

Vetterli, M., and Kovačević, J., *Wavelets and Subband Coding,* Prentice-Hall, 1995.

Ziv, J., and Lempel, A., "A universal algorithm for sequential data compression," *IEEE Trans. Information Theory,* 23, pp. 337–343, 1977.

Performance Evaluation and Monitoring

Implementing a network is a major task that requires the design of specific hardware and software. When designing a new network, it is desirable to be able to predict its level of performance before implementing it. This chapter briefly reviews the methods used by network engineers to predict performance.

These methods can be classified into three groups: *mathematical models, simulations,* and *emulations.* They are discussed in the following sections. In addition to performance prediction methods, we will discuss *performance monitoring* methods, which are an important aspect of *network management.*

Except for the first section, this chapter requires a level of mathematical sophistication higher than that required by the rest of the text. Only the results of the queueing analysis are given here. Derivations can be found in Appendix B.

9.1 Monitoring, SNMP, CMOT, and RMON

How can a network administrator evaluate the level of performance of a network? How can such an evaluation guide modifications to the network?

To evaluate the performance of a network, the administrator must monitor its behavior. Some aspects of the behavior can be monitored by adding software to specific nodes of the network. Other aspects require specialized monitoring hardware. For instance, a computer on an Ethernet network can be programmed to collect statistics about the traffic of packets it sees on the network. Such statistics can be used to calculate the load on the network, i.e., the fraction of the network throughput used by the traffic. The network administrator can use this load information to determine when the network is approaching saturation.

More detailed statistics can also be measured by software. For instance, it is possible to *filter* the packets by examining their headers. Using such filtering, the network administrator can determine the amount of traffic due to electronic mail or to FTP, for instance. Traffic can also be classified by pair (source, destination). Many vendors sell *network analyzers* and network management software that perform this type of monitoring and that provide the network administrator with a user-friendly interface and tools for computing various traffic statistics.

By adding specialized monitoring hardware to the network interfaces, the network administrator is able to collect some information that cannot be monitored by software alone. For instance, some hardware failures, such as cable rupture or loose connections can be detected and identified with specialized monitoring equipment. Monitoring the number of collisions faced by a packet in an Ethernet interface requires a specific monitoring of the interface algorithm that is not possible with all interface boards. Measuring the delay of packets between nodes makes it necessary to add time stamps to the packets and for the nodes to synchronize their clocks by exchanging time-stamped control packets.

Hardware failure indications (*alarms*) are clearly valuable. Statistical monitoring of information is not as straightforward to use. The network administrator can use traffic information to identify when the network has to be modified. One modification that can relieve an Ethernet network that is approaching its capacity is to partition it into subnetworks connected by bridges. The subnetworks should connect nodes that communicate frequently in a way that limits the traffic between different subnetworks. The network administrator can select the subnetworks by analyzing the traffic statistics for the different source/destination pairs.

The collection of network information is facilitated when the various devices use standardized formats and protocols. Two classes of network management standards were developed: CMOT (*common management information services and protocol over TCP/IP*) and SNMP (*simple network management protocol*). Many vendors provide SNMP products, whereas fewer implementations of CMOT are in operation. These two standardization efforts result from recommendations of the *Internet Activities Board.*

Both SNMP and CMOT maintain images of the network in a database called *management information base* (MIB). The consistency between the image and the network is maintained by exchanging messages. When a management application modifies attributes of an object in the database, messages are sent to the physical device represented by the object to modify its corresponding attributes. Conversely, the database can learn the attributes of the physical devices by polling the devices periodically or by having the devices send relevant information about selected events.

In SNMP, a management application sends the commands *get, set,* and *get-next* to objects in the MIBs of managed devices. These commands are formatted in ASN.1 and are sent using UDP/IP. Some implementations of SNMP send these commands directly over LLC. This connectionless exchange of commands and responses is used to set and read attributes of managed objects. The managed objects can also signal events to the management application.

CMOT is a protocol suite for the implementation of the ISO *common management information protocol* (CMIP). CMOT provides connection-oriented services between a management application and managed objects. The services are standardized as *common management information services* (CMIS). The application layer contains a *common management information service element* (CMISE). The messages are exchanged over TCP/IP and UDP/IP. CMIP provides more complex naming and applications than SNMP. For instance, CMIP permits management applications to retrieve large numbers of data by linked replies that are not possible under SNMP. Also, CMIP can address multiple objects by type, value, and relative location without having to name them individually, as SNMP requires. Moreover, *system management functions* (SMFs) are being standardized for CMIS. The SMFs provide sophisticated management services that are not available under SNMP.

The remote monitoring of a network results in the collection of statistics about the traffic and network operations. The statistics of RMON that should be collected are standardized in RFCs 1757 and 2021.

9.1.1   Monitoring Summary

- A network manager can monitor the performance of a network by using specialized software and possibly some dedicated hardware.
- Alarms identify malfunctions, and traffic statistics can be used to reconfigure the network to improve its performance.
- SNMP is a connectionless management system.
- CMOT is a connection-oriented protocol suite for implementing CMIP.
- RMON is a standardized classification of the statistics that should be collected about the operations of a network. RMON is specified in RFCs.

9.2   Models and Analysis

Some simplified models of computer systems can be analyzed mathematically. The insight provided by the analytical results can be very valuable, even if the methods do not permit the evaluation of detailed models of real systems. The most commonly used models are *queueing systems*. The most widely used results are reviewed in this section.

9.2.1   A FIFO Queue

Consider a buffer equipped with a transmitter. The transmitter operates at rate R bps. If the packets arrive at the buffer far enough apart, then the buffer is empty every time a new packet arrives. In this situation, the delay that a packet faces in the buffer is equal to its transmission time. The transmission time of a packet of L bits is equal to L/R seconds.

For instance, if $R = 1$ Mbps and every packet has exactly 1000 bits, then the transmission time of a packet is equal to 1 ms. Assume that the arrival times of packets are always separated by at least 1 ms, then the delay of each packet in the buffer is equal to 1 ms. That is, if a packet enters the buffer at time T, its last bit leaves the buffer at time $T + 1$ ms. In this discussion we assume implicitly that a packet arrives instantaneously at the buffer, which is a standard simplification.

Consider once again our example but assume that the time between successive arrivals of packets may be less than 1 ms. For instance, assume that the first packet arrives at time 0 and that the second packet arrives at time 0.4 ms. Suppose the transmitter sends the packets in the order that they arrive. In this case, the last bit of the first packet leaves the buffer at time 1 ms. At that time, the second packet is in the buffer and starts being transmitted. The last bit of the second packet leaves the buffer at time 2 ms. Since the second packet arrived at time 0.4 ms, it spent 1.6 ms in the buffer. During the first 0.6 ms, from its arrival time 0.4 ms to the start of its transmission, the second packet was waiting its turn to be transmitted. This delay is called the *queueing delay*. The total delay of a packet in a buffer is equal to its queueing delay plus its transmission time.

Although the transmission time of a packet depends only on its length and on the transmission rate, its queueing delay depends on the arrival times of the packets at the buffer. That is, the queueing delay depends on the traffic or load that the buffer faces. The queueing delays increase with the load because the buffer gets congested: it fills up when the transmitter cannot keep up with the arriving packets.

In most network links, the arrival times of packets are rather unpredictable. However, by measuring the arrival times on many links one can construct mathematical models that are representative of typical situations. These mathematical models capture the irregular fluctuations of interarrival times. Similar models are used for the packet lengths. Given the models for arrival times and for packet lengths, techniques from queueing theory can predict the statistics of the queueing delays and of the total delays in the buffers.

9.2.2 *M/M/1 Queue*

Consider a buffer equipped with a transmitter. Assume that packets arrive at the buffer at times $0 < T_1 < T_2 < T_3 < \cdots$ and that the transmission times of the successive packets, which are transmitted on a first come–first served basis, are $S_1, S_2, S_3, \ldots$.

This system is called an *M/M/1* queue if:

- The random *interarrival times* $T_1, T_2 - T_1, T_3 - T_2, \ldots$ are independent and identically distributed with

$$P\{T_{n+1} - T_n \geq t\} = \exp\{-\lambda t\}, t \geq 0. \tag{9.1}$$

  That is, the interarrival times are *exponentially distributed with rate* $\lambda > 0$. In that case, the arrival times $\{T_n\}$ are said to form a *Poisson process with rate* λ.

- The transmission times $\{S_n\}$ are independent and exponentially distributed with rate μ. Equivalently, the packet lengths are independent and exponentially distributed with mean μ^{-1} (in bits) where R is the transmission rate (in bits per second). That is,

$$P\{S_n \geq t\} = \exp\{-\mu t\}, t \geq 0. \tag{9.2}$$

In the notation M/M/1, the first M indicates that the arrival process is memoryless and the second M that the service times are memoryless. It so happens that the only memoryless distribution is exponential. The 1 in M/M/1 indicates that the queue has a single server.

Under the M/M/1 assumptions, if one denotes by x_t the number of packets that are either in the buffer or being transmitted at time $t > 0$, we show in Appendix B that, *in statistical equilibrium,*

$$P\{x_t = n\} = \rho^n(1 - \rho), n \geq 0, \text{ if } \rho := \frac{\lambda}{\mu} < 1. \tag{9.3}$$

By statistical equilibrium we mean that the system has reached steady state in the sense that the probability that the queue length takes given values does not change with time. This does not mean, of course, that the queue length stops evolving. This notion of equilibrium is similar to what happens when one injects gas into an empty bottle. After a while, the distribution of the gas molecules stabilizes even though the molecules keep on moving. The likelihood of finding a given number of molecules in some section of the bottle approaches some constant value.

If $\lambda \geq \mu$, then the average number of packets in the buffer is infinite. That is, the queue builds up over time, without bound.

If $\lambda < \mu$, we conclude from (9.3) that the average *queue length* is given by

$$L := E\{x_t\} = \sum_{n=0}^{\infty} n P\{x_t = n\} = \frac{\rho}{1-\rho} = \frac{\lambda}{\mu - \lambda}. \tag{9.4}$$

As an illustration of (9.4), consider the following situation. Packets arrive at a buffer with transmission rate $R = 1$ Mbps. The packets have random lengths L_n that are independent and exponentially distributed with mean length equal to $\alpha^{-1} := 1000$ bits. The arrival times of the packets form a Poisson process with rate $\lambda = 800$ packets/s. We want to calculate the average number of packets in the buffer, including the packet possibly being transmitted. To apply (9.4), the key observation is that the packet transmission times are $S_n = L_n/R$ and are therefore independent and exponentially distributed with mean $\mu^{-1} = 1$ ms/packet. Indeed,

$$P\{S_n \geq t\} = P\left\{\frac{L_n}{R} \geq t\right\} = P\{L_n \geq Rt\} = \exp\{-\alpha Rt\} = \exp\{-\mu t\}$$

with $\mu = \alpha R = 10^{-3} \times 10^6 = 10^3$. The third equality in the derivation above uses the assumption that L_n is exponentially distributed with rate α. Using (9.4) we then conclude that the average number of packets in the buffer is equal to $L = \lambda/(\mu - \lambda) = 800/(1000 - 800) = 4$ packets.

Next we use formula (9.4) to derive the *average delay T* per packet through the system. The average delay T is related to the average queue length by *Little's result*

$$L = \lambda T. \tag{9.5}$$

Using this result, we conclude that

$$T = \frac{1}{\mu - \lambda} \tag{9.6}$$

for the M/M/1 queue.

Little's result (9.5) applies to a large class of queueing systems. A simple interpretation of that result can be given for first come–first served queueing systems. If a typical customer spends T time units, on the average, in the system, then the number of customers left behind by that typical customer is equal to λT, on the average. Indeed, the customers left behind are those who arrived during the T time units spent by the typical customer in the system. Thus, a typical customer leaves λT customers behind upon leaving the network. This must be the average number L of customers in the system. Thus, $L = \lambda T$.

We apply formula (9.5) to our previous numerical example, with $\lambda = 800$ packets/s and $\mu^{-1} = 1$ ms/packet. We find in that case that

$$T = \frac{1}{\mu - \lambda} = \frac{1}{1000 - 800} = 5 \text{ ms/packet.}$$

This value is consistent with our previous conclusion that there are 4 packets on average in the buffer when a new packet arrives and each packet takes 1 ms to be transmitted.

Little's result holds for systems that are not necessarily first come–first served, as the following argument explains. Say that each customer pays one unit of cost per unit of time spent in the system. Then the average amount paid per unit of time by all the customers is equal to the average number L of customers in the system at any given time. On the other hand, each customer pays a total of T, on the average, since T is the average time that a customer spends in the system. Therefore, since customers arrive at rate λ, customers pay $\lambda \times T$ per unit of time. Hence (9.5).

9.2.3   *Application to Statistical Multiplexing*

The above results for the M/M/1 queue can be used to appreciate the gain achieved by statistical multiplexing, when compared to time-division multiplexing. Consider one transmission line with transmission rate R. A number N of traffic streams of packets, each with rate λ, have to be transmitted along that line. Assume that the arrival times of the packets form independent Poisson processes (with rate λ) and that the packet lengths are independent and exponentially distributed with mean L.

Time-division multiplexing consists of dividing the capacity of the transmitter into N channels with rate R/N. Each traffic stream can then be modeled as an M/M/1 queue with arrival rate λ and transmission rate $\mu := R(LN)^{-1}$ (in packets per second). The average delay per packet is given by formula (9.6):

$$T = \frac{1}{\mu - \lambda}. \tag{9.7}$$

Statistical multiplexing consists of buffering the packets coming from the N streams into a single buffer and transmitting them one at a time. The arrival process into the single buffer now has rate $N\lambda$. (The process can be shown to be Poisson.) This system can now be modeled as an M/M/1 queue with arrival rate $N\lambda$ and with transmission rate $RL^{-1} = N\mu$ packets per second. The average delay is now given by

$$\frac{1}{N\mu - N\lambda} = \frac{T}{N}. \tag{9.8}$$

Thus, the average delay per packet for the statistical multiplexing system is a fraction $1/N$ of the delay for the time-division multiplexing.

As an illustration of this result, imagine that you are designing a wireless data network. The transmitters can implement a wireless channel with rate 128 kbps. There are 10 users that transmit packets that have independent and exponentially distributed lengths with average value 1000 bits. Each user gets packets to transmit according to a Poisson process with average rate 10 packets per second. In one design, the 128-kbps channel is divided equally among the 10 users, so that each user gets a dedicated 12.8-kbps channel.

In another design, the users all share the 128-kbps channel. Using formula (9.7), one finds that the average delay T per packet in the first design is $T = 1/(\mu - \lambda)$ with $\mu =$ (12.8 kbps)/(1000 bits/packet) $= 12.8$ packets/s and $\lambda = 10$ packets/s. That is,

$$T = \frac{1}{12.8 - 10} \approx 360 \text{ ms.}$$

Using (9.8) one sees that the average delay per packet in the second design is equal to $T/10 \approx 36$ ms.

9.2.4   Networks of M/M/1 Queues

Most networks have more than one buffer. In this section we explain simple formulas that are used to estimate the average delay per packet in a network with multiple buffers. We start by examining the network shown in Figure 9.1. Three streams of packets go through this network. We assume that the arrival streams are Poisson processes with the rates indicated in the figure. We also assume that the service times at the three buffers are independent and exponentially distributed with the rates indicated.

Remarkably, we can analyze this network "one queue at a time." That is, one can show that the average number of packets in each buffer is given by formula (9.5). Specifically, for $i = 1, 2, 3$, the average number L_i of packets in buffer i is equal to

$$L_i = \frac{\lambda_i}{\mu_i - \lambda_i}. \tag{9.9}$$

In this formula, λ_i is the average rate of packets that go through buffer i. As Figure 9.1 indicates, $\lambda_1 = \gamma_1 + \gamma_2$, $\lambda_2 = \gamma_1 + \gamma_2 + \gamma_3$, and $\lambda_3 = \gamma_1 + \gamma_3$.

The plausibility argument that we gave for Little's result for non-FIFO queues extends to networks. Precisely, one can show that the average delay T per packet in the network is related to the average number L of packets in the network by the formula $L = \gamma T$, where γ is the average total arrival rate of packets into the network. For the network of Figure 9.1, we find that

$$L = L_1 + L_2 + L_3 \quad \text{and} \quad \gamma = \gamma_1 + \gamma_2 + \gamma_3. \tag{9.10}$$

As a numerical illustration, assume that $\mu_1 = \mu_2 = \mu_3 = 1000$ packets/s and that $\gamma_1 =$

FIGURE 9.1

A network with three buffers.

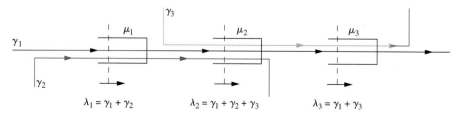

$\gamma_2 = \gamma_3 = 300$ packets/s. Using (9.9) we find that

$$L_1 = \frac{\lambda_1}{\mu_1 - \lambda_1} = \frac{600}{1,000 - 600} = 1.5 \text{ packets.}$$

Similarly, we find that $L_2 = 9$ packets and $L_3 = 1.5$ packets. Accordingly, from (9.10), $L = 12$ packets and, since $L = \gamma T$, the average delay T per packet is equal to $T = L/\gamma = 12/(900 \text{ packets/s}) \approx 13.3$ ms.

The result that we applied to the example above holds for more general networks. We show in Appendix B that the average delay T per packet in a network (in seconds) is given by the following formula:

$$T = \frac{1}{\gamma} \sum_{j=1}^{J} \frac{\lambda_j}{\mu_j - \lambda_j} \tag{9.11}$$

where γ is the average rate of flow into the network, λ_j is the average rate of flow through queue j, and μ_j is the average transmission rate of queue j. All these rates are in packets per second, and the λ_j are obtained by solving the equations that express the conservation of the flows through the nodes.

The assumptions required for the formula (9.11) to be valid are:

1. The arrival streams from outside into the network form independent Poisson processes.
2. The packet transmission times at all the queues are independent and exponentially distributed.

In practice, assumption 1 may be verified. However, assumption 2 cannot be. Indeed, the successive transmission times of a given packet into the various nodes are all proportional to the packet length, and so they cannot be independent. Nevertheless, simulation experiments show that formula (9.11) provides a reasonable estimate for the average delay per packet in store-and-forward packet switched networks.

9.2.5   *M/G/1 Queues*

So far we have assumed that the packet lengths are exponentially distributed. Measurements show that this model is not quite correct. Rather, packets tend to be split into short email and control packets and long files generated by file transfers. Because of this fact, we need to investigate the impact of the packet length distribution on the behavior of the queues.

Consider once again the M/M/1 queue but assume that the successive packet transmission times S_n are independent and have some common distribution that is not necessarily exponential. The resulting system is called the M/G/1 queue. In the notation M/G/1, the first M indicates that the arrival process is memoryless, i.e., that it has exponentially distributed interarrival times and is thus Poisson. The G indicates that the service times have a general distribution. The 1 recalls that there is a single server.

We show in Appendix B that the average delay per packet through the system is given by

$$T = \frac{1}{\mu} + \frac{\lambda E\{S_n^2\}}{2(1 - \rho)} = \frac{1}{\mu} + \frac{\lambda(\sigma^2 + \mu^{-2})}{2(1 - \rho)} \tag{9.12}$$

where μ^{-1} is the average transmission time, $\rho = \lambda/\mu$, and σ^2 is the variance of the transmission times. That is, $\sigma^2 = E\{(S_n - \mu^{-1})^2\}$. Formula (9.12) shows that the average delay increases with the variance of the packet length.

For instance, if the transmission times are exponentially distributed with mean μ^{-1}, then $\sigma^2 = \mu^{-2}$, so that (9.12) becomes (9.6). As another example, if all the packet lengths are identical, then $\sigma^2 = 0$ and the average delay of the packets is the delay through an M/M/1 queue times $(1 - \rho/2)$.

As yet another example, assume that the packet transmission times S_n are such that

$$P(S_n = a) = p = 1 - P(S_n = b)$$

where $p \in (0, 1)$ and $0 < a < b$ are two given durations. That is, the packets have only two possible lengths. In this case, $\mu^{-1} = pa + (1 - p)b$ and $E\{S_n^2\} = pa^2 + (1 - p)b^2$. Accordingly, we find that

$$T = \frac{1}{pa + (1 - p)b} + \frac{\lambda(pa^2 + (1 - p)b^2)}{2[1 - \lambda(pa + (1 - p)b]}.$$

For instance, if $\lambda = 300$ packets/s, $p = 0.8$, $a = 0.8$ ms, and $b = 10$ ms, one finds that $T \approx 17.4$ ms.

9.2.6   A Word of Caution

The result of (9.12) shows that we must be careful when assuming some distribution for packet lengths. That result shows that the average values of the service times and interarrival times are not sufficient to predict average delays. The variance of the service times has any important impact on the average delay.

A similar caution holds for the packet arrival process. For instance, if we assume that packets arrive exactly every λ^{-1} time units at a queue and that their service times are constant and equal to $\mu^{-1} < \lambda^{-1}$, then every packet arrives after the previous ones have been transmitted, so that no queueing takes place. As a consequence, the delay per packet is equal to a transmission time, i.e., to μ^{-1}. Comparing this to (9.12), we confirm that the distributions of interarrival times and of service times strongly influence delays.

One should, therefore, be very wary of hasty assumptions, even when they are convenient. It may be very tempting to assume that some processes are Poisson so as to be able to carry out some analysis. (This is the assumption that we made when analyzing the CSMA/CD and ALOHA protocols in Chapter 4.) However, network engineers should not trust blindly the results of an analysis based on such arbitrary assumptions. If a more precise analysis is not feasible and if the assumptions cannot be validated, then it is less misleading to admit that fact and to resort to a careful simulation than to make arbitrary assumptions for the sake of convenience.

Careful measurements of traffic streams show that simple models can capture their behavior over reasonably long periods of time even though they are inaccurate over longer periods of observations. In addition, analytical studies and simulations show that these models predict the delays and backlogs in buffers satisfactorily. However, the suitable models are more complex than Poisson processes and independent service times. The upshot of these studies is that analysis provides powerful and insightful tools to examine the operation of networks. These tools enable us to understand how various aspects of the networks affect their performances. For instance, the simple M/G/1 model teaches us the importance of the variance of the service time. More complete analytical studies add to such qualitative understanding and inform our measurement and simulation methodologies and experiments.

9.2.7 Queues with Vacations

Consider the following variation on the M/G/1 queue. Assume that once the transmitter has emptied the queue it is turned off for some random time, called a *vacation epoch*. Assume also that the successive vacation durations are independent, are independent of the past evolution of the system, and have a common distribution.

With these assumptions it can be shown that the average delay per packet is given by

$$T = T_0 + \frac{\nu}{2} E\{V^2\} \tag{9.13}$$

where T_0 is the average delay through an M/G/1 queue without vacations, V denotes a typical vacation duration, and ν^{-1} is the average vacation duration.

For instance, if the vacations have a fixed duration equal to A, then (9.13) shows that $T = T_0 + A/2$. This particular case is not too surprising. Indeed, a packet that arrives into the queue may join other packets whose transmission has been delayed by half a vacation time $A/2$, on average, while waiting for the transmitter to be turned on again.

Interestingly, if the vacation epochs are not all constant, then (9.13) shows that their impact on the average delay is larger than the average vacation duration. (Mathematically, $E\{V^2\} > (E\{V\})^2 = \nu^{-2}$, so that $T > T_0 + \nu^{-1}/2$.) Intuitively, packets are more likely to be delayed by longer vacations since these affect more packets than shorter vacations.

The queue with vacations is a prelude to our discussion of a buffer that shares the transmitter with other buffers in "Cyclic-Service Systems" below.

9.2.8 Priority Systems

Consider transmission systems with two types of packets: voice and data. Assume that the arrivals of the packets form two independent Poisson processes with rate λ_1 for the voice packets and λ_2 for the data packets. Assume that the transmission times of the voice packets are all independent and are distributed as S_1 and that the transmission times of the data packets are independent and distributed as S_2.

The voice packets are given *priority* over the data packets. That is, whenever a packet transmission is completed, the transmitter starts the transmission of a voice packet if there is one in the buffer, and the transmission of a data packet, if any, otherwise.

We show in Appendix B that the average delay per voice packet is given by $\mu_1^{-1} + W_1$ and the average delay per data packet is given by $\mu_2^{-1} + W_2$ where $\mu_i^{-1} = E\{S_i\}$ for $i = 1, 2$ and

$$W_1 = \frac{\sum_{i=1}^2 \lambda_i E\{S_i^2\}}{2(1 - \rho_1)} \quad \text{and} \quad W_2 = \frac{\sum_{i=1}^2 \lambda_i E\{S_i^2\}}{2(1 - \rho_1)(1 - \rho_1 - \rho_2)} \tag{9.14}$$

where $\rho_i := \lambda_i \mu_i^{-1}$ for $i = 1, 2$.

For instance, let $\lambda_1 = 300$, $\lambda_2 = 10$, $S_1 \equiv 2$ ms, and S_2 be exponentially distributed with mean 20 ms. We find $E\{S_2^2\} = 8 \times 10^{-4}$ s^2. Substituting in (9.14), we obtain $W_1 = 11.5$ ms and $W_2 = 57.5$ ms. Using (9.14) once again, you can verify that reversing the priorities would give $W_1 = 28.75$ ms and $W_2 = 5.75$ ms. Comparing these two reversed priorities demonstrates clearly the effect of the priority on the average queueing time.

9.2.9 Cyclic-Service Systems

Consider the model of a token ring network illustrated in Figure 9.2. In this model, packets arrive at each station as a Poisson arrival process with the same rate λ. The packet transmission times are independent and are distributed as the random variable S. The token travels around the ring until it is captured by a station that has a nonempty queue. At that time, that station can transmit one packet. After transmission, it releases the token. This is the protocol that we discussed in Section 4.3. The token travel time around the ring, when it is not captured by any station, is equal to R.

The problem is to calculate the average delay per packet. It has been shown that this average delay is equal to $E\{S\} + W = \mu^{-1} + W$ where

$$W = \frac{N\lambda E\{S^2\} + R[1 + \rho N^{-1}]}{2[1 - \rho - \lambda R]}. \tag{9.15}$$

In this expression, $\rho := N\lambda\mu^{-1}$. This formula is valid provided that $\lambda R < 1 - \rho$; otherwise, the system is unstable.

Note that if R (the token walk time) is negligible, then this formula yields

$$W = \frac{N\lambda E\{S^2\}}{2(1 - \rho)},$$

which gives the same queueing time as that in an M/G/1 queue. This result should not be surprising, since the system then behaves exactly as an M/G/1 queue, except that it is not necessarily first come–first served. This modification cannot change the average number in the system and, therefore, cannot affect the average delay, by Little's result.

The stability condition $\lambda R < 1 - \rho$ can be written as $E\{NS + R\} < \lambda^{-1}$. This condition can be explained by observing that when all the queues are nonempty, the average

FIGURE 9.2

Token ring network. The token travel time is R when the queues are empty.

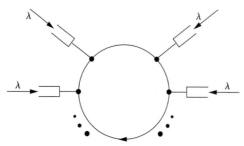

time between successive visits to a station by the token is $E\{NS + R\}$. This average time should be less than the average interarrival time λ^{-1} at a station.

9.2.10   *Model Summary*

- The M/M/1 queue is a single-server queue with Poisson arrivals and exponential service times. The average queue length is $\lambda/(\mu - \lambda)$, and the average delay is $1/(\mu - \lambda)$.
- The average delay through a network of queues is given by (9.11).
- The average delay through an M/G/1 queue is given by (9.12). The average delay increases with the variance of the service times. When the server goes on vacation, the average delay increases by the second moment of the vacations divided by twice their mean value.
- We stated the values of the average delay in queues with priorities and in a symmetric cyclic-service system.

9.3   Simulation

A computer *simulation* is a process that mimics the evolution of a physical system. By tracing that evolution we can determine quantities about the evolution of the physical system. We will discuss *time-driven* and *event-driven* simulations of an M/M/1 queue. We will then examine the *regenerative simulation* method and comment on simulation packages.

9.3.1   *Time-Driven Simulation*

Our objective is to simulate the evolution of an M/M/1 queue. Recall that such a queue is characterized by independent and exponentially distributed interarrival times with rate λ and by independent and exponentially distributed transmission times with rate μ.

Let $\epsilon > 0$ and $t > 0$. Assume that we have simulated the M/M/1 queue up to time t and that the arrival times up to t were $0 < T_1 = t_1 < \cdots < T_n = t_n < t$. The probability that an arrival will occur during the time interval $(t, t + \epsilon)$ is approximately equal to $\lambda\epsilon$. Similarly, the probability that a transmission in progress at time t is completed by time $t + \epsilon$ is approximately equal to $\mu\epsilon$.

This leads to the following simulation procedure. We divide the time into intervals of duration ϵ. Assume that we know x_t. We then simulate two independent coin flips. A flip of the first coin yields head with probability $\lambda\epsilon$ and tail with probability $1 - \lambda\epsilon$. A flip of the second coin yields head with probability $\mu\epsilon$ and tail with probability $1 - \mu\epsilon$. The interpretation is that the first coin flip yields head to represent that an arrival occurs between time t and time $t + \epsilon$ and that the second coin flip yields head to represent that a service in progress at time t (when $x_t > 0$) terminates between time t and time $t + \epsilon$. The value of $x_{t+\epsilon}$ is then given by

$$x_{t+\epsilon} = (x_0 - C_2)^+ + C_1$$

where C_i takes the value 1 when the flip of coin number i yields head and the value 0 otherwise (for $i = 1, 2$). The notation a^+ means the maximum of a and 0. We repeat this procedure to produce $x_{t+2\epsilon}$, and so on.

It remains to explain how the coin flipping experiments can be simulated. As you probably know, most programming languages and a large number of pocket calculators have a function *random*. When invoked, this function returns a random number that is uniformly distributed in $[0, 1]$, i.e., a random number equally likely to take any value in that interval. Moreover, every call to this function is supposed to provide a new random number that is independent of the previous ones. The computers and calculators usually generate these random numbers by computing a regression of the form

$$U_{n+1} = \frac{X_{n+1}}{M} \qquad \text{with} \qquad X_{n+1} = \{a \times X_n + b\} \bmod M$$

where X_0, a, b, and M are well chosen. A suitable choice of these numbers produces a sequence $\{X_1, X_2, \ldots\}$ that is difficult to distinguish from a sequence of independent random numbers uniformly distributed over $[0, 1]$.

Thus, if we choose $0 < p < 1$ and if we define Z_n to be the random variable that takes the value 1 if $U_n < p$ and takes the value 0 otherwise, then the value Z_n behaves almost as independent random variables that take the value 1 with probability p and the value 0 otherwise. This permits the simulation of coin flipping experiments.

The disadvantage of a time-driven simulation is that it may spend a lot of time generating time intervals during which no action takes place. This may result in a very slow simulation.

9.3.2   Event-Driven Simulation

We now describe a different simulation method for the M/M/1 queue. Assume that x_0 is known and let $\{U_n, n \geq 1\}$ be independent and uniformly distributed over $[0, 1]$. Define

$$\tau = -\frac{1}{\lambda} \log U_1 \qquad \text{and} \qquad \sigma = -\frac{1}{\mu} \log U_2. \tag{9.16}$$

You can verify that τ and σ are independent and exponentially distributed with rates λ and μ, respectively. Thus, τ is distributed as the first arrival time T_1 and σ is distributed as the first service time S_1.

The simulation of the queue length x_t then proceeds as follows. If $\tau < \sigma$, we define

$$x_t = x_0 \text{ for } t \in [0, \tau) \text{ and } x_\tau = x_0 + 1.$$

The interpretation is that the first arrival occurs at time τ before the first service is completed. If $\tau > \sigma$, we define

$$x_t = x_0 \text{ for } t \in [0, \sigma) \text{ and } x_\sigma = (x_0 - 1)^+.$$

The interpretation is that the completion of the first service occurs at time σ, before the first arrival.

The simulation then proceeds by repeating the previous construction. This simulation method simulates the successive transitions of x_t. It jumps from event to event.

9.3.3   Regenerative Simulation

The evolution of an M/M/1 queue, or of a network of such queues, can be shown to start afresh every time that the system becomes empty. That is, if τ_n denotes the nth time that the system becomes empty, then the evolutions

$$\{x_t, t \in [\tau_n, \tau_{n+1})\}, n \geq 1$$

are independent and identically distributed.

As a consequence, functions of those "cycles" are also independent and identically distributed. For instance, if D_n is the sum of the delays of the packets that arrived during the cycle $[\tau_n, \tau_{n+1})$ and if K_n is the number of packets that arrived during that cycle, then the random pairs (D_n, K_n) for $n \geq 1$ are independent and have a common distribution. It follows that

$$\frac{D_1 + \cdots + D_n}{n} \to E\{D_1\}, \tag{9.17}$$

by the *strong law of large numbers* (see Appendix A), and similarly for the K_n. As a consequence,

$$\frac{D_1 + \cdots + D_n}{K_1 + \cdots + K_n} \to \frac{E\{D_1\}}{E\{K_1\}}, \tag{9.18}$$

and this limit is then the average delay per packet. Moreover, we can use the *central limit theorem* (see Appendix A) to construct confidence intervals for the estimates. This idea to decompose the evolution into cycles that are independent and identically distributed is called *regenerative simulation*.

This approach cannot be used for complex systems that take a very long time to "regenerate." Calculations show that a network of 40 M/M/1 queues with $\mu_i = 2\lambda_i$ for each queue and with $\gamma = 200$ packets per second spends 170 years between two successive returns to the empty state. A simulation on a fast computer would require a few weeks between regenerations.

9.3.4 Simulation Packages

There are many simulation packages available for simulating communication networks such as OPNET, BONES, and Ptolemy (see References). These packages provide the user with a convenient interface and a large number of facilities for setting up the model of the network to be simulated and for analyzing the results of the simulation.

These simulation tools are useful for testing modifications of protocols, sizing networks, performing case studies on network modifications, and for evaluating original designs. Such simulation environments use discrete-event or time-driven simulations.

9.3.5 Simulation Summary

- A simulation of a discrete-event system, such as an M/M/1 queue, can be either *time-driven* or *event-driven*.
- A time-driven simulation generates the state of the system every multiple of a small time unit.
- An event-driven simulation generates the successive transitions of the system.
- The *regenerative* simulation method enables us to analyze the output of a simulation by decomposing it into independent and identically distributed cycles.

Summary

- Monitoring enables the network manager to evaluate the level of performance of a network. The monitoring information is used to identify corrective actions to be taken and also modifications of the network to improve its performance.
- SNMP and CMOT are management protocol suites being adopted as standards.
- RMON and RMON2 specify the statistics that should be collected about the operations of a network.
- Queueing theory enables one to predict the delays through queues and networks. We discussed the M/M/1 queue and its networks and the M/G/1 queue with and without vacations and priorities.
- Simulations are used by network engineers to evaluate network performance when analytical results are not available.

Problems

Problems with a * are somewhat more challenging. Those with a ^{c} are based on material in a complement. Problems with a ^{1} are borrowed from the first edition.

1. Consider two buffers in series. The first buffer is equipped with a transmitter with rate R_1 bps. The second transmitter has rate R_2. Packets arrive at the first transmitter as a Poisson process with rate λ. The packet lengths are independent and exponentially distributed with mean μ^{-1}.
    a. First assume $R_1 < R_2$. Argue that packets do not queue in the second buffer. Compute the average delay per packet through the two buffers.
    b. Second, assume $R_1 > R_2$. What is now the average delay per packet?
    c. Finally, compare the results with those that correspond to the model of the system as a network of M/M/1 queues.

^{1}2. Consider an M/G/1 queue with arrival rate λ and mean service time μ^{-1}. Show that the probability that the queue is empty is equal to $1 - \lambda\mu^{-1}$. (*Hint:* Apply Little's result to the server. The number of customers with the server is equal to 1 if the queue is nonempty or to 0 otherwise. As a consequence, the average number of customers with the server is equal to the probability that the queue is nonempty.)

^{1}3. Consider an M/M/1 queue with arrival rate λ and service rate μ. What is the average time between two successive returns to the empty state? [*Hint:* Denote that time by A. Since the queue remains empty for λ^{-1} units of time, on the average, every time that it becomes empty, the fraction of time that the queue is empty is equal to $\lambda^{-1}[\lambda^{-1} + A]$. Since this is the probability that the queue is empty at an arbitrary time, using (9.3) allows one to determine A.]

^{1}4. Formula (9.15) shows that

$$W \approx \frac{R}{2(1 - \lambda R)}$$

in the case where $S \approx 0$. Try to explain this value. [*Hint:* A typical customer has to wait for a residual token travel time (equal to $R/2$ on the average), plus X token travel times, if X denotes the number of customers found upon arrival in the queue. Little's result states that $E\{X\} = \lambda W$.]

[1]5. Is it better to use one server at rate 1 or two servers at rate $\frac{1}{2}$? To address this question, assume that the rate of arrivals in an M/M/1 queue is λ and that the service rate is 1. Show that the mean delay is $T_1 = 1/(1-\lambda)$. Instead of using one server, use two, but each with half the service rate. The arrival rate remains the same. Compute the delay T_2 for the latter case. (*Hint:* The latter system is an M/M/2 queue. Consult Problem 4 of Appendix B.)

[1]6. Is it better to form one or two lines in a counter with two employees? The second system of the previous problem models the situation in which customers arrive at rate λ in a bank with two employees, each working at rate $\frac{1}{2}$, and form one single line and are serviced in an FIFO fashion. Now consider the situation in which there are two lines, one for each employee, and a customer joins one of them at random. Compute the delay T_3 in this case. Make a plot of T_1, T_2 (see Problem 5 above), and T_3 as a function of λ.

[1]7. Argue that the second model of Problem 6 is a model for the time-division multiplexing of two channels and that the first model of Problem 5 is a model for the statistical multiplexing of the same channels. Conclude that statistical multiplexing is superior to time-division multiplexing even when the arrival streams are random.

[1]8. Packets arrive at rate 1 Mbs according to a Poisson process. Each message can either follow path 1, consisting of a single M/M/1 queue with service rate 30 Mbps, or path 2, consisting of two M/M/1 queues in series with service rates 10 and 20 Mbs, respectively. A static algorithm is a controller that sends a packet to path 1 with probability p or to path 2 with probability $1 - p$. (A more sophisticated algorithm would actually look at the queue sizes before making any decision.) We are given the constraint that the average delay on path 2 should not exceed 0.10 μs. What is the minimum allowable value of p?

[1]9. The interarrival times of packets in a communication link are random variables $T_1, T_2, \ldots$ that are independent with a common distribution, which unfortunately is unknown. For flow control purposes, one is interested in knowing the arrival rate and the standard deviation of the interarrival time. Explain how to estimate these quantities from real-time observations.

10. Consider the network of Figure 9.3.

FIGURE 9.3

A network with two parallel queues.

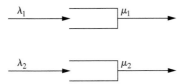

a. We want to choose the positive values of μ_1 and μ_2 that minimize the average delay per packet in the network subject to $\mu_1 + a\mu_2 \leq C$. The interpretation is that $\mu_1 + a\mu_2 \leq C$ is the cost of installing link 1 with average transmission rate μ_1 packets/s and $a\mu_2$ is the cost of installing link 2. This is an example of capacity assignment problem.

b. For fixed values of μ_1 and μ_2, can you find the nonnegative values of λ_1 and λ_2 subject to $\lambda_1 + \lambda_2 = \lambda$ that minimize the average delay per packet in the network. The interpretation is that one choose the routing of the packets.

c. Can you combine part *a* and part *b* of the problem?

d. Can you generalize the problem to more than two queues?

References

Ben-Artzi (1990) contains a clear discussion of SNMP and CMOT. Case (1989) is a more detailed presentation of SNMP. Chapter 3 in Bertsekas (1992) is a nice introduction to performance evaluation. Kleinrock (1975) is also highly recommended. Walrand (1988) is a more advanced presentation of the theory. Varaiya (1996) discusses the performance evaluation of circuit-switched networks such as the telephone network, datagram networks such as Internet, and virtual circuit networks such as ATM.

For information about simulation environments, consult the following Web pages: Ptolemy project (http://ptolemy.eecs.berkeley. edu/), BONES (http://www.cadence.com/ alta/products/hhtml/designer/main.html), OPNET (http://www.mil3.com/products/modeler/ network.html).

Bibliography

Ben-Artzi, A., Chanda, A., and Warrier, U., "Network management of TCP/IP networks: Present and future," *IEEE Network Magazine*, pp. 35–43, July 1990.

Bertsekas, D., and Gallager, R., *Data Networks*, Prentice-Hall, 1992.

Case, J., Fedor, M., Schoffstall, M., and Davin, C., "A simple network management protocol (SNMP)," *Network Working Group RFC 1098*, April 1989.

Kleinrock, L., *Queueing Systems*, vol. 1, John Wiley & Sons, 1975.

Varaiya, P., and Walrand, J., *High-Performance Communication Networks*, Morgan Kaufmann, 1996.

Walrand, J., *An Introduction to Queueing Networks*, Prentice-Hall, 1988.

Probability

Uncertainty affects the operations and the design of communication networks in a fundamental way. The retransmission protocols and the statistical multiplexing strategies used in packet-switched networks are designed to handle the unpredictable traffic and transmission errors. *Probability theory* is one method for quantifying uncertainty. That theory is useful because of its predictive power. Probability theory has been used successfully to predict and to optimize the performance of computer systems, communication networks, manufacturing plants, and many other engineering systems. In this appendix, we explain the few results from probability theory that we need in this text and we illustrate them with examples.

In Section A.1, we define *probability* and *random variable*. The *expected value* of a random variable is defined in Section A.2. Section A.3 explains the definition of *independence* of random variables. The *regenerative method* for computing some mean values is explained in Section A.4.

A.1 Probability and Random Variables

We do not give a formal definition of probability or of a random variable. Let us just say that a random variable is a function of the outcome of a random experiment and that the probabilities characterize how likely different outcomes are. The following examples illustrate what we have in mind.

A *roulette wheel* has N slots that are equally likely to be selected by spinning the wheel. The N slots are marked with $n \leq N$ different numbers $\{x_1, x_2, \ldots, x_n\}$. Number x_i appears on N_i of the N slots, for $i = 1, \ldots, n$. Thus,

$$\sum_{i=1}^{n} N_i := N_1 + \cdots + N_n = N.$$

The odds that the number x_i is selected by a roulette wheel spin are N_i to N. We say that the *probability* that the number selected by the spin is x_i is equal to

$$p_i = \frac{N_i}{N} \text{ for } i = 1, 2, \ldots, n.$$

Thus, the actual number selected is unpredictable. Let us denote it by X. By definition, X is a function of the outcome of a random experiment. In this example, X takes the value x_i if the outcome of the random experiment, i.e., the slot selected by spinning the wheel, happens to be one of the N_i slots marked with the value x_i. A function of the outcome of a random experiment is called a *random variable*. We define

$$P\{X = x_i\} := p_i.$$

Thus, $P\{X = x_i\}$ is the probability that $X = x_i$. Notice that

$$0 \leq p_i \leq 1 \text{ for all } i \text{ and } \sum_{i=1}^{n} p_i = 1.$$

For any set A of values that X can take, we define

$$P\{X \in A\} := \sum_{x_i \in A} p_i.$$

This last expression means that the probability that X takes a value in A is defined as the sum of the probabilities p_i of all the values x_i in A. The justification is that if $A = \{x_1, \ldots, x_k\}$, then a value in A is chosen by the spin if it selects one out of the $N_1 + \cdots + N_k$ slots marked $x_1, x_2, \ldots,$ or x_k. The probability that this occurs is equal to $(N_1 + \cdots + N_k)/N$, which is seen to be equal to

$$p_1 + \cdots + p_k = \sum_{x_i \in A} p_i$$

since the indices i such that $x_i \in A$ are $1, 2, \ldots, k$.

The roulette wheel has only a finite number of slots. Our next example examines a random experiment that has an infinite number of possible outcomes. The experiment consists of counting the photons Y that hit a specific photodetector in a time interval of 1 minute. By repeating the experiment many times, one finds that the fraction of experiments when $Y = n$ is equal to

$$p_n = \frac{\alpha^n}{n!} e^{-\alpha}, \ n \geq 0 \tag{A.1}$$

where α is some positive number. (By definition, $n!$—called *n factorial*—is equal to 1 when $n = 0$ and to $1 \times 2 \times 3 \times \cdots \times n$ when n is a positive integer.) The number p_n is called the *probability* that $Y = n$. You can verify that

$$0 \leq p_n \leq 1, \text{ for all } n \text{ and } \sum_{n=0}^{\infty} p_n = 1$$

by using the definition of the exponential

$$e^{\alpha} = \sum_{m=0}^{\infty} \frac{\alpha^m}{m!}. \tag{A.2}$$

As before, for any set A, we define

$$P\{Y \in A\} := \sum_{n \in A} p_n.$$

By definition, a random variable with the probabilities (A.1) is called a *Poisson random variable* with parameter α.

So far, our random variables admitted values in a countable set. That is, we could enumerate the possible values as $x_1, x_2, x_3, \ldots$. The random variable that we examine next takes arbitrary values in $[0, \infty)$. We buy a lightbulb and we are told by the manufacturer that the lifetime X of that bulb is such that

$$P\{X > t\} = e^{-\lambda t}, t \geq 0 \tag{A.3}$$

where λ is a positive real number. A random variable such that (A.3) holds is said to be *exponentially distributed* with rate λ. From (A.3) we can calculate that

$$P\{X \in (t, t + \epsilon]\} \approx \lambda \epsilon e^{-\lambda t}, \text{ for } 0 \leq t \text{ and } 0 \leq \epsilon \ll 1. \tag{A.4}$$

Indeed, since the set $(t, t + \epsilon]$ is the difference between the set (t, ∞) and the set $(t + \epsilon, \infty)$, we conclude that the left-hand side of (A.4) is equal to

$$P\{X > t\} - P\{X > t + \epsilon\} = e^{-\lambda t} - e^{-\lambda(t+\epsilon)} = \{1 - e^{-\lambda \epsilon}\}e^{-\lambda t}$$

and (A.4) then follows from (A.3) by using $e^{-\lambda \epsilon} \approx 1 - \lambda \epsilon$. In general, when X is a random variable that takes values in $(-\infty, +\infty)$ and

$$P\{X \in (t, t + \epsilon)\} \approx f(t)\epsilon, \text{ for } t \in (-\infty, +\infty) \text{ and } 0 \leq \epsilon \ll 1,$$

we say that the function $f(.)$ is the *probability density function* (p.d.f.) of the random variable X. Thus, the probability density function of an exponentially distributed random variable with rate λ is

$$f(t) = \begin{cases} \lambda e^{-\lambda t}, & \text{for } t \geq 0 \\ 0, & \text{for } t < 0. \end{cases} \tag{A.5}$$

So far, we have defined a random variable as the outcome of a random experiment and we have examined three examples. We now turn to a more subtle question that is of central importance in probability theory. That question is how we should take available information into account when calculating probabilities. The answer is provided by the notion of *conditional probability*. We start our discussion of conditional probability with a simple example from which we derive a general definition. We then apply this definition to a few more examples.

Consider the roulette wheel discussed at the beginning of this section. We are told that the outcome of the spin is such that $X \in \{x_2, x_4, x_5\}$. We want to find the probability that $X \in \{x_1, x_2, x_4, x_6\}$ given that information. Since we know that $X \in \{x_2, x_4, x_5\}$, the slot selected by the spin must be one of the $N_2 + N_4 + N_5$ slots marked x_2, x_4, or x_5. Also, these slots are equally likely to have been selected and X takes a value in $\{x_1, x_2, x_4, x_6\}$ if it takes one of the values x_2 or x_4, i.e., if the slot selected by the spin is one of $N_2 + N_4$ among the $N_2 + N_4 + N_5$ equally likely slots known to have been selected. Thus, the probability that $X \in \{x_1, x_2, x_4, x_6\}$ given that $X \in \{x_2, x_4, x_5\}$ is given by

$$\frac{N_2 + N_4}{N_2 + N_4 + N_5}.$$

By dividing both numerator and denominator by N, the above probability can be expressed as

$$\frac{P\{X \in \{x_2, x_4\}\}}{P\{X \in \{x_2, x_4, x_5\}\}}.$$

Therefore, if one denotes by A the event $X \in \{x_1, x_2, x_4, x_6\}$ and by B the event $X \in \{x_2, x_4, x_5\}$, one sees from the above discussion that the *conditional probability of A given B* is equal to

$$P[A|B] := \frac{P\{A \text{ and } B\}}{P\{B\}}. \tag{A.6}$$

We adopt the above formula as the definition of conditional probability. Let us see how we can apply that formula to two other examples. As a first example, let Y be a Poisson random variable with parameter α [see (A.1)]. Then

$$P[Y = 1|Y > 0] = \frac{P\{Y = 1 \text{ and } Y \geq 0\}}{P\{Y \geq 0\}} = \frac{P\{Y = 1\}}{1 - P\{Y = 0\}} = \frac{\alpha e^{-\alpha}}{1 - e^{-\alpha}}.$$

As another example, let X be the exponential lifetime of a light bulb [see (A.3)]. We want to estimate the probability that a light bulb that has been on for a seconds will survive for at least t more seconds. That is, we want to calculate $P[X > a + t|X > a]$. Using (A.3) and (A.6) we find

$$P[X > a + t|X > a] = \frac{P\{X > a + t\}}{P\{X > a\}} = \frac{e^{-\lambda(a+t)}}{e^{-\lambda a}}$$
$$= e^{-\lambda t} = P\{X > t\}.$$

This calculation reveals the *memoryless property* of the exponential distribution: a light bulb that has been on for some time is equally likely as a new light bulb to survive for t more seconds. In other words, an old light bulb is exactly as good as a new one (assuming that the lifetime is exponentially distributed).

A.2 Expectation

Intuitively, the *expected value* (or *average* or *mean value*) of a random variable is the arithmetic mean of the values observed when the random experiment is replicated many times. For instance, consider the roulette wheel and denote by X_m the number selected by the mth spin of the wheel. The arithmetic mean of the values selected by one's spinning the wheel N times is

$$\frac{X_1 + X_2 + \cdots + X_N}{N}. \tag{A.7}$$

We expect that the fraction of times that a spin selects the number x_i is close to p_i. That is, we expect that the number x_i is selected about $p_i N$ times during the N experiments. Consequently, the numerator of (A.7) is approximately equal to

$$(p_1 N)x_1 + (p_2 N)x_2 + \cdots + (p_n N)x_n,$$

so that we expect the ratio (A.7) to be approximately equal to

$$p_1x_1 + p_2x_2 + \cdots + p_nx_n.$$

We use this intuitive discussion to define the expected value of a random variable. Specifically, let X be a random variable with possible values $\{x_i, i \geq 1\}$. Assume that the x_i are real numbers. (X is then called a *real-valued random variable*.) The *expected value* $E\{X\}$ of X is defined as *the sum of the values of X weighted by their probabilities*, i.e., by

$$E\{X\} = \sum_{i=1}^{\infty} x_i P\{X = x_i\}. \tag{A.8}$$

We illustrate this definition with a few simple examples. By definition, a random variable X is *geometrically distributed with parameter* $p \in [0, 1]$ if

$$P\{X = n\} = (1 - p)p^{n-1}, \text{ for } n = 1, 2, 3, \ldots. \tag{A.9}$$

These probabilities sum to one since

$$\sum_{m=0}^{\infty} a^m = \frac{1}{1 - a}, \text{ for } a \in (-1, +1). \tag{A.10}$$

A simple way to verify (A.10) is to denote the sum by A and to observe that

$$A = \sum_{m=0}^{\infty} a^m = 1 + \sum_{m=1}^{\infty} a^m = 1 + a \sum_{m=0}^{\infty} a^m = 1 + aA,$$

from which (A.10) follows. We use (A.9) and (A.8) to calculate the expected value of X. We find

$$E\{X\} = \sum_{n=1}^{\infty} n(1 - p)p^{n-1} = \frac{1}{1 - p}. \tag{A.11}$$

To verify the last equality, denote the sum by S and observe that

$$S - pS = \sum_{n=1}^{\infty} n(1 - p)p^{n-1} - \sum_{n=0}^{\infty} n(1 - p)p^n$$

$$= \sum_{n=0}^{\infty} (n + 1)(1 - p)p^n - \sum_{n=0}^{\infty} n(1 - p)p^n$$

$$= (1 - p) \sum_{n=0}^{\infty} p^n = 1.$$

As another example, let us calculate the expected value of a Poisson random variable X with parameter α [see (A.1)]. We find

$$E\{X\} = \sum_{n=0}^{\infty} n \frac{\alpha^n}{n!} e^{-\alpha} = \alpha \sum_{n=0}^{\infty} \frac{\alpha^{n-1}}{(n - 1)!} e^{-\alpha} = \alpha \sum_{m=0}^{\infty} \frac{\alpha^m}{m!} e^{-\alpha} = \alpha. \tag{A.12}$$

As yet another example, consider a random variable X that always takes the same value μ (some real number). Thus, $P\{X = \mu\} = 1$. In that case, one says that X is constant and one may denote X by μ. It follows from the definition (A.8) that

$$E\{\mu\} = \mu$$

since there is only one term in the sum (A.8) which corresponds to the value μ and its probability 1.

We conclude our discussion of expected value by examining a random variable that takes values in a continuous set. Let X be an exponentially distributed random variable with rate λ. We want to calculate $E\{X\}$. We argue that, for very small ϵ,

$$P\{X \in (n\epsilon, (n+1)\epsilon]\} \approx f(n\epsilon)\epsilon$$

where $f(.)$ is the p.d.f. of X defined in (A.5). Thus, X can be approximated by a random variable that takes the discrete values $\{0, \epsilon, 2\epsilon, 3\epsilon, \ldots\}$ with the probabilities given above. Consequently, from (A.8),

$$E\{X\} \approx \sum_{n=0}^{\infty} (n\epsilon)\{f(n\epsilon)\epsilon\}.$$

This approximation becomes better as ϵ becomes smaller and the above sum approaches

$$E\{X\} = \int_0^\infty xf(x)dx,$$

by definition of the integral. We conclude that

$$E\{X\} = \int_0^\infty xf(x)dx = \int_0^\infty x\lambda e^{-\lambda x}dx = \lambda^{-1}. \tag{A.13}$$

The above example can be generalized, of course, to a real-valued random variable X with p.d.f. $f(.)$ to show that its expected value is given by

$$E\{X\} = \int_{-\infty}^\infty xf(x)dx, \tag{A.14}$$

provided that this integral exists.

In many applications, we are led to consider functions of random variables. We explain how the expected value of such a function can be calculated. The method is very simple. Let X be a random variable and $H(X)$ a real-valued function of X. Then $Y = H(X)$ is a real-valued random variable. (It is a function of the outcome of a random experiment since X is such a function.) We want to calculate $E\{H(X)\}$. Assume that X takes the values $\{x_i, i \geq 1\}$ with probabilities $p_i = P\{X = x_i\}$. Assume further that the function $H(.)$ is such that the values $H(x_i)$ are distinct for different i's. Then Y takes the value $H(x_i)$ with probability p_i. Indeed, Y takes the value $H(x_i)$ if and only if X takes the value x_i, which occurs with probability p_i. Hence, by definition (A.8) of expected value,

$$E\{H(X)\} = \sum_{i=1}^{\infty} H(x_i)p_i. \tag{A.15}$$

In general, if the values $H(x_i)$ are not all distinct, a minor perturbation of the function H (after the hundredth decimal place, say) can make them distinct, so that (A.15) must again hold. (See Problem 2 for another argument.)

Often we need to calculate the mean value of sums of random variables. Let (X, Y) be a function of some random experiment. Assume that both X and Y take real values.

Consequently, $aX + bY$ is a real-valued random variable for arbitrarily chosen real numbers a and b. We show that

$$E\{aX + bY\} = aE\{X\} + bE\{Y\}. \tag{A.16}$$

To perform the calculation, let us assume that the pair (X, Y) takes the value (x_i, y_i) with probability p_i for $i \geq 1$. By applying (A.15) to the function $H(X, Y) = aX + bY$, we find

$$E\{aX + bY\} = \sum_{i=1}^{\infty}(ax_i + by_i)p_i = a\sum_{i=1}^{\infty}x_i p_i + b\sum_{i=1}^{\infty}y_i p_i = aE\{X\} + bE\{Y\},$$

as claimed. To derive the last equality, we use

$$E\{X\} = \sum_{i=1}^{\infty}x_i p_i \text{ and } E\{Y\} = \sum_{i=1}^{\infty}y_i p_i,$$

which can be verified by applying (A.15) to the function $H(X, Y) = X$ and to $H(X, Y) = Y$, respectively.

A.3 Independence

Independence is probably the most fertile concept in probability theory. Intuitively, two random variables are *independent* if knowing the value of one does not provide information about the value of the other. We define this concept precisely in this section, and we explore some of its consequences.

We start with a simple example. Consider two roulette wheels, one in Las Vegas and one in Atlantic City. The wheel in Las Vegas has N equally likely slots, and N_i of those slots are marked with the number x_i, for $i = 1, 2, \ldots, n$. The wheel in Atlantic City has M equally likely slots, and M_j of those slots are marked with the number y_j, for $j = 1, 2, \ldots, m$. Thus,

$$P\{X = x_i\} = \frac{N_i}{N} \quad \text{and} \quad P\{Y = y_j\} = \frac{M_j}{M}.$$

Assume that both wheels are spun simultaneously. Denote the values selected by the two wheels by X (in Las Vegas) and Y (in Atlantic City). If we consider spinning the two wheels as a single random experiment, then intuition suggests that there are NM equally likely outcomes, corresponding to N possible slots selected by the first wheel and to M for the second wheel. To every slot of the first wheel correspond M possible slots for the second wheel. Out of these NM possibilities, $N_i M_j$ result in $X = x_i$ and $Y = y_j$. Hence,

$$P\{X = x_i \text{ and } Y = y_j\} = \frac{N_i M_j}{NM} = \left(\frac{N_i}{N}\right)\left(\frac{M_j}{M}\right),$$

so that

$$P\{X = x_i \text{ and } Y = y_j\} = P\{X = x_i\}P\{Y = y_j\}, \tag{A.17}$$

for $i = 1, 2, \ldots, n$ and $j = 1, 2, \ldots, m$.

We adopt (A.17) as a *definition* of independence. Specifically, two random variables X and Y with possible values $\{x_i, i \geq 1\}$ and $\{y_j, j \geq 1\}$ are said to be *independent* when (A.17) holds. More generally, two random variables X and Y are independent if

$$P\{X \in A \text{ and } Y \in B\} = P\{X \in A\}P\{Y \in B\} \text{ for all sets } A \text{ and } B \qquad \text{(A.18)}$$

and the random variables $\{X_n, n \geq 1\}$ are independent if

$$P\{X_1 \in A_1, X_2 \in A_2, \ldots, X_n \in A_n\} = P\{X_1 \in A_1\} \cdots P\{X_n \in A_n\}, \qquad \text{(A.19)}$$

for all $n \geq 1$ and all sets A_i.

Note that it is generally not true that the random variables $\{X_n, n \geq 1\}$ are independent if X_i and X_j are independent for all $i \neq j$. In this latter case, one says that the random variables are *pairwise independent*. Thus, pairwise independence does not imply independence of a set of random variables. (See Problem 3.)

As an illustration of independence, consider the following experiment. One is given a coin that, when flipped, yields *head* (H) with probability $p \in (0, 1)$ and yields *tail* (T) otherwise. Denote by X the number of times that the coin must be flipped before yielding the first T. We want to calculate $P\{X = n\}$ for $n \geq 1$. In our description of the experiment, we implicitly assume that the successive outcomes of the coin flips are independent random variables. Thus, if one denotes by Y_n the outcome of the nth flip, one has

$$\begin{aligned} P\{X = n\} &= P\{Y_1 = \text{H}, \ldots, Y_{n-1} = \text{H}, Y_n = \text{T}\} \\ &= P\{Y_1 = \text{H}\} \cdots P\{Y_{n-1} = \text{H}\}P\{Y_n = \text{T}\} \\ &= p^{n-1}(1 - p), n \geq 1 \end{aligned}$$

where the second equality follows from (A.19). This shows that X is geometrically distributed with parameter p [see (A.9)]. In particular, $E\{X\} = (1 - p)^{-1}$, by (A.11).

You probably expect that functions of independent random variables are independent. Let us show that this is indeed the case. Let X and Y be two independent random variables and $f(.), g(.)$ be two arbitrary functions. To show that $f(X)$ and $g(Y)$ are independent, we must show that

$$P\{f(X) \in A \text{ and } g(Y) \in B\} = P\{f(X) \in A\}P\{g(Y) \in B\},$$

for all sets A and B. Now, $f(X)$ takes values in A if and only if X takes values in some specific set C. That set C is the set of values x such that $f(x) \in A$. Similarly, $g(Y)$ takes values in B if and only if Y takes values in some set D. Hence,

$$\begin{aligned} P\{f(X) \in A \text{ and } g(Y) \in B\} &= P\{X \in C \text{ and } Y \in D\} \\ &= P\{X \in C\}P\{Y \in D\} \\ &= P\{f(X) \in A\}P\{f(Y) \in B\}. \end{aligned}$$

In this derivation, the first and third equalities follow from the definition of the sets C and D and the second equality follows from the independence of X and Y.

We need the following result about the expectation of the product of independent random variables. The result states that

$$E\{XY\} = E\{X\}E\{Y\} \text{ whenever } X \text{ and } Y \text{ are independent.} \qquad \text{(A.20)}$$

We prove this result when X takes the values $\{x_i, i \geq 1\}$ with probabilities p_i and Y takes the values $\{y_j, j \geq 1\}$ with probabilities q_j. Then (X, Y) takes the value (x_i, y_j) with probability $p_i q_j$ (by independence). Consequently, (A.15) implies that

$$E\{XY\} = \sum_{i,j} x_i y_j p_i q_j$$

This sum can be computed by first summing over j and then summing over i. We find

$$E\{XY\} = \sum_i [\sum_j x_i y_j p_i q_j] = \sum_i [x_i p_i \sum_j y_j q_j] = \sum_i [x_i p_i E\{Y\}]$$
$$= E\{Y\} \sum_i x_i p_i = E\{Y\} E\{X\},$$

as we set out to prove.

Two very useful results, which we do not derive here, are known for independent and identically distributed random variables $\{X_n, n \geq 1\}$. The first result is the *strong law of large numbers* that states that

$$Y_n := \frac{X_1 + X_2 + \cdots + X_n}{n} \to E\{X_1\} \text{ as } n \to \infty.$$

That is, for large n, the arithmetic mean Y_n of n independent random variables with the same distribution is close to the expected value of these random variables.

The second result, the *central limit theorem,* states that

$$\{Y_n - E\{X_1\}\}\sqrt{n} \approx N(0, \sigma^2) \text{ for } n \gg 1.$$

In this statement, $N(0, \sigma^2)$ denotes a Gaussian random variable with mean zero and variance σ^2, where σ^2 is the variance of each X_n. By definition,

$$X = N(0, \sigma^2) \text{ if } P\{X \in (x, x + dx)\} = \frac{1}{\sqrt{2\pi\sigma^2}} e^{-x^2/2\sigma^2} dx \text{ for } -\infty < x < \infty.$$

This theorem says that the difference between the arithmetic mean Y_n and the expected value $E\{X_1\}$ is distributed as a Gaussian random variable divided by $\sqrt{n}$, for large n.

A.4 Regenerative Method

The purpose of this section is to explain a method for calculating mean values of random variables that arise frequently in models of communication networks. The method replaces complicated calculations with elementary algebra.

To introduce the method, we need to develop a preliminary result about a costly gambling experiment defined in two phases. In the first phase, we flip a coin. If the outcome is H (which happens with probability $p \in (0, 1)$), then we go to Las Vegas to spin the roulette wheel X defined before. If the outcome of the coin flip is T, then we go to Atlantic City to spin the roulette wheel Y. We would like to calculate the average earnings ultimately obtained by playing the randomly chosen wheel. To perform this calculation, we denote the final earnings by Z. Additionally, we define V as the random variable that takes the value

1 when the outcome of the coin flip is H and the value 0 otherwise. With these definitions, we can write the ultimate earnings Z as

$$Z = VX + (1 - V)Y,$$

since $Z = X$ when $V = 1$ and $Z = Y$ when $V = 0$. Using the independence of V, X, and Y, we conclude from (A.20) and (A.16) that

$$E\{Z\} = E\{V\}E\{X\} + E\{1 - V\}E\{Y\} = pE\{X\} + (1 - p)E\{Y\}.$$

Summarizing, we have seen that if

$$Z = \begin{cases} X, & \text{with probability } p \\ Y, & \text{with probability } 1 - p \end{cases} \tag{A.21}$$

then

$$E\{Z\} = pE\{X\} + (1 - p)E\{Y\}. \tag{A.22}$$

We illustrate how the above result can be applied to the analysis of a communication system. A transmitter sends packets to a receiver. With probability $p \in (0, 1)$, the transmission is unsuccessful (contains errors) and it has to be repeated. We want to calculate the expected number of transmissions that are necessary until the packet is successfully transmitted. One way to solve the problem is to calculate the distribution of the number X of necessary transmissions. Using our discussion of the coin flipping experiment, we can verify that

$$P\{X = n\} = p^{n-1}(1 - p), n = 1, 2, \ldots.$$

From this expression we can derive that

$$E\{X\} = \frac{1}{1 - p}.$$

It turns out that this method is difficult to use in more complicated situations. We explain another approach.

The random variable X is equal to 1 with probability $1 - p$ if the first transmission is successful. With probability p, the number X is equal to $1 + Y$ where Y is distributed as X. Indeed, with probability p, there is a first unsuccessful transmission, and then the problem starts afresh so that the number of remaining transmissions, after the first unsuccessful one, is statistically equivalent to the random variable X. Thus, we can write that

$$X = \begin{cases} 1 & \text{with probability } 1 - p \\ 1 + Y & \text{with probability } p. \end{cases}$$

Note that Y is independent of the outcome of the first transmission. Consequently, from (A.22),

$$E\{X\} = (1 - p)1 + pE\{1 + Y\} = (1 - p) + p[1 + E\{Y\}].$$

Now, $E\{Y\} = E\{X\}$ since Y and X are statistically equivalent. It follows that

$$E\{X\} = (1 - p) + p + pE\{X\} = 1 + pE\{X\},$$

so that, solving for $E\{X\}$, we find

$$E\{X\} = \frac{1}{1 - p}.$$

In this simple example, the economy of the method may not be very apparent. Notice, however, that no complicated summation was needed. The power of the method is more evident in the next examples.

The diagram shown in Figure A.1 summarizes how the random variable X behaves. The figure indicates two points: the Start and Stop of the transmission. The arrows are labeled with an expression of the form $[D, P]$. The meaning of D is the duration of the corresponding transition. P is the probability of the transition. Thus, the arrow corresponding to Success indicates that the transmission takes one unit of time (time is measured here in transmissions) and that it is successful with probability $1 - p$. Similarly, the arrow corresponding to the Error takes one unit of time and has probability p. If the transmission is a success, then the packet transmission is completed. If not, it restarts. From this diagram, one sees that if X is the total number of transmissions, then

$$E\{X\} = (1 - p)1 + p[1 + E\{X\}].$$

Consider now the following variation of the previous problem. A successful transmission takes T units of time. If it is unsuccessful, then it takes $T + \alpha$ time units to realize that the transmission was not successful. (For instance, α could be the time taken to receive a negative acknowledgment.) We want to calculate the average time taken by a successful transmission. The diagram corresponding to this problem is indicated in Figure A.2. To solve the problem, we denote by τ the duration until the first successful transmission. Arguing as in the previous example, we can write that

$$E\{\tau\} = (1 - p)T + p[T + \alpha + E\{\tau\}].$$

Solving for $E\{\tau\}$ gives

$$E\{\tau\} = \frac{T + p\alpha}{1 - p}.$$

The same method applies to more complicated situations. Say that a packet must be transmitted via three successive links. Each transmission takes T seconds and is successful with probability $1 - p$. If the first, second, or third transmission is not successful, then the transmitter finds out that the transmission is not successful after $T + \alpha, T + 2\alpha$, or $T + 3\alpha$ time units, respectively. The diagram corresponding to this system is indicated in Figure A.3. Denoting again by τ the total transmission time until the packet first reaches

FIGURE A.1

Transmitter model.

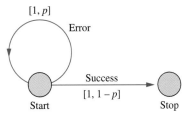

FIGURE A.2

Transmission durations.

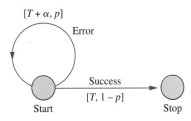

FIGURE A.3

Multistep transmission.

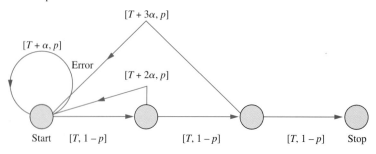

the destination without error, we find

$$E\{\tau\} = p(T + \alpha + E\{\tau\}) + (1 - p)[T + T_1]$$
$$T_1 = p(T + 2\alpha + E\{\tau\}) + (1 - p)[T + T_2]$$
$$T_2 = p(T + 3\alpha + E\{\tau\}) + (1 - p)T.$$

These equations can be solved to obtain $E\{\tau\}$.

A.5 Complement 1: Channel Coding

You should read "Source Coding Theory" in Section 8.10 before this complement.

How fast can we transmit bits over a telephone line or over an optical fiber or between two microwave antennas? A better question is how fast we can transmit the bits reliably. Indeed, it is plausible that if we are willing to receive incorrectly a fraction of the bits, then we should be able to send them faster, a little like trying to throw balls in a basket.

As soon as we try to make the question precise we run into an apparent contradiction. Indeed, no matter what communication equipment we use, it is impossible to guarantee that any single bit is received correctly. Thermal noise, electromagnetic interference, or other sources of noise have some positive probability of corrupting any given bit. Moreover, we can also argue that any finite set of bits also has some positive probability of being all corrupted. Consequently, it seems that the only possible answer to our question is that the maximum rate at which we can transmit bits reliably is 0: We can not transmit bits reliably at any rate.

Asking the right question is the key to basic research, and making the question precise and relevant is the essential step to interesting research. In fact, it is rarely the first step because formulating the question correctly requires understanding important aspects of the situation.

Shannon asked the precise question and answered it. The precise question is "What is the maximum rate at which one can transmit bits with an arbitrarily small probability of errors?" Say that that rate is C bps. The meaning of C is that for any given ϵ with $0 < \epsilon \le 1$ one can transmit bits at any rate less than C bps with the fraction of the bits that are not received correctly at most ϵ. The specific procedure that we use depends on ϵ and this is why we cannot transmit the bits fully reliably. In particular, the delay before the receiver gets the bits increases without bound as ϵ gets closer to 0.

To explore this question, Shannon constructs the simplest model of a communication channel. His model is a channel that corrupts a bit with probability p and transmits it correctly otherwise, with $0 \le p \le 1$. Moreover, the channel corrupts the bits independently of one another. How can we transmit reliably through this channel? The trick is to transmit a group of n bits—called a *codeword*—for some large n. We know that a fraction close to p of the bits of that codeword is corrupted. Say that $p = 0.1$ and that $n = 1000$. We send a codeword of 1000 bits and about 100 bits are corrupted. Say that we choose a number of codewords of 1000 bits that differ from one another by at least 220 bits. A codeword with 100 bits corrupted is still more similar to that codeword than to any other codeword since it differs from any other codeword by at least 120 bits. This key idea is called *block coding*. Assume that there are 2^{nR} codewords of n bits that differ by at least 220 bits from one another. Then we can send "reliably" nR bits that number these codewords by sending n bits through the channel. We say that the reliable rate of transmission is then R bits per unit of time (or nR bits every n units of time).

How do we calculate an achievable rate R? Here we need another breakthrough idea. A tempting approach is to try to calculate how many codewords of n bits differ by at least $n(p + \epsilon)$ bits from one another. This calculation is not easy and we could spend many nights without getting the answer. Shannon's approach is surprising and striking in how it simplifies the calculations. Here is how it goes. Say we want to choose $K = 2^{nR}$ groups of n bits. Instead of searching for codewords that differ sufficiently from each other, let's just pick these codewords randomly. To pick the first codeword, we simply flip a fair coin n times. Say that the first codeword is $\mathbf{Y}^1$. Thus, $\mathbf{Y}^1$ is a string of n 0s and 1s. We repeat the procedure for the other codewords $\mathbf{Y}^2, \ldots, \mathbf{Y}^K$. You might argue that some of these codewords might not differ enough from each other so that this is not a particularly clever way of choosing them. Shannon's insight is that we should not worry: Typically, we have found good codewords! That is, the outputs of the channel that correspond to these different codewords are likely to remain far apart. In fact, these outputs are far apart as long as we do not try to pick too many codewords. The calculation that we perform at the end of this section shows that all this is fine as long as $R \le 1 - H$ where

$$H = -p \log p - (1 - p) \log(1 - p) \tag{A.23}$$

is the binary entropy that we studied in Section 8.10. The idea of picking the codewords at random is called a *random coding argument*.

Summarizing, we can send one of 2^{nR} groups of n bits and be able to identify which codeword was sent after the channel alters some of the bits. Sending one of 2^{nR} codewords amounts to delivering nR bits of information: the index number of the codeword that was sent. Hence, we can send nR bits by transmitting n symbols through the channel. The transmission rate is then nR bits in n units of time, or R bits per unit of time. We call the maximum value of R the *Shannon capacity* of the channel and we denote it by C. Thus, for the simple channel model that we analyzed,

$$C = 1 - H = 1 + p \log_2 p + (1 - p) \log_2(1 - p). \tag{A.24}$$

In practice, channels are more complicated than the simple model that Shannon used initially. He showed that the same ideas apply to much more complicated channels. Since his work in the late 1940s, many extensions have been developed and practical coding methods with efficient decoding algorithms are now available and are implemented in modems and other communication link components.

Derivation of (A.24)

To transmit information through the noisy channel we choose $K = 2^{nR}$ sequences of n bits. These sequences are codewords for K different messages. We give that list of sequences to the receiver. To send message i, we send the ith sequence of n bits. The receiver gets a corrupted version of that sequence and looks in the list to find the most likely sequence that was sent: the sequence that resembles the received sequence the most if $p < 0.5$. We want to show that the receiver guesses correctly with a probability very close to 1 as long as K is not too large. Specifically, we show that the probability can be made arbitrarily close to 1 by choosing n large enough as long as $R < 1 - H$ where H is given in (A.23).

To show this result, we choose the K sequences—the K codewords—that we denote by $\mathbf{Y}^1, \ldots, \mathbf{Y}^K$ by flipping a fair coin nK times. When we send the codeword $\mathbf{Y}^i$ through the channel, we can write the sequence of n bits at the output of the channel as

$$\mathbf{Z} = \mathbf{Y}^i + \mathbf{X}$$

where $\mathbf{X} = (X_1, X_2, \ldots, X_n)$ and the X_m are independent with $P(X_m = 1) = p = 1 - P(X_m = 0)$ for $1 \leq m \leq n$ and where the addition is modulo 2 and without carry. Indeed, this addition means that the channel modifies the mth input when $X_m = 1$, with probability p, which is precisely how the channel was defined.

Note that the bits at the output are all independent and are equally likely to be 0 or 1, by symmetry. Thus, the output is equally likely to be any of the 2^n sequences in $\{0, 1\}^n$. From the results in Section 8.10 we know that $\mathbf{X}$ is equally likely to be any of the 2^{nH} sequences in some set $\mathbf{S}$. Consequently, for any specific input codeword $\mathbf{Y}^i$, the output $\mathbf{Z}$ is equally likely to be any one of the 2^{nH} sequences in the set $\mathbf{S}(i) := \{\mathbf{Y}^i + \mathbf{x}, \mathbf{x} \in \mathbf{S}\}$. One possible decoding rule for the receiver is then to list the sets $\mathbf{S}(1), \ldots, \mathbf{S}(K)$ and to decide that message i was sent if $\mathbf{Z} \in \mathbf{S}(i)$.

Let's send the codeword $\mathbf{Y}^j$ through the channel. The resulting output $\mathbf{Z}$ is equally likely to be any of the 2^{nH} sequences in $\mathbf{S}(j)$. The probability that this sequence $\mathbf{Z}$ happens to be in $\mathbf{S}(i)$ is essentially 1 if $i = j$. However, if $i \neq j$, that probability is equal to

$$P(\mathbf{Z} \in \mathbf{S}(i)) = \frac{|S(i)|}{|\{0, 1\}^n|} = \frac{2^{nH}}{2^n} = 2^{n(H-1)}$$

where by $|\mathbf{A}|$ we denote the number of elements in a set $\mathbf{A}$. To see why the above equality holds, recall that $\mathbf{Z}$ is equally likely to be any sequence in $\{0, 1\}^n$ so that the probability that it actually is one of the sequences in $\mathbf{S}(i)$ is given by the ratio of the number of sequences in $\mathbf{S}(i)$ over the total number of sequences in $\{0, 1\}^n$.

The receiver makes the wrong guess for the message j that was sent if the output $\mathbf{Z}$ happens to be in one of the $K - 1$ sets $\mathbf{S}(i)$ for some $i \neq j$. Using the above equality we can bound that probability of error α as follows:

$$\alpha = P(\mathbf{Z} \in \mathbf{S}(i) \text{ for some } i \in \{1, \ldots, K\} \text{ with } i \neq j) \leq K \cdot 2^{n(H-1)}.$$

To derive this bound we use the fact that the probability of the union of sets is less than the sum of the probabilities of those sets.

Let us choose $K = 2^{nR}$. We then find that the above probability α of error is at most

$$2^{nR} 2^{n(H-1)} = 2^{n[R-(1-H)]}.$$

This probability of error goes to zero as $n \to \infty$ as long as

$$R < C := 1 - H,$$

as we wanted to prove.

Summarizing, the above argument hinges on the observation that the channel noise is not arbitrary but instead falls in a relatively small set $\mathbf{S}$. As a result, each codeword produces an output that falls in a set of size $|\mathbf{S}| = 2^{nH}$. Since the outputs have 2^n equally likely values, the probability that the output that corresponds to a codeword j falls in the set that corresponds to another specific codeword i is $2^{nH}/2^n = 2^{n(H-1)}$. Thus, if there are 2^{nR} codewords, the probability that the receiver makes the wrong guess is at most $2^{nR} \times 2^{n(H-1)}$ and it goes to 0 as $n \to \infty$ as long as $R < C = 1 - H$. The outputs are uniformly distributed in $\{0, 1\}^n$ because the codewords are chosen randomly. These ideas extend to more complicated channels.

Summary

- Probability theory quantifies uncertainty. It is useful because of its predictive power.
- The first step in the definition of a random variable is to consider a random experiment and to assign numbers between 0 and 1 to certain sets of possible values of the outcome. The number associated with a set is called its *probability*.
- A random variable is a function of the outcome of a random experiment. One can define the probability that the random variable takes value in a given set by considering the set of outcome values that correspond to that occurrence. (Recall the roulette wheel example: The probability that $X = x_i$ is the probability that the spin selects one of the N_i slots marked x_i.)
- The independence of random variables embodies the intuition that knowing the values of some of them does not give any information about the others.
- The interpretation of the *expected value* of a random variable is the average value that is to be observed over a large number of replications of the random experiment. We learned how to calculate the expected value of (1) a function of a random variable, (2) the sum of random variables, and (3) the product of independent random variables. We stated the strong law of large numbers and the central-limit theorem.
- The *regenerative method* is used for calculating the expected value of random variables that occur frequently in models of communication networks.

Problems

1. *Card game.* One is given a perfectly shuffled deck of 52 cards. Four cards are drawn out of the deck. What is the probability that they are all diamonds? What is the probability that they are four aces? What is the probability that they are four consecutive spades?

2. *Expectation of functions.* Give a derivation of (A.15) without assuming the values $H(x_i)$ are distinct. (*Hint:* Denote all the possible values of $H(X)$ by $\{y_j\}$ and let A_j be the set of indices i such that $H(x_i) = y_j$. Then

$$E\{H(X)\} = \sum_j y_j \sum_{i \in A_j} p_i = \sum_j [\sum_{i \in A_j} p_i H(x_i)] = \sum_i p_i H(x_i).$$

The second equality comes from observing that $y_j = H(x_i)$ when i is in A_j.)

3. *Independence.* Consider the following example. Let X and Y be two independent random variables with $P\{X = 0\} = P\{X = 1\} = P\{Y = 0\} = P\{Y = 1\} = \frac{1}{2}$. Define $Z = X + Y \pmod 2$. (Thus, $Z = 0$ if $X = Y$ and $Z = 1$ otherwise.) Show that $\{X, Y, Z\}$ are pairwise independent but that they are not independent.

4. *Multiple access.* In a multiple-access system, N stations compete for the time slots of a common channel. The slots have duration T. In any given time slot, every station attempts to transmit with probability $p \in (0, 1)$, independently of the others. What is the probability that a single station tries to transmit in a given slot? What is the expected time until a single station transmits?

5. *Reliability.* A physical link consists of N segments. The link is "up" if and only if all the segments are up. Moreover, the segments are up each with probability $p \in (0, 1)$, independently of each other, and are "down" otherwise. What is the probability that the link is up? Consider now two computers that are connected by two independent links. The first link is as described above, and the second one is similar but is made up of M segments that are up, each with probability $q \in (0, 1)$, independently of each other. The connection is working if and only if one of the two links is up. Find the probability that the connection is working. Generalize to an arbitrary topology (a mesh of independent segments).

6. *Branching.* Generalize the branching result (A.22) to an arbitrary number of alternatives. For instance, say that one first tosses a die. The toss yields i with probability p_i and, in that case, one spins a roulette wheel with some real-valued random outcome X_i. Denote the outcome of the randomly chosen wheel by Z. Show that

$$E\{Z\} = \sum_i p_i E\{X_i\}.$$

7. *Random motion.* Consider a random motion of an object on the set $\{1, 2, \ldots, N\}$. The rules of evolution are as follows. If the object is in position $i \in \{1, 2, \ldots, N\}$ at time n, then a die is tossed to determine the position at time $n + 1$. The die tossed in position i is such that the next position is j with probability $P(i, j)$ for $j \in \{1, 2, \ldots, N\}$. (One assumes that $\sum_{j=1}^{N} P(i, j) = 1$ for $1 \le i \le N$.) Denote by $A(i)$ the average time that

it takes the object to reach the position N starting from position i. Explain why it must be that

$$A(i) = 1 + \sum_{j \neq N} P(i, j) A(j).$$

(*Hint:* Denote by X_i the random time to reach N starting from i and explain why $X_i = 1 + X_j$ with probability $P(i, j)$ with $X_N = 0$. Use the ideas explained in the branching result.)

8. *Regenerative method.* A packet has to be sent from node A to node D via nodes B and C. The transmission proceeds as follows. First, A transmits the packet to B. This transmission is always successful and takes T units of time. Second, B sends the packet to C. This is successful with probability $1 - \epsilon$ and takes T time units. If the transmission from B to C is unsuccessful, then B finds out that the transmission is incorrect after $2T$ time units. It then repeats the transmission until the first success. Third, C sends the packet to D. This takes T time units and is successful with probability $1 - \epsilon$. If it is not successful, then A finds out after $3T$ time units, and A must then repeat the whole process. Find the mean time needed until D first gets a successful packet.

References

There are many excellent texts on probability theory. Hoel (1971) is an easy introduction. Bremaud (1988) is a more sophisticated, yet very readable, text. For the material in the complement, read Shannon (1948).

Bibliography

Bremaud, P., *An Introduction to Probabilistic Modeling,* Springer Verlag, 1988.

Hoel, P. G., Port, S. C., and Stone, C. J., *An Introduction to Probability Theory,* Houghton Mifflin Company, 1971.

Shannon, C., "A mathematical theory of communication," *Bell System Technical Journal,* 27, pp. 379–423 and 623–656, 1948.

Queues and Networks of Queues

The objective of this appendix is to provide some details on the queueing theory results used in Chapter 9. We limit our discussion to some of the simplest results of the theory. The reader should consult the references for a more complete exposition.

Queueing theory is the study of models of service systems in which tasks wait to be processed. The objectives of the theory are to predict the *delays* faced by tasks before their processing is completed and also the *backlog* of tasks waiting to be processed. The theory is used to design the telephone network, computer systems, manufacturing plants, and computer networks.

A *queue* is a service facility equipped with a waiting room. Using the standard terminology, *customers* arrive at the queue where they are required to spend some time, called a *service time,* with a *server.* There may be more than one server, and customers may have to wait for an available server. The customers leave the queue once they have received their required amount of service time.

The simplest queueing model is the M/M/1 *queue* defined in Section 9.1. Its analysis is based on the theory of *Markov chains,*, which we explain in Section B.1. We analyze *networks of* M/M/1 *queues* using similar tools in Section B.2. The average delay in many queueing systems can be derived from *Little's result* [formula (9.5)]. In Section B.3 we explain how to derive the average delay in an M/G/1 *queue,* in an M/G/1 *queue with vacations,* and in *priority systems* using Little's result.

B.1 Markov Chains and M/M/1 Queues

Markov chains are a class of models of random evolution. These models are useful for two reasons. First, they are general enough to model many physical systems. Second, the theory of Markov chains provides numerical methods for evaluating the performance measures of these models.

A *Markov chain* is a random process $\mathbf{x} = \{x_t, t \geq 0\}$ that describes a random evolution in a countable set $\mathbf{X}$. The interpretation of this definition is that x_t denotes the *state* of some system at time t. The set $\mathbf{X}$ is called the *state space*. The elements of $\mathbf{X}$ are denoted by $x, y, z, \ldots$ and are called *states*. The Markov chain specifies the random law of motion of x_t in $\mathbf{X}$. Given that $x_t = x$ and the values of $\mathbf{x}$ up to time t, the conditional probability (see Section A.2) that $x_{t+\epsilon} = y$ is equal to

$$P\{x_{t+\epsilon} = y | x_t = x; x_s, s < t\} = \begin{cases} q(x, y)\epsilon + o(\epsilon) \text{ for } y \neq x, \\ 1 - \sum_{z \neq x} q(x, z)\epsilon + o(\epsilon) \text{ for } y = x. \end{cases} \quad \text{(B.1)}$$

In this expression, $o(\epsilon)$ denotes a function of $\epsilon > 0$ such that $o(\epsilon)/\epsilon \to 0$ as $\epsilon \to 0$ and $\{q(x, y), x \neq y \in \mathbf{X}\}$ are given nonnegative real numbers with

$$\sum_{y \neq x} q(x, y) < \infty \text{ for all } x \in \mathbf{X}.$$

Thus, x_t jumps from x to $y \neq x$ with probability $q(x, y)\epsilon$ in $\epsilon \ll 1$ time unit. The numbers $q(x, y)$ are called the *transition rates* of $\mathbf{x}$. Notice that the motion of $\mathbf{x}$ after time t depends on its motion up to time t only through the position at time t. That is, the system modeled by $\mathbf{x}$ is *memoryless:* It forgets how it got to its present value. Equivalently, one can consider that x_t contains all the information about the past evolution that is relevant for predicting the future.

One important question about $\mathbf{x}$ is whether it approaches some statistical equilibrium, or steady state. That is, does the probability $P\{x_t = x\}$ approach a constant value (depending on x but not on t) as $t \to \infty$? This question is of interest because if the answer is affirmative, it tells us the likelihood of finding the system in any given state, and that information can be used to derive measures of performance of the system. For instance, if $\mathbf{x}$ models a communication network, then the limiting probabilities determine how likely it is for the system to be congested. To discuss the evolution of $P\{x_t = x\}$, we first define

$$\pi_t(x) := P\{x_t = x\} \text{ for } x \in \mathbf{X}. \quad \text{(B.2)}$$

Using (B.1), we find that

$$\pi_{t+\epsilon}(y) = \sum_{x \neq y} \pi_t(x)[q(x, y)\epsilon + o(\epsilon)] + \pi_t(y)[1 - \sum_{x \neq y} q(y, x)\epsilon + o(\epsilon)].$$

This equation expresses that $x_{t+\epsilon}$ is equal to y either when x_t is equal to some $x \neq y$ and $\mathbf{x}$ jumps from x to y in ϵ time units or when x_t is equal to y and $\mathbf{x}$ remains equal to y for the next ϵ time units. We can rewrite the above equation by subtracting $\pi_t(y)$ from both sides, dividing by ϵ, and letting ϵ go to zero. This calculation gives

$$\frac{d}{dt}\pi_t(y) = \sum_x \pi_t(x)q(x, y) \quad \text{(B.3)}$$

where we defined

$$q(x, x) := -\sum_{y \neq x} q(x, y).$$

The equation (B.3) describes the evolution of $\pi_t(.)$. Let us attempt to find an *invariant* solution, i.e., a solution that does not depend on t. To do this, we assume that $\pi_t(x) \equiv \pi(x)$ for $x \in \mathbf{X}$. If $\pi_t(x) \equiv \pi(x)$, then (B.3) becomes

$$0 = \sum_x \pi(x)q(x, y). \quad \text{(B.4)}$$

Using the definition of $q(x, x)$, we can also write these equations as

$$\pi(y) \sum_{x \neq y} q(y, x) = \sum_{x \neq y} \pi(x) q(x, y) \text{ for } x \in \mathbf{X}. \tag{B.5}$$

These equations are called the *balance equations*. They express the equality of the rate of transitions leaving y (the left-hand side) with the rate of transitions entering state y from another state. Summarizing, the above calculations show that if the probability distribution of x_t does not depend on t, then it must satisfy (B.5). The system is said to be in *steady state* when the probability distribution of x_t does not depend on t.

Two questions arise at this point. First, how many solutions do the equations (B.5) admit? Second, if the system is not initially in steady state, can one expect it to approach it as time goes by? The answers are as follows. The Markov chain is said to be *irreducible* when it can reach any state from any other state, possibly by making many jumps. If the Markov chain $\mathbf{x}$ is irreducible, then (B.5) admits *at most* one solution. Moreover, if it has one solution π, then $\pi_t(x)$ approaches $\pi(x)$ as $t \to \infty$, for all $x \in \mathbf{X}$.

Let us examine an example to illustrate how we can use these results. Consider the Markov chain defined by $\mathbf{X} = \{0, 1\}$, $q(0, 1) = \lambda$, $q(1, 0) = \mu$ where λ and μ are two positive real numbers. We summarize this definition by the *transition diagram* in Figure B.1. This diagram indicates the possible states and the transition rates. The balance equations (B.5) are

$$\pi(0)\lambda = \pi(1)\mu \qquad \text{and} \qquad \pi(1)\mu = \pi(0)\lambda.$$

These two equations are redundant and are not sufficient to determine the invariant distribution π. We know that π must satisfy the equation

$$\sum_{x \in \mathbf{X}} \pi(x) = 1 \tag{B.6}$$

since π is a probability distribution. Thus, $\pi(0) + \pi(1) = 1$. Using this equation together with the balance equations, we conclude that

$$\pi(0) = \frac{\mu}{\lambda + \mu} \qquad \text{and} \qquad \pi(1) = \frac{\lambda}{\lambda + \mu}.$$

The vector π is the unique invariant probability distribution. We also know that $\pi_t(0) \to \pi(0)$ and that $\pi_t(1) \to \pi(1)$, since the Markov chain is irreducible.

We are now ready to apply our knowledge of Markov chains to the analysis of the M/M/1 queue that we defined at the beginning of Section 9.2. We show that the queue length $\mathbf{x} = \{x_t, t \geq 0\}$ of an M/M/1 queue is a Markov chain. To do this, we have to show that (B.1) holds. Assume that the queue length has been observed up to time $t > 0$ and that $x_t = n > 0$ (see Figure B.2). The probability that, given the observed evolution, $x_{t+\epsilon} = n + 1$ is the probability that the interarrival time in progress at time t completes in $(t, t + \epsilon)$ and that the service does not complete during the same interval. By observing the past evolution, one knows when the interarrival time started and also when the service started. Since these random variables are exponentially distributed, they are memoryless (see Section A.1). That is, given the past evolution of the queue length, the probability that the interarrival time terminates in the next ϵ time units is equal to $\lambda\epsilon + o(\epsilon)$. Also, the

FIGURE B.1

Transition diagram.

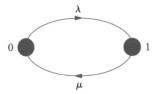

FIGURE B.2

Evolution of M/M/1 *queue length.*

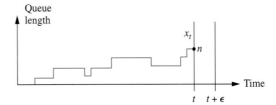

FIGURE B.3

Transition diagram of M/M/1 *queue.*

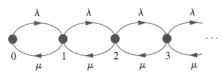

probability that the service time in progress at time t terminates in the next ϵ time units is equal to $\mu\epsilon + o(\epsilon)$.

Consequently, the probability that $x_{t+\epsilon} = n + 1$ given the evolution up to time t is equal to

$$\{\lambda\epsilon + o(\epsilon)\} \times \{1 - \mu\epsilon - o(\epsilon)\} = \lambda\epsilon + o(\epsilon).$$

(We used the fact that terms of the form $o(\epsilon)o(\epsilon)$, $o(\epsilon) + o(\epsilon)$, $o(\epsilon)\epsilon$, and ϵ^2 are all $o(\epsilon)$.) We write the above result as

$$P[x_{t+\epsilon} = n + 1 | x_t = n; x_s, s < t] = \lambda\epsilon + o(\epsilon). \tag{B.7}$$

Similar arguments (see Problem 2) enable us to show that:

$$P[x_{t+\epsilon} = n - 1 | x_t = n; x_s, s < t] = \begin{cases} \mu\epsilon + o(\epsilon) \text{ if } n > 0 \\ 0 \text{ if } n = 0 \end{cases} \tag{B.8}$$

This shows that **x** is a Markov chain with the following transition rates:

$$q(n, n + 1) = \lambda \text{ and } q(n + 1, n) = \mu \text{ for } n \geq 0$$
$$q(m, n) = 0 \text{ for } |m - n| > 1. \tag{B.9}$$

These relations are summarized by the transition diagram shown in Figure B.3.

The M/M/1 queue is an irreducible Markov chain since the queue length can reach any value from any other. Thus, if the balance equations admit a solution, then that solution is unique and is the limiting distribution of the queue length. The balance equations (B.5) become [with (B.9)]

$$\pi(0)\lambda = \pi(1)\mu$$
$$\pi(1)\{\lambda + \mu\} = \pi(0)\lambda + \pi(2)\mu$$
$$\pi(2)\{\lambda + \mu\} = \pi(1)\lambda + \pi(3)\mu$$

and so on. To solve these equations, subtract $\pi(1)\mu = \pi(0)\lambda$ (obtained from the first equation) from the second equation. The result is $\pi(1)\lambda = \pi(2)\mu$. Subtracting $\pi(2)\mu = \pi(1)\lambda$ from the third equation then gives $\pi(2)\lambda = \pi(3)\mu$. Continuing in this way shows that

$$\pi(n)\lambda = \pi(n+1)\mu \text{ for } n \geq 0. \tag{B.10}$$

These equations imply that

$$\pi(n+1) = \rho\pi(n) \text{ for } n \geq 0 \text{ with } \rho := \frac{\lambda}{\mu}.$$

Hence, $\pi(n) = \rho^n \pi(0)$. This formula shows that the balance equations admit a solution if and only if $\rho < 1$, i.e., $\lambda < \mu$. In that case, choosing $\pi(0)$ so that the numbers $\{\pi(n), n \geq 0\}$ sum to 1 gives

$$\pi(n) = (1-\rho)\rho^n \text{ for } n \geq 0. \tag{B.11}$$

Note that the equations (B.10) admit a simple interpretation. They express the equality of the rate of transitions of x_t from $A := \{0, 1, \ldots, n\}$ to $A^c := \{n+1, n+2, \ldots\}$ and the rate of transitions of x_t from A^c to A in equilibrium. (See Problem 3.)

B.2   Networks of M/M/1 Queues

Our objective in this section is to explain the formula for the average delay per customer in a network of M/M/1 queues. We consider the network shown in Figure B.4. The circles represent queues. Customers arrive from outside as independent Poisson processes, and they follow a random path in the network. There are J queues. For $i = 1, \ldots, J$, queue i has exponential service times with rate μ_i. When a customer leaves queue i, it is sent to queue j with probability r_{ij} for $j = 1, \ldots, J$ and it leaves the network otherwise, i.e., with

FIGURE B.4

Model of a network of queues.

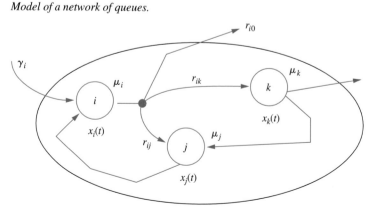

probability

$$r_{i0} := 1 - \sum_{j=1}^{J} r_{ij}.$$

Also, customers arrive from outside as independent Poisson processes, with rate γ_i into queue i for $i = 1, 2, \ldots, J$.

Denote by λ_i the average rate of flow of customers through queue i, assuming that the network is in steady state. The claim is that these rates must solve the following *flow conservation equations:*

$$\lambda_i = \gamma_i + \sum_{j=1}^{J} \lambda_j r_{ji} \text{ for } i = 1, \ldots, J. \qquad (B.12)$$

To see why the equations (B.12) hold, note that the traffic through queue i is composed of the traffic entering the queue from outside (with rate γ_i) and of the traffic coming from the other queue. For $j \neq i$, node j sends to queue i a fraction r_{ji} of the rate λ_j of the traffic that goes through it. Hence, (B.12).

The network is said to be *open* if every customer eventually leaves. For such networks, it can be shown that the equations (B.12) admit a unique solution for every given vector $(\gamma_1, \ldots, \gamma_J)$. (For closed networks, see Problem 6.)

For $t \geq 0$, we define $x_t := (x_t^1, \ldots, x_t^J)$ where x_t^i is the number of customers in queue i at time t, including the one being served, if any. The process $\mathbf{x} = \{x_t, t \geq 0\}$ is a Markov chain because the interarrival times and the service times are memoryless. The transition rates are easily computed (see Figure B.5). For instance, let $y = (x^1, \ldots, x^{i-1}, x^i + 1, x^{i+1}, \ldots, x^J)$. That is, y is obtained from the vector x by adding one customer in queue i. Thus, $y = x + e_i$ where $e_i := (0, \ldots, 1, \ldots, 0)$ is the unit vector in direction i in $\{0, 1, 2, \ldots\}^J$. A transition from x to y occurs when a customer arrives in queue i from outside. Given the past evolution of $\mathbf{x}$ up to time t, this external arrival into queue i occurs with rate γ_i. Hence,

$$q(x, x + e_i) = \gamma_i \text{ for } i = 1, 2, \ldots, J.$$

As another example, let $y = x - e_i + e_j$ where we assume that $x^i > 0$ and $i \neq j$. A transition from x to y corresponds to a customer finishing service at node i (with rate μ_i) and being sent to queue j upon leaving queue i (which occurs with probability r_{ij}). Thus,

FIGURE B.5

Transition rates for network of M/M/1 *queues.*

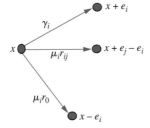

the probability that the transition from x to y occurs in ϵ time units is equal to $\{\mu_i \epsilon + o(\epsilon)\} r_{ij}$. Hence,

$$q(x, x - e_i + e_j) = \mu_i r_{ij} \text{ whenever } x^i > 0.$$

Similarly, one finds that

$$q(x, x - e_i) = \mu_i r_{i0} \text{ whenever } x^i > 0.$$

We want to calculate the invariant distribution of **x.** To do this, we denote by $(\lambda_1, \ldots, \lambda_J)$ a solution of (B.12). Assume that

$$\rho_i := \frac{\lambda_i}{\mu_i} < 1 \text{ for } i = 1, 2, \ldots, J. \tag{B.13}$$

We claim that a solution of the balance equations of **x** is given by

$$\pi(x^1, \ldots, x^J) = \Pi_{i=1}^{J} \{(1 - \rho_i) \rho_i^{x^i}\} \text{ for } (x^1, \ldots, x^J) \in \{0, 1, 2, \ldots\}^J. \tag{B.14}$$

The most elegant known proof of (B.14) is as follows. (You may want to skip this proof on a first reading.) Define, for $1 \le i \ne j \le J$,

$$\gamma_i' := \lambda_i r_{i0}, \, r_{ij}' := \frac{\lambda_j r_{ji}}{\lambda_i}, \text{ and } r_{i0}' = \frac{\gamma_i}{\lambda_i}. \tag{B.15}$$

Consider the network with J nodes defined as before except that the parameters γ_i, r_{ij}, and r_{i0} are replaced by γ_i', r_{ij}', and r_{i0}', respectively. Denote by $q'(x, y)$ the transition rates of this new network. These transition rates are obtained from the original rates by making the suitable parameter substitutions. For instance,

$$q'(x, x - e_i + e_j) = \mu_i r_{ij}' \text{ whenever } x^i > 0.$$

One can then verify that

$$\pi(x) q(x, y) = \pi(y) q'(y, x) \text{ for all } x \ne y. \tag{B.16}$$

As an example of verification of (B.16), let $x^i > 0$ and $y = x - e_i + e_j$. Then,

$$q(x, y) = \mu_i r_{ij} \quad \text{and} \quad q'(y, x) = \mu_j r_{ji}' = \frac{\mu_j \lambda_i r_{ji}}{\lambda_j}$$

where the expression for $q'(y, x)$ is obtained by observing that the transition from y to x occurs when a customer is served in node j and is sent to node i. Thus, for this choice of x and y, (B.16) becomes

$$\pi(x) \mu_i r_{ij} = \pi(y) \frac{\mu_j \lambda_i r_{ij}}{\lambda_j}. \tag{B.17}$$

This equality is verified by using (B.14), which implies that

$$\pi(y) = \pi(x) \frac{\rho_j}{\rho_i} = \pi(x) \frac{\lambda_j \mu_i}{\mu_j \lambda_i},$$

which yields (B.17). The other possible pairs (x, y) in (B.16) are verified similarly. Moreover, the new network is such that

$$\sum_{y \ne x} q(x, y) = \sum_{y \ne x} q'(x, y). \tag{B.18}$$

To see this, observe that the left-hand side of (B.18) is the rate of transitions out of state x in the original network while the right-hand side is that rate in the new network. Therefore,

$$\sum_{y \neq x} q(x, y) = \sum_i \gamma_i + \sum_{\{i : x^i > 0\}} \mu_i \qquad \text{and} \qquad \sum_{y \neq x} q'(x, y) = \sum_i \gamma_i' + \sum_{\{i : x^i > 0\}} \mu_i.$$

Using (B.12) you can verify that $\sum_i \gamma_i = \sum_i \gamma_i'$, so that (B.18) holds. The balance equations are immediate consequences of (B.16) and (B.18). To see this, sum (B.16) over $x \neq y$. This gives

$$\sum_{x \neq y} \pi(x) q(x, y) = \pi(y) \sum_{x \neq y} q'(y, x) \tag{B.19}$$

$$= \pi(y) \sum_{x \neq y} q(y, x) \tag{B.20}$$

where the last equality follows from (B.18). The equations (B.19) are the balance equations (B.5). It turns out that by proving (B.16) and (B.18), we also proved that the new network behaves as the original network *reversed in time*.

We finally come to our goal: deriving the formula for the average delay per customer in the network. The key to the derivation is *Little's result, $L = \lambda T$*, which we discussed in Chapter 9. Recall that this result states that the average number of customers in a queueing system is equal to the arrival rate multiplied by the average time spent by each customer in the system.

To calculate the average delay T per customer in the network, we first calculate the average number L of customers in the network and the arrival rate λ. The arrival rate λ is equal to the sum of the arrival rates γ_i. Thus,

$$\lambda = \gamma := \sum_i \gamma_i.$$

The average number of customers in the network is equal to the sum of the average number of customers in the J queues. That is,

$$L = \sum_{i=1}^{J} L_i$$

where L_i is the average number of customers in queue i. To calculate L_i, we use (B.14), which shows that, in steady state, the queue length at node i is distributed as the queue length in an M/M/1 queue with arrival rate λ_i and service rate μ_i. Thus, L_i is equal to the average queue length of an M/M/1 queue with parameters λ_i and μ_i. Using (9.4), the formula for the average delay in an M/M/1 queue, we obtain

$$L_i = \frac{\lambda_i}{\mu_i - \lambda_i}.$$

Finally, we find

$$T = \frac{1}{\gamma} \sum_{i=1}^{J} L_i = \frac{1}{\gamma} \sum_{i=1}^{J} \frac{\lambda_i}{\mu_i - \lambda_i},$$

which is the desired formula.

B.3 Average Delays

This section explains the formulas for the average delay in an M/G/1 queue. The derivations are based on Little's result. Consider first an M/G/1 queue and assume that every customer in the queue pays at rate R when his or her remaining service time is equal to R. A graph of this cost rate is illustrated in Figure B.6. The total cost paid by this customer is equal to the integral of the rate over time, i.e., to

$$SQ + \frac{S^2}{2} \tag{B.21}$$

where S denotes the service time of the customer and Q his or her queueing delay before the service actually starts. The service time S of a customer and his or her queueing time Q are independent. As a consequence, the expected cost paid by each customer, the average value of (B.21), is equal to

$$C := \frac{E\{Q\}}{\mu} + \frac{E\{S^2\}}{2} \tag{B.22}$$

where we used $E\{SQ\} = E\{S\}E\{Q\} = \mu^{-1}E\{Q\}$, because of the independence of Q and S. It follows that the customers pay at rate λC since each customer pays C on the average and λ customers go through the queue per unit of time.

At a given time t, the customers pay at a rate equal to the sum of the remaining service times of all the customers in the queue. The queue being first come–first served, this sum is equal to the queueing time of a customer who would enter the queue at time t. Let us denote this "virtual" queueing time by Q^*.

We show below that Q^* has the same distribution as Q. As a consequence, the average rate at which customers pay is equal to $E\{Q\}$. We just showed that this average rate is also equal to λC. Consequently,

$$E\{Q\} = \lambda C = \lambda \left(\frac{E\{Q\}}{\mu} + \frac{E\{S^2\}}{2} \right).$$

Solving this equation for $E\{Q\}$ yields

$$W := E\{Q\} = \frac{\lambda E\{S^2\}}{2(1 - \rho)},$$

and this result enables us to calculate the average total delay T since $T = W + \mu^{-1}$.

It remains to show that Q^* is distributed as Q. That is, we want to show that the waiting time of a customer who would enter the queue in steady state at time t is distributed as the

FIGURE B.6

Cost rate for a typical customer.

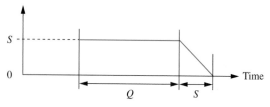

waiting time of a typical customer. This equality is due to a stronger result which states that *Poisson arrivals see time averages* (PASTA). This statement means that customers who arrive as a Poisson process at a queue in steady state find the queue with its invariant distribution. To show PASTA, we denote the state of the queue by x_t and we assume that the queue is in steady state with $P\{x_t = x\} = \pi(x)$. Denote the arrival times of the customers by $\{T_n, n \geq 1\}$ and assume that these arrival times form a Poisson process with rate λ. We want to show that, given that there is an arrival in $(t, t + \epsilon)$, x_t has the distribution π. This shows that the distribution just before an arrival is the invariant distribution. Intuitively, this result is plausible because the state of the queue does not affect the arrival rate. To derive the result formally, we must calculate

$$P[x_t = x | T_n \in (t, t + \epsilon) \text{ for some } n]. \tag{B.23}$$

By the definition of conditional probability (A.6), the quantity (B.23) is equal to

$$\frac{P\{x_t = x \text{ and } T_n \in (t, t + \epsilon) \text{ for some } n\}}{P\{T_n \in (t, t + \epsilon) \text{ for some } n\}}. \tag{B.24}$$

The numerator of this fraction is equal to

$$P\{x_t = x\}P\{T_n \in (t, t + \epsilon) \text{ for some } n\}$$

since the Poisson process is memoryless and the occurrence of an arrival in $(t, t + \epsilon)$ is therefore independent of whatever happened before time t, and in particular of x_t. The preceding expression implies that

$$P[x_t = x | T_n \in (t, t + \epsilon) \text{ for some } n] = \pi(x),$$

as was to be shown.

We now turn our attention to an M/G/1 queue with vacations. We prove formula (9.13) for the average delay through the queue. Assume that the server pays the queue at rate R when that server is on vacation with a residual vacation time equal to R. The total cost paid during a vacation with duration V is then equal to $E\{V^2\}/2$. We show below that the rate of vacations is equal to

$$\frac{1 - \rho}{E\{V\}}. \tag{B.25}$$

Consequently, the server pays with the average rate

$$(1 - \rho)\frac{E\{V^2\}}{2E\{V\}}.$$

Assume now that a customer with a residual service time equal to R pays with rate R. The average total cost paid by a customer is then equal to

$$E\{QS\} + \frac{E\{S^2\}}{2},$$

as in the M/G/1 queue. As a consequence, the average rate at which the customers pay is equal to

$$\lambda\left[E\{QS\} + \frac{E\{S^2\}}{2}\right].$$

It follows that the server and the customers together pay with the total rate

$$\lambda[E\{QS\} + \frac{E\{S^2\}}{2}] + (1-\rho)\frac{E\{V^2\}}{2E\{V\}}.$$

Now, the rate at which the server and customers pay at any given time t is the sum of the residual vacation time and the remaining waiting times of all the customers in the queue. This sum is equal to the waiting time of a customer who would enter the queue at time t. Since Poisson arrivals see time averages, it follows that the average value of this "virtual waiting time" is equal to $E\{Q\}$, as in the analysis of the M/G/1 queue. Hence,

$$W = E\{Q\} = \lambda\left[E\{QS\} + \frac{E\{S^2\}}{2}\right] + (1-\rho)\frac{E\{V^2\}}{2E\{V\}}.$$

Using $E\{QS\} = E\{Q\}/\mu$ and solving for $E\{Q\}$ leads to

$$E\{Q\} = \frac{1}{1-\rho}\frac{\lambda E\{S^2\}}{2} + \frac{E\{S^2\}}{2},$$

which implies, since $T = E\{Q\} + \mu^{-1}$,

$$T = T_0 + \frac{E\{V^2\}}{2E\{V\}} = T_0 + \frac{v}{2}E\{V^2\}.$$

This last formula is precisely (9.13).

It remains to show that the rate of vacations is given by (B.25). To derive that result, we assume that the server pays at a unit rate when serving customers. The server then pays at a rate equal to the probability that it is serving. That rate can be computed as the product of the average cost per customer paid by the server times the rate at which customers are served. This computation shows that the probability that the server is serving must be equal to $\rho = \lambda/\mu$. Thus, the server is on vacation a fraction $1-\rho$ of the time. Say that the vacation rate is equal to α. Since the average duration of a vacation is equal to $E\{V\}$, it follows that the average time spent on vacation per unit of time is equal to $\alpha E\{V\}$. Thus,

$$1 - \rho = \alpha E\{V\},$$

which shows that α is indeed given by (B.25).

We conclude our discussion of average delays by deriving (9.14), the average delays through a queue with priorities. The first step is to show that if V denotes the residual service time of a customer in service at an arbitrary time, then

$$E\{V\} = \frac{1}{2}\sum_{i=1}^{2}\lambda_i E\{S_i^2\}. \tag{B.26}$$

To show (B.26), we assume that a customer pays at rate R when the customer is in service with a residual service time R. The total cost paid by a customer with service time S is then equal to $S^2/2$. Now, S is equal to S_1 with probability $\lambda_1(\lambda_1 + \lambda_2)^{-1}$ and to S_2 with probability $\lambda_2(\lambda_1 + \lambda_2)^{-1}$. As a consequence, the expected cost per customer is given by the following expression [see (A.22)]:

$$(\lambda_1 + \lambda_2)^{-1}\sum_{i=1}^{2}\frac{\lambda_i E\{S_i^2\}}{2}.$$

Multiplying this expression by the rate $(\lambda_1 + \lambda_2)$ of customers gives the rate at which customers pay. But this rate must be equal to $E\{V\}$, by definition of V. Hence, (B.26).

Consider now a customer of class 1 and denote her waiting time (before service) by Q_1. Then

$$E\{Q_1\} = E\{V\} + E\{N_1\}E\{S_1\}$$

where N_1 denotes the typical number of customers of class 1 in the queue (not in service). Indeed, the customer of class 1 must wait for the residual service time V and for the completion of the service times of all the customers of class 1 upon her arrival. Little's result also gives

$$E\{N_1\} = \lambda_1 E\{Q_1\}.$$

Hence,

$$W_1 := E\{Q_1\} = E\{V\} + \rho_1 E\{Q_1\},$$

so that

$$W_1 = \frac{E\{V\}}{1 - \rho_1},$$

which is the first formula in (9.14).

The case of class 2 is similar. One finds that

$$E\{Q_2\} = E\{V\} + E\{N_1\}E\{S_1\} + E\{N_2\}E\{S_2\} + \lambda_1 E\{Q_2\}E\{S_1\} \qquad \text{(B.27)}$$

where Q_2 and N_2 were defined analogously to Q_1 and N_1. To understand (B.27), notice that if one calls C the customer of class 2, then C must wait for V time units, then for the completion of the service of the N_1 class 1 customers in the queue when C arrived, then for the completion of the N_2 customers of class 2 when C arrived, and then for the service of the $\lambda_1 E\{Q_2\}$ customers of class 1 who arrived during the waiting time Q_2 of C in the queue. Combining (B.27), the value of $E\{Q_1\}$ and $E\{N_2\} = \lambda_2 E\{Q_2\}$ gives the second formula in (9.14) for $W_2 := E\{Q_2\}$.

Summary

- A *Markov chain* $\mathbf{x} = \{x_t, t \geq 0\}$ is a model of random evolution in a countable set $\mathbf{X}$. The process $\mathbf{x}$ jumps from x to $y \neq x$ at rate $q(x, y)$, independently of how it got to x. If $\mathbf{x}$ is *irreducible* and if the *balance equations* admit a solution π (the *unique* invariant distribution of $\mathbf{x}$), then $P\{x_t = x\}$ converges to $\pi(x)$ as t increases. Thus, $\mathbf{x}$ approaches *steady state*.
- The length of an M/M/1 queue is an irreducible Markov chain. It admits an invariant distribution if and only if the arrival rate is less than the service rate. From the invariant distribution, one can calculate the average queue length and the average delay per customer is then obtained from *Little's result*.
- The vector of queue lengths in an *open network of M/M/1 queues* is also an irreducible Markov chain. Its invariant distribution exists if and only if the *flow conservation equations* show that the average rate of flow through each queue is less than the service

rate of the queue. Moreover, in that case, at any given time, the queue lengths of the network in steady state are independent and are distributed as M/M/1 queues with arrival rates given by the flow conservation equations. (This property holds even though one can show that the arrival processes at the queues in the network are generally not Poisson.) From this result, one can calculate the average number of customers in the network, and the average delay is then derived from Little's result.

· The average delay through an M/G/1 queue with or without vacations or priorities can be derived by using Little's result.

Problems

1. *M/M/1/N queue.* Consider a queue defined as the M/M/1 queue except that it can hold at most N customers (including the one in service). Customers arrive as a Poisson process with rate λ. Customers who arrive when there are N customers in the queue are rejected and never come back. This type of queue is called an M/M/1/N queue. Show that the queue length x_t is a Markov chain with the transition diagram indicated in Figure B.7. Write and solve the balance equations.

2. *Exponential distribution.* Let σ and τ be two independent exponentially distributed random variables with rates λ and μ, respectively. Show that $\min\{\sigma, \tau\}$ is exponentially distributed with rate $\lambda + \mu$. Show that

$$P\{\sigma \leq \tau\} = \frac{\lambda}{\lambda + \mu}.$$

3. *Balance equations.* The point of this exercise is to elaborate on the simplification that occurs in the solution of the balance equations of the M/M/1 queue. It turns out, as we explain below, that the same simplification occurs for more general systems. Let $\mathbf{x} = \{x_t, t \geq 0\}$ be a Markov chain with state space $\mathbf{X}$ and with transition rates $\{q(x, y), x \neq y \in \mathbf{X}\}$. Assume that $\{\pi(x), x \in \mathbf{X}\}$ is an invariant distribution of $\mathbf{x}$. Show that if A is a subset of $\mathbf{X}$ and if $A^c := \mathbf{X} - A$ denotes the complement of A, then

$$\sum_{x \in A} \sum_{y \in A^c} \pi(x)q(x, y) = \sum_{x \in A^c} \sum_{y \in A} \pi(x)q(x, y).$$

The interpretation of this identity is that the rate of jumps leaving A must be equal to the rate of jumps entering A when $\mathbf{x}$ is in steady state. [*Hint:* Use the balance equations (B.5).]

4. *M/M/s queue.* Consider a queue with Poisson arrivals with rate λ and with exponentially distributed service times with rate μ. The queue has $s \geq 1$ servers.

FIGURE B.7

Transition diagram of M/M/1/N *queue.*

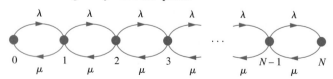

Thus, when there are n customers in the queue, $\min\{n, s\}$ of them are being served and the time until the next service completion is, in view of Problem 2, exponentially distributed with rate $\mu \min\{n, s\}$. Conclude that the queue length x_t is a Markov chain with rates

$$q(n, n + 1) = \lambda \quad \text{and} \quad q(n + 1, n) = \mu \min\{n + 1, s\} \text{ for } n \geq 0.$$

Write and solve the balance equations.

5. *Transient behavior.* In matrix notation, with π_t being the row vector with components $\pi_t(y)$, the equations (B.3) read

$$\frac{d}{dt}\pi_t = \pi_t Q$$

where Q is the matrix whose entry (x, y) is equal to $q(x, y)$. Formally at least, the solution of this equation is

$$\pi_t = \pi_0 e^{Qt} \text{ where } e^{Qt} := \sum_{n=0}^{\infty} \frac{(Qt)^n}{n!}.$$

This solution turns out to be correct when the entries of Q are bounded. Apply these results to the example of Figure B.1. Show that

$$\pi_t(0) = \pi_0(0)\frac{\mu + \lambda e^{-(\lambda+\mu)t}}{\lambda + \mu} + \pi_0(1)\frac{\mu - \mu e^{-(\lambda+\mu)t}}{\lambda + \mu} \text{ for } t \geq 0.$$

Verify that $\pi_t(0) \to \mu(\lambda + \mu)^{-1}$.

6. *Closed network of M/M/1 queues.* Consider the network of M/M/1 queues of Section B.3 but assume that it is *closed*. That is, no customer can leave and no customer can enter. Thus, $r_{i0} = 0$ and $\gamma_i = 0$ for all i. Show that (B.14) is again an invariant distribution with the λ_i's given as a solution of (B.12). (*Hint:* The proof given in the open case still applies without modification.)

References

Elements of queueing theory are introduced in Bertsekas (1992). Kelly (1979) is an excellent reference for product-form networks. Kleinrock (1975) is also highly recommended. Walrand (1988) explains analysis, design, and control methods for queueing networks.

Bibliography

Bertsekas, D., and Gallager, R., *Data Networks,* Prentice-Hall, 1992.
Kelly, F. P., *Reversibility and Stochastic Networks,* John Wiley & Sons, 1979.
Kleinrock, L., *Queueing Systems,* vol. 1, John Wiley & Sons, 1975.
Walrand, J., *An Introduction to Queueing Networks,* Prentice-Hall, 1988.

Communication Principles

This appendix is intended for readers who do not have a background in communication theory. It provides some details about the basic ideas of the theory.

In Section C.1, we explain the concept of *frequency spectrum*. Section C.2 discusses some modulation and demodulation techniques. Section C.3 explains the phase-locked loop used in demodulation and for synchronizing receivers. *Nyquist's sampling theorem* is discussed in Section C.4.

C.1 Frequency Spectrum

Telecommunication engineers usually think about signals in terms of their *frequency spectrum*. Let $x(.) = \{x(t), -\infty < t < +\infty\}$ be a function of t. The function takes real or complex values. In the early 1800s, the French mathematician Jean Baptiste Fourier suggested that almost every such function $x(.)$ can be written as

$$x(t) = \int_{-\infty}^{+\infty} X(f)e^{j2\pi ft}\, df. \tag{C.1}$$

The right-hand side of (C.1) expresses $x(.)$ as a sum of *complex exponentials* $e^{j2\pi ft}$. Recall that $e^{j2\pi ft} = \cos(2\pi ft) + j\sin(2\pi ft)$. Thus, (C.1) expresses $x(.)$ as a sum of sine waves. The expression (C.1) shows that $x(.)$ can be described by specifying the coefficient $X(f)$ of the complex exponential $e^{j2\pi ft}$ for each f. The values $\{X(f), -\infty < f < +\infty\}$ constitute the *frequency spectrum* (or *Fourier transform*) of $x(.)$. The two equivalent representations of $x(.)$, by specifying the value of $x(t)$ for each t or the value of $X(f)$ for each f, are illustrated in Figure C.1.

In some cases, the integral in (C.1) gets replaced by a sum. For instance, if $x(.)$ is *periodic* with period T, i.e., if $x(t) = x(t+T)$ for all t, then

$$x(t) = \sum_{n=-\infty}^{\infty} c_n e^{j(2\pi n/T)t}, \text{ with } c_n = \frac{1}{T}\int_{-T/2}^{+T/2} x(t)e^{-j(2\pi n/T)t}\, dt. \tag{C.2}$$

FIGURE C.1

The function x(.) and its frequency spectrum.

That is, a periodic function with period T can be written as a sum of sine waves whose frequencies are multiples of $1/T$.

Let us examine a few examples of this decomposition of functions into sine waves. First, consider $x(t) = \cos(2\pi f_1 t)$. We know the identity $e^{j\theta} = cos(\theta) + j\sin(\theta)$. Using this identity, we derive $\cos(\theta) = (e^{j\theta})/2 + (e^{-j\theta})/2$ and, therefore,

$$\cos(2\pi f_0 t) = \frac{1}{2}e^{j2\pi f_0 t} + \frac{1}{2}e^{-j2\pi f_0 t}. \tag{C.3}$$

This expression is a particular case of (C.2). Another example, which we used to derive Figure 7.7, is the decomposition of the *square wave*

$$q(t) = \begin{cases} 1, & \text{if } \dfrac{4n-1}{4}T \le t < \dfrac{4n+1}{4}T \text{ for some integer } n \\ 0, & \text{otherwise} \end{cases}$$

into sine waves as

$$q(t) = \frac{1}{2} + \frac{2}{\pi}\cos\left(\frac{2\pi}{T}t\right) - \frac{2}{3\pi}\cos\left(\frac{6\pi}{T}t\right) + \frac{2}{5\pi}\cos\left(\frac{10\pi}{T}t\right) - \cdots.$$

The coefficients $1/2, 2/\pi, -2/3\pi, \ldots$ are calculated by using the expressions given in (C.2) for the coefficients c_n. Another useful example is the identity

$$\sum_{n=-\infty}^{\infty} \delta(t - nT) = \frac{1}{T}\sum_{n=-\infty}^{\infty} e^{j(2\pi n/T)t}. \tag{C.4}$$

In this identity, $\delta(t)$ denotes the *Dirac impulse,* which is defined by

$$\int_{-\infty}^{\infty} x(t)\delta(t - s)\,dt = x(s) \text{ for } -\infty < s < +\infty \tag{C.5}$$

whenever the function $x(.)$ is continuous at s. That is, the Dirac impulse *samples* the signal $x(.)$. The Dirac impulse is the mathematical idealization when $\epsilon \downarrow 0$ of the function $\delta_\epsilon(.)$ defined by

$$\delta_\epsilon(t) = \begin{cases} 1/\epsilon \text{ for } -\epsilon/2 < t < +\epsilon/2 \\ 0, \text{ otherwise} \end{cases} \tag{C.6}$$

From this definition you see that

$$\int_{-\infty}^{\infty} x(t)\delta_\epsilon(t - s)\,dt = \frac{1}{\epsilon}\int_{s-\epsilon/2}^{s+\epsilon/2} x(t)\,dt$$

is the average value of $x(.)$ in the interval $(s - \epsilon/2, s + \epsilon/2)$. As $\epsilon \downarrow 0$, this average value approaches $x(s)$ when the function $x(.)$ is continuous at s. This observation justifies the definition (C.5). To understand the identity (C.4), note that the left-hand side defines a periodic function

$$g(t) := \sum_{n=-\infty}^{\infty} \delta(t - nT)$$

with period T. Consequently, we can use the formula (C.2) to write $g(t)$ as a sum of complex exponentials. Using the formula given in (C.2) for the coefficients c_n, we find

$$c_n = \frac{1}{T} \int_{-T/2}^{+T/2} g(t)e^{-j(2\pi n/T)t} \, dt.$$

Now, for t in the interval $[-T/2, +T/2]$ of integration, $g(t) = \delta(t)$ since the other terms vanish. Indeed, $\delta_\epsilon(t - s)$ is nonzero only in a small interval around s, and the same is true for $\delta(t)$. Therefore, all the terms in $g(t)$ are equal to zero in $[-T/2, +T/2]$, except the term $\delta(t)$. Consequently, we get

$$c_n = \frac{1}{T} \int_{-T/2}^{+T/2} \delta(t)e^{-j(2\pi n/T)t} \, dt = \frac{1}{T},$$

where the last equality follows from (C.5) and from $e^0 = 1$. This proves (C.4).

Thus, Fourier tells us that essentially any function $x(.)$ is a sum of sine waves. The restriction *essentially* refers to some mathematical conditions that are satisfied by all the functions that we consider in this text.

C.2 Modulation and Demodulation

Modulation is a technique central to communication systems. One form of modulation, called *amplitude modulation*, is explained next. First, consider the signal

$$x(t) = Ae^{j2\pi f_1 t}, \, t \in (-\infty, +\infty).$$

Since this signal is a single complex exponential with frequency f_1, we can represent its spectrum as shown in the left part of Figure C.2.

Now, consider the new signal $y(t) = x(t)\cos\{2\pi f_0 t\}$ obtained by multiplying the original signal $x(t)$ by a sine wave. Using (C.3), we find that

$$y(t) = \frac{A}{2}e^{j2\pi(f_1+f_0)t} + \frac{A}{2}e^{j2\pi(f_1-f_0)t}.$$

FIGURE C.2

Left: frequency spectrum of $x(t) = Ae^{j2\pi f_1 t}$; right: frequency spectrum of $x(t)\cos\{2\pi f_0 t\}$.

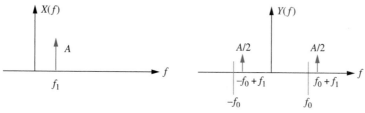

Assuming that $f_0 > f_1$, one sees that $y(t)$ now consists of two complex exponentials: one with the frequency $f_0 + f_1$ and the other with the frequency $f_0 - f_1$. The spectrum of $y(t)$ is shown in the right part of Figure C.2.

Consider now a signal $x(t)$ that is the sum of complex exponentials with frequencies in the range $[-f_1, f_2]$. Arguing as in our previous example, we conclude that $y(t) = x(t) \cos\{2\pi f_0 t\}$ is now a sum of sine waves at frequencies in the range $[f_0 - f_1, f_0 + f_1]$ and in $[-f_0 - f_1, -f_0 + f_1]$. In fact, the amplitudes of the various frequencies are half of what they were in the original signal. Figure C.3 illustrates the transformation of the spectrum of a signal when it is multiplied by a sine wave.

The operation of multiplying a signal by a sine wave is called the *amplitude modulation* of the sine wave by the signal. Thus, we have seen that amplitude modulation provides a simple way of shifting the spectrum of a signal.

Some radio transmitters use amplitude modulation. The laws of electromagnetic theory are such that a radio antenna is efficient only when its length is of the order of the wavelength of the signal being transmitted. The wavelength of a signal is equal to the speed of light $(3 \times 10^8$ meters per second) divided by the frequency of the signal. For instance, a voice signal has a typical frequency of 1000 Hz. This corresponds to an electrical signal with a wavelength equal to $3 \times 10^8/10^3 = 300$ km. Thus, without modulation, an antenna would have to be around 100 kilometers long to be efficient. This is clearly not feasible. If we modulate a sine wave at 100 MHz with the voice signal, the signal to be transmitted has a wavelength equal to $3 \times 10^8/10^8 = 3$ meters, for which a short antenna is efficient.

A few different methods can be used to *demodulate* a signal. We explain next *synchronous demodulation*. The signals are the same as in Figure C.3. That is, $x(t)$ is a sum of sine waves with frequencies in the range $[-f_1, f_1]$ and $y(t) = x(t) \cos\{2\pi f_0 t\}$. We assume that $0 < f_1 < f_0$. The objective of demodulation is to recover $x(t)$ from $y(t)$. We first compute

$$z(t) = y(t) \cos\{2\pi f_0 t\}.$$

We then *filter* the signal $z(t)$ to eliminate all the complex exponentials with frequencies not in $[-f_1, f_1]$. A filter is a device that amplifies different frequencies differently. We use a filter that does not amplify at all the frequencies outside the range $[-f_1, f_1]$ but that does not modify the frequencies within that range. Denote by $v(t)$ the result of this filtering. The claim is that $v(t) = x(t)/2$. The proof of this claim is as follows. From the definitions of

FIGURE C.3

Modification of frequency spectrum by modulation.

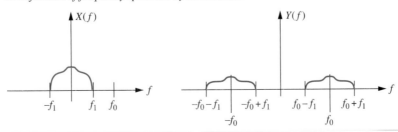

$y(t)$ and $z(t)$, we get

$$z(t) = x(t)[\cos\{2\pi f_0 t\}]^2 = x(t)\frac{1 + \cos\{4\pi f_0 t\}}{2} \tag{C.7}$$

$$= \frac{x(t)}{2} + \frac{x(t)}{2}\cos\{4\pi f_0 t\} \tag{C.8}$$

The first part in the last expression, $x(t)/2$, goes through the filter unchanged. The claim is that the complex exponentials in $x(t)\cos\{4\pi f_0 t\}$ have frequencies larger than f_1. To see this, note that this signal is the sine wave at the frequency $2f_0$ modulated by $x(t)$. The spectrum of that signal is therefore the same as that shown in Figure C.3 with f_0 replaced by $2f_0$. It follows that the second term in (C.8) does not go through the filter and that the output of the filter is indeed equal to $x(t)/2$.

The above synchronous demodulation method requires that the receiver be able to generate $\cos\{2\pi f_0 t\}$. The oscillator that generates $\cos\{2\pi f_0 t\}$ in the receiver must be kept in phase with the received signal $y(t)$. This requirement is achieved by a circuit called a *phase-locked loop*. We explain the operations of that circuit in the next section.

C.3 Phase-Locked Loop

The objective of the phase-locked loop is to match the phase of a local oscillator with that of a received signal. The receiver has an oscillator that operates at a frequency approximately equal to f_0. The oscillator frequency is slightly adjustable by means of an input voltage. To be specific, say that the frequency of the oscillator increases when input voltage V is positive and decreases when V is negative (see Figure C.4). Denote the output of the oscillator by $y(t)$ and write it as

$$y(t) = \cos\{2\pi f_0 t + \theta(t)\}$$

where $\theta(t)$ is some phase value that changes slowly.

The signal received is $x(t) = \sin\{2\pi f_0 t + \phi\}$ where ϕ is an unknown phase shift due to the propagation delay.

Assuming that $\theta(t) \approx \theta \approx \phi$, we find that

$$x(t)y(t) \approx \frac{1}{2}\sin\{4\pi f_0 t + \phi + \theta\} + \frac{1}{2}\sin\{\phi - \theta\}$$

FIGURE C.4

Phase-locked loop.

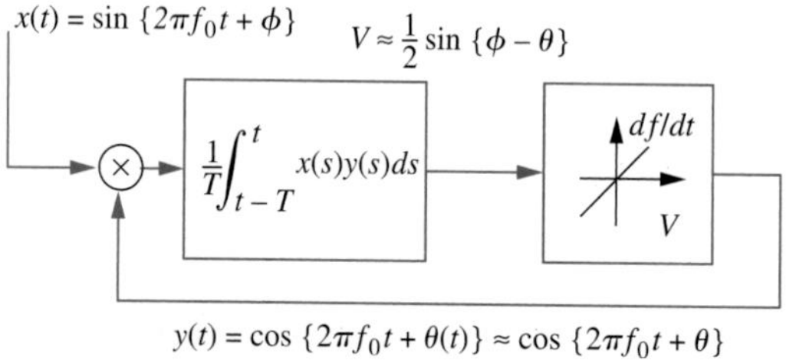

$x(t) = \sin\{2\pi f_0 t + \phi\}$ $V \approx \frac{1}{2}\sin\{\phi - \theta\}$

$\frac{1}{T}\int_{t-T}^{t} x(s)y(s)ds$

df/dt

V

$y(t) = \cos\{2\pi f_0 t + \theta(t)\} \approx \cos\{2\pi f_0 t + \theta\}$

and the average value V of this expression is then such that

$$V = \frac{1}{T} \int_{t-T}^{t} x(s)y(s)ds \approx \frac{1}{2} \sin \{\phi - \theta\}.$$

We claim that the circuit adjusts the local oscillator so that $\theta(t)$ stays close to ϕ and V stays close to 0. To verify this claim, assume that $0 < \phi - \theta < \pi/2$. Then $V > 0$ and the frequency of the local oscillator tends to increase, which makes $\theta - \phi$ increase, thereby reducing $\phi - \theta$. A similar argument holds for $0 \le \theta - \phi < \pi/2$. Thus, the circuit "locks" the phase of the local oscillator to that of the received signal.

C.4   Nyquist's Sampling Theorem

Nyquist discovered that a signal $x(t)$ with no energy at frequencies larger than f_{max} can be reconstructed *exactly* from the values of its samples $\{x(nT), n = 0, \pm 1, \pm 2, \ldots\}$, provided that $T < 1/(2f_{max})$. This fact is called *Nyquist's sampling theorem*.

This result makes the digital transmission of audio and video signals possible by sampling and quantizing the signals. Nyquist's sampling theorem follows from the identity (C.4) as we show next. Using (C.4), we see that

$$\sum_{n=-\infty}^{\infty} x(t)\delta(t - nT) = \frac{1}{T} \sum_{n=-\infty}^{\infty} x(t)e^{j(2\pi n/T)t}. \qquad (C.9)$$

Now, $\delta(.)$ is the limit of $\delta_\epsilon(.)$ as $\epsilon \downarrow 0$, and (C.6) shows that

$$x(t)\delta_\epsilon(t - nT) \approx x(nT)\delta_\epsilon(t - nT),$$

provided that $x(.)$ is approximately constant in a small interval around nT. Consequently, you can expect, and this can be justified, that

$$x(t)\delta(t - nT) = x(nT)\delta(t - nT).$$

Using this expression, we can write (C.9) as

$$\sum_{n=-\infty}^{\infty} x(nT)\delta(t - nT) = \frac{1}{T} \sum_{n=-\infty}^{\infty} x(t)e^{j(2\pi n/T)t}. \qquad (C.10)$$

We claim that one can recover $x(.)$ from the right-hand side of (C.10) when T is small enough. Since the left-hand side of (C.10) depends only on the samples of $x(.)$ every multiple of T, this claim proves Nyquist's theorem. To prove the claim, we observe that each term

$$x(t)e^{j(2\pi n/T)t}$$

has the same frequency spectrum as $x(.)$, except that the frequency spectrum is shifted by the frequency n/T. Consequently, the frequency spectrum of the right-hand side of (C.10) is the sum over $n \in \{\ldots, -2, -1, 0, 1, 2, \ldots\}$ of a copy of the frequency spectrum of $x(.)$ shifted by n/T. That frequency spectrum is shown in Figure C.5 which shows that if $\frac{1}{T} > 2f_{max}$, then the replications of the frequency spectrum of $x(.)$ every $1/T$ do not overlap.

It is then possible to recover the frequency spectrum of $x(.)$ by filtering the signal (C.10) so as to keep only the frequencies in the range $(-f_{max}, +f_{max})$. The diagram in Figure C.6 illustrates the recovery of $x(.)$ from its samples $\{x(nT), n = 0, \pm 1, \pm 2, \ldots\}$.

FIGURE C.5

Spectrum of signal (C.10).

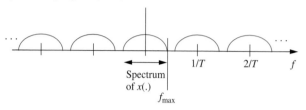

FIGURE C.6

Recovery of x(t) from its samples.

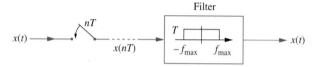

Summary

- Most signals can be written as a sum of complex exponentials. The coefficients of these exponentials constitute the *frequency spectrum* of the signal.
- *Modulation* is used to modify the spectrum of a signal. Modulation enables one to transmit electromagnetic waves efficiently with short antennas.
- The *phase-locked loop* maintains the phase of a local oscillator very close to that of a received signal.
- According to *Nyquist's sampling theorem,* when a signal has no energy above some frequency f_{max}, it suffices to specify $2 f_{max}$ *samples* of the signal per second in order to describe it completely.

References

Couch (1983), Cooper (1986), Proakis (1995), and Lee (1988) are recommended texts on communication theory.

Bibliography

Cooper, G. R., and McGillem, C. D., *Modern Communications and Spread Spectrum,* McGraw-Hill, 1986.

Couch, L. W., *Digital and Analog Communication Systems,* Macmillan, 1983.

Lee, E. A., and Messerschmitt, D. G., *Digital Communication,* Kluwer, 1988.

Proakis, J. G., *Digital Communications,* 3rd ed., McGraw-Hill, 1995.

APPENDIX D

References

Aggrawal, S., Courcoubetis, C., and Wolper, P., "Adding liveness properties to coupled finite state machines," *ACM TOPLAS,* April 1990.

Ahuja, V., *Network and Internet Security,* Academic Press, 1996.

Alles, A., "ATM Internetworking" Cisco Systems, http://www.cisco.com/, May 1995.

Bellman, R., "On a routing problem," *Quart. Appl. Math.,* vol. 16, pp. 87–90, 1958.

Ben-Artzi, A., Chanda, A., and Warrier, U., "Network management of TCP/IP networks: Present and future," *IEEE Network Magazine,* pp. 35–43, July 1990.

Bertsekas, D., and Gallager, R., *Data Networks,* Prentice-Hall, 1987.

Bertsekas, D., and Gallager, R., *Data Networks,* Prentice-Hall, 1992.

Bremaud, P., *An Introduction to Probabilistic Modeling,* Springer Verlag, 1988.

Case, J., Fedor, M., Schoffstall, M., and Davin, C., "A simple network management protocol (SNMP)," *Network Working Group RFC 1098,* April 1989.

Comer, D., and Stevens, D. (for vols. 2 and 3): *Internetworking with TCP/IP,* vols. 1–3, Prentice-Hall, 1991–93.

Cooper, G. R., and McGillem, C. D., *Modern Communications and Spread Spectrum,* McGraw-Hill, 1986.

Couch, L. W., *Digital and Analog Communication Systems,* Macmillan, 1983.

De Prycker, M., *Asynchronous Transfer Mode,* 2nd ed., Ellis Horwood, 1993.

Dijkstra, E. W., "A note on two problems in connection with graphs," *Numerische Mathematik,* vol. 1, pp. 269–271, 1959.

Feynman, R. P., *Lectures on Computation,* edited by J. G. Hey and R. W. Allen, Addison Wesley, 1996.

Göhring, H-G, and Kauffels, F-J, *Token Ring—Principles, Perspectives and Strategies,* Addison-Wesley, 1992.

Har' El and Kurshan, R. P., "Software for analytical development of communications protocol," *AT&T Tech. J.,* pp. 45–59, January/February, 1990.

Hegering, H-G, and Läpple, A., *Ethernet: Building a Communications Infrastructure,* Addison-Wesley, 1993.

Henry, P., "Lightwave primer," *IEEE Journal of Quantum Electronics,* QE-21, no. 12, pp. 1862–1879, 1985.

Hoel, P. G., Port, S. C., and Stone, C. J., *An Introduction to Probability Theory,* Houghton Mifflin Company, 1971.

Holtzmann, G., *Design and Validation of Computer Protocols,* Prentice-Hall, 1991.

Huitema, C., *Routing in the Internet,* Prentice-Hall, 1995.

Jacobson, V., "Congestion avoidance and control," *Proc. SIGCOMM '88 Symposium,* pp. 314–329, August 1988.

Jain, R., "A delay-based approach for congestion avoidance in interconnected heterogeneous computer networks," *ACM Computer Communication Review,* 19, pp. 56–71, October 1989.

Johnson, H. W., *Fast Ethernet—Dawn of a New Network,* Prentice-Hall, 1996.

Kaufman, C., Perlman, R., and Speciner, M., *Network Security—Private Communication in a Public World,* Prentice-Hall, 1995.

Kelly, F. P., *Reversibility and Stochastic Networks,* John Wiley & Sons, 1979.

Kershenbaum, Aaron, *Telecommunications Network Design Algorithms,* McGraw-Hill, 1993.

Kleinrock, L., *Queueing Systems,* vol. 1, John Wiley & Sons, 1975.

Kleinrock, L., editor, *Realizing the Future; The Internet and Beyond,* National Academy Press, 1994.

Le Boudec, J-Y, "The asynchronous transfer mode: A tutorial," *Computer Networks and ISDN Systems,* 24, 279–309, 1992.

Lee, E. A., and Messerschmitt, D. G., *Digital Communication,* Kluwer, 1988.

LeGall, D., "MPEG: A video compression standard for multimedia applications," *Communications of the ACM,* 34, no. 4, pp. 46–58, April 1991.

Lin, S., and Costello, D. J., *Error Control Coding,* Prentice-Hall, 1983.

McEliece, R. J., Rodemich, E. R., and Cheng, J.-F., "The Turbo decision algorithm," *33rd Allerton Conference on Communication, Control, and Computing,* October 1995.

McGraw G., and Felten, E. W., *Java Security,* John Wiley & Sons, 1997.

Menezes, A. J., van Oorschot, P. C., and Vanstone, S., editors, *Handbook of Applied Cryptography,* CRC Press, 1997.

Metcalfe, R. M., and Boggs, D. R., "Ethernet: distributed packet switching for local computer networks," *Comm. ACM,* 19, 395–404, 1976.

Palais, J. C., *Fiber Optic Communications,* 2nd ed., Prentice-Hall, 1988.

Partridge, C., *Gigabit Networking,* Addison-Wesley, 1994.

Peterson, Larry L., and Davie, Bruce S., *Computer Networks: A Systems Approach,* Morgan Kaufmann, 1996.

Preneel, B., and van Oorschot, P. C., "MDx-MAC and building fast MACs from hash functions," *Proc. Crypto '95, LNCS* 963, D. Coppersmith, ed., Springer-Verlag, 1–14, 1995.

Prim, R. C., "Shortest connection networks and some generalizations," *Bell System Technical Journal,* vol. 36, pp. 1389–1401, 1957.

Proakis, J. G., *Digital Communications,* 3rd ed., McGraw-Hill, 1995.

Rényi, A., *A Diary on Information Theory,* John Wiley & Sons, 1984.

RFC 2200, IAB, Postel, J., editor, Internet Official Protocol Standards, June 1997.

Shannon, C., "A mathematical theory of communication," *Bell System Technical Journal,* 27, pp. 379–423 and 623–656, 1948.

Stevens, W. Richard, *TCP/IP Illustrated,* vols. 1 and 2, Addison-Wesley, 1994.

Tanenbaum, A. S., *Computer Networks,* 3rd ed., Prentice-Hall, 1996.

Varaiya, P., and Walrand, J., *High-Performance Communication Networks,* Morgan Kaufmann, 1996.

Vetterli, M., and Kovačević, J., *Wavelets and Subband Coding,* Prentice-Hall, 1995.

Walrand, J., *An Introduction to Queueing Networks,* Prentice-Hall, 1988.

Ziv, J., and Lempel, A., "A universal algorithm for sequential data compression," *IEEE Trans. Information Theory,* 23, pp. 337–343, 1977.

Index